endorsed for
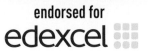

SALTERS HORNERS
A level Physics

2

THIRD EDITION

ALWAYS LEARNING

PEARSON

Published by **Pearson Education Limited, 80 Strand, London WC2R 0RL**

www.pearsonschoolsandfecolleges.co.uk

Copies of official specifications for all Edexcel qualifications may be found on the website: www.edexcel.com

Text © University of York Science Education Group, 2015

Designed by Elizabeth Arnoux, Pearson Education Ltd

Typeset by Wearset Ltd, Boldon, Tyne and Wear

Original illustrations © Pearson Education

Illustrated by Pantek Arts, Maidstone, Kent; Wearset Ltd, Boldon, Tyne and Wear

Cover design by Juice Creative

Picture research by Caitlin Swain

Cover photo © Shutterstock/cherezoff

The rights of the authors to be identified as authors of this work have been asserted by them in accordance with the Copyright, Designs and Patents Act 1988.

First edition published 2000
Second edition published 2008
This edition published 2015
19 18 17 16 15
10 9 8 7 6 5 4 3 2 1

British Library Cataloguing in Publication Data
A catalogue record for this book is available from the British Library

ISBN 978 1 447 99099 4

Printed in Slovakia by Neografia

Websites
Pearson Education Limited is not responsible for the content of any external internet sites. It is essential for tutors to preview each website before using it in class so as to ensure that the URL is still accurate, relevant and appropriate. We suggest that tutors bookmark useful websites and consider enabling students to access them through the school/college intranet.

From the Publisher

In order to ensure that this resource offers high-quality support for the associated Pearson qualification, it has been through a review process by the awarding body. This process confirms that this resource fully covers the teaching and learning content of the specification or part of a specification at which it is aimed. It also confirms that it demonstrates an appropriate balance between the development of subject skills, knowledge and understanding, in addition to preparation for assessment.

Endorsement does not cover any guidance on assessment activities or processes (e.g. practice questions or advice on how to answer assessment questions), included in the resource nor does it prescribe any particular approach to the teaching or delivery of a related course.

While the publishers have made every attempt to ensure that advice on the qualification and its assessment is accurate, the official specification and associated assessment guidance materials are the only authoritative source of information and should always be referred to for definitive guidance.

Pearson examiners have not contributed to any sections in this resource relevant to examination papers for which they have responsibility.

Examiners will not use endorsed resources as a source of material for any assessment set by Pearson.

Endorsement of a resource does not mean that the resource is required to achieve this Pearson qualification, nor does it mean that it is the only suitable material available to support the qualification, and any resource lists produced by the awarding body shall include this and other appropriate resources.

Picture Credits

The publisher would like to thank the following for their kind permission to reproduce their photographs:

(Key: b-bottom; c-centre; l-left; r-right; t-top)

akg-images Ltd: John Hios 174; **Alamy Images:** Arterra Picture Library 71t, BSIP SA 6tl, Dan Leeth 240t, David Wall 183, Erik Tham 76r, Eye Ubiquitous 13, GL Archive 291b, 310, GL Archive 291b, 310, Image Source 44tl, 93tr, Inga Spence 182l, Jake Lyell 172, John Boud 38, John Henshall 190, Keystone Pictures USA 313, Kirsty McLaren 219, Lifestyle pictures 148, Oleg Kozlov 63, Paul Carstairs 188, Paul Rapson 51, Richard Cooke 2l, Robert Morris 3, Roger Bamber 4l, sciencephotos 153c, 291t, sciencephotos 153c, 291t, Septemberlegs 257t, SFL Travel 52, Steven May 77, Universal Images Group 185, VintageCorner 277t, World History Archive 270l, World Pictures 4r, Zoonar GmbH 273b; **CERN Geneva:** 1995 CERN 110, 2006 CERN 144, 146, 159b, 160t, 160b, 2008 CERN 106, 2014 CERN 159t, 2015 CERN 107; **Corbis:** 13/Philip and Karen Smith/Ocean 114 (a), Bettmann 257, Colin Garratt/Milepost 92 1/2 7, Digital Art 108, Heritage Images 270r, Kevin Fleming 112, Roger Ressmeyer 203; **ESA:** Rosetta/ NAVCAM – CC BY-SA IGO 3.0 280; **Fotolia.com:** IRC 212, Joe Gough 76bl, spacekris 273t, underworld 76tl; **Getty Images:** Bloomberg 9, Bryn Lennon 152b, Cultura RM/Joseph Giacomin 173, John Giustina/Photodisc 114 (k), Michael Steele 114 (g), Pool/ Pool 152, Ryan Kelly 206, Science & Society Picture Library 2r, 46tl, Science & Society Picture Library 2r, 46tl, Time Life Pictures 300, Universal Images Group/Contributor 253c; **GNU Free Documentation License: Wikipedia:** 210; **http://mindsetsonline. co.uk/Site/Home:** 191; **ITER Organization:** 272; **NASA:** Atlas Image [or Atlas Image mosaic] courtesy of 2MASS/UMass/ IPAC-Caltech/NASA/NSF 317, /CXC/SAO; Optical: Detlef Hartmann; Infrared: NASA/JPL-Caltech 319 (f), /CXC/University of Toronto/M. Durant et al; Optical: DSS/Davide De Martin 305, ESA 327, ESA/Hubble Heritage Team 296, ESA, HEIC and The Hubble Heritage Team (STScI/AURA) 239l, ESA, J. Hester (Arizona State University) 240b, ESA, M. Robberto (Space Telescope Science Institute/ESA) and the Hubble Space Telescope Orion Treasury Project Team 288, JPL-Caltech 303, JPL Public/JPL NASA 304, LASCO, SOHO Consortium, NRL, ESA 242l, Marshall Space Flight Center 114 (j), Reto Stöckli, Nazmi El Saleous, and Marit Jentoft-Nilsen 125t, SDO 269t, SDO/AIA 242r, WMAP Science Team 319 (d); **National Trust Photo Library:** Chris Lacey 277b; **Rex Features:** 114 (b), David Crump/Daily Mail 46bl; **Science & Society Picture Library:** Science Museum 99; **Science Photo Library Ltd:** 126b, 136, 247, 253tr, 263, 276, 126b, 136, 247, 253tr, 263, 276, Andrew Lambert Photography 72, 139, 189, Babak Tafreshi 248, C. Carreau/European Space Agency 245, Celestial Image Picture Co. 307b, David A. Hardy 257b, David Parker 71c, 109, 116, David Parker 71c, 109, 116, Dirk Wiersma 268, Dr Rudolph Schild 71b, Emilio Segre Visual Archives/American Institute Of Physics 252l, 253br, 307t, 307c, Emilio Segre Visual Archives/American Institute Of Physics 252l, 253br, 307t, 307c, Emilio Segre Visual Archives/American Institute Of Physics 252l, 253br, 307t, 307c, Emilio Segre Visual Archives/American Institute Of Physics 252l, 253br, 307t, 307c, GIPhotostock 142, Harvard College Observatory 252r, John Sanford 249, Juan Carlos Casado (starryearth. com) 243, Julian Baum 319 (e), Library of Congress 186, Michael Abbey 260, NASA 123, NASA/ESA/STSCI/J. Hester & P. Scowen, ASU 238, NASA/ESA/STSCI/R.Williams, HDF Team 239r, Outlast Technologies 218, Physics Dept., Imperial College 269b, Royal Astronomical Society 286l, Stanford Linear Accelerator Center 151; **Shutterstock.com:** Aleksey Stemmer 255cr, FreshPaint 44r, Hal_P. 87l, Ivan Smuk 217, ra3rn 65, Sideways Design. 87r, Stuart Jenner 62, Tupungato 46r; **Stable Micro Systems:** 114 (f); **The Kobal Collection:** Paramount/Bad Robot 119; **TopFoto:** Henrietta Swan Leavitt 255b; **University of York:** Algol, a sculpture by Harry Mercer (1968), Goodricke College, University of York 286cr; **University of York Science Education Group:** 22cl, 64c, 67, 83, 90, 204, 213; **www.vivarail.co.uk:** 43, 114 (h), 23

Cover images: *Front:* **Shutterstock.com:** cherezoff

All other images © Pearson Education

Picture Research by: Caitlin Swain

Acknowledgements
We are grateful to the following for permission to reproduce copyright material:

Figures
Figure 3.4 on page 110 from https://proj-cngs.web.cern.ch/proj-cngs/GeneralDescriptionVe/images/fig01.jpg, CERN, with permission; Figure 3.57 on page 159 from http://livefromcern-archive.web.cern.ch/livefromcern-archive/antimatter/history/historypictures/LHC-drawing-half.jpg, CERN, with permission; Figure 4.27a on page 191 is available from Mindsets (UK) Ltd www.mindsetsonline.co.uk

Text
Extract on page 11 from https://www.gov.uk/government/uploads/system/uploads/attachment_data/file/312249/the-highway-code-typical-stopping-distances.pdf, Contains public sector information licensed under the Open Government Licence v3.0; Extract on page 145 from The ATLAS Simulation Infrastructure, *The ATLAS collaboration Journal: Journal of Instrumentation, 2014, Volume 9, Number 08, Page P08009 DOI: 10.1088/1748-0221/9/08/P08009*, http://link.springer.com/article/10.1140/epjc/s10052-010-1429-9, Attribution 4.0 International (CC BY 4.0)

Many people from schools, colleges, universities, industries and the professions have contributed to the Salters Horners Advanced Physics (SHAP) project and the preparation of SHAP course materials.

Authors of this edition

Jonathan Allday (The Royal Hospital School, Ipswich)
Christina Astin (The King's School, Canterbury)
Paul Cooney (John Leggott College, Scunthorpe)
Ian Francis (Bath Academy, Somerset)
Jan Hatherell (Hardenhuish School, Wiltshire)
Neil Jaques (Bancroft's School, Essex)
David Neal (Barton on Humber, Lincs)
Chris Pambou (City & Islington College, London)
Siôn Peters-Flynn (Grimsby Institute Group, Lincs)
Nicky Robinson (Reigate College, Surrey)
David Swinscoe (City & Islington College, London)
Carol Tear (York)

Project director and general editor

Elizabeth Swinbank (University of York)

Acknowledgements

We would like to thank the following for their help and support with the preparation of this edition.

Marc Bloch (Pearson)
Hovan Catchatoor (Reigate College)
Joanna MacDonald (University of York)
William D Phillipps (Leek)
Robin Millar (University of York)

Authors of previous editions

This third edition is based on previous editions and incorporates the work of the following authors:

Jonathan Allday	Alasdair Kennedy	Tony Sherborne
Chris Butlin	Bob Kibble	Richard Skelding
Steven Chapman	Paul Lee	Sandy Stephens
Steve Cobb	Maureen Maybank	Wendy Swarbrick
Tony Connell	Averil Macdonald	Elizabeth Swinbank
Howard Darwin	Graham Meredith	David Swinscoe
Nick Fisher	David Neal	Carol Tear
Frances Green	Chris Pambou	Clare Thomson
Greg Hughes	Kerry Parker	Nigel Wallis
Neil Jaques	David Sang	

Sponsors

We are grateful to the following for sponsorship that has continued to support the Salters Horners Advanced Physics project after its initial development and has enabled the production of this edition.

The Worshipful Company of Horners
The Worshipful Company of Salters

Advisory Committee for the initial development

Prof. Frank Close
Prof. Cyril Hilsum FRS

Prof. Robin Millar
Prof. Sir Derek Roberts FRS

HOW TO USE THIS BOOK

Context-led study

This book covers the second year of a two-year A level course. SHAP is a context-led approach to the Pearson Edexcel AS and A level Physics specification.

Each chapter starts by looking at particular situations in which physics is used or studied, and then develops the physics you need to learn to explore this 'context'.

We have tried to select contexts to give you some idea of how physics can help improve people's lives, how physics is used in engineering and technology, and how physics research extends our understanding of the physical world at a fundamental level. These will show you just some of the many physics-related careers and further study that might be open to you in the future.

Within each chapter, you will develop your knowledge and understanding of one or more areas of physics. In later chapters, you will meet many of these ideas again – in a completely different context – and develop them further. In this way, you will gradually build up your knowledge and understanding of physics and learn to apply key principles of physics to a variety of contexts. Each chapter includes the following features:

Main text

This presents the context of each teaching unit and explains the relevant physics, as you need it.

Within the main text, some words are printed in **bold**. These are key terms relating to the physics. We suggest that you make your own summary of these terms (and others if you wish) as you go along. Then you can refer back to your summary when you revisit a similar area of physics later in the course and when you revise for exams.

There are references to relevant websites in this book. We have made the links available on our website at www.pearsonhotlinks.co.uk. Search for Salters Horners A level Physics or ISBN 978 1 447 990994.

Activities

The text refers to many Activities. These include practical work, the use of ICT, reading, writing, data handling and discussion. Some Activities are best carried out with one or more other students, while others are intended for you to do on your own. For some activities, there are handout sheets giving further information, details about apparatus and so on.

> ### ⚙ ACTIVITY 13 PLOTTING LINES OF EQUIPOTENTIAL
> Use conducting paper to plot lines of equipotential in various two-dimensional fields.
>
> Alternatively, use a single-flame probe to find the equipotential surfaces between parallel plates and other arrangements. (You can use the equipment from Activity 12 but just light one of the probes and connect the earth of the electroscope to one of the plates.)

Questions

There are two types of Questions in this book:

- Some are intended for you to do as you go along, as a self-check or perhaps for homework; the answers to these questions appear at the end of each chapter.

- Questions on the whole chapter are intended as a summary to what you have learned in each chapter: your teacher will provide the answers to these questions.

Once you have had a go at a question, check your answer. If you have gone wrong, use the answer (and the relevant part of the book chapter) to help you sort out your ideas. Working in this way is not cheating; rather, it helps you to learn.

MATHS REFERENCE

Manipulating powers on a
calculator
See Maths note 1.4

Maths notes

Maths references in the margin will direct you to the Maths notes at the end of the book, which are to help you with the maths needed in physics. This may involve calculations, rearranging equations, plotting graphs and so on. You will probably have covered most of what is required at GCSE, but you may not be used to applying it in physics. The Maths notes summarise the key maths ideas that you need and show you how to apply them to situations in physics.

PRACTICAL SKILLS
REFERENCE

Recording data
See Practical skills note 3.2

Practical skills references

These references in the margin direct you to guidance notes on practical skills at the end of the book. The notes set out a framework for carrying out practical work and give guidance on particular aspects of such work. They draw attention to points that you need to consider when preparing for and carrying out the task and when writing a report.

STUDY NOTE

For a summary of information
about motion see TRA
Additional Sheet 1.

Study notes

These notes in the margin are intended to help you get to grips with the physics. For example, they indicate links with other parts of the course.

Link the learning

At the end of each chapter there is the opportunity to revisit and consolidate the physics you have already learned. The questions and activities will help you connect different aspects of your learning from across the course and apply your knowledge to new contexts. A table of achievements shows exactly what material you've covered in the chapter, so you can recap what you've learned at glance.

Getting the most from your ActiveBook

Your ActiveBook is the perfect way to personalise your learning as you progress through your A level Physics course. You can:

● access your content online, anytime, anywhere;

● use the inbuilt highlighting and annotation tools to personalise the content and make it really relevant to you;

● search the content quickly.

Highlight tool

Use this to pick out key terms or topics so you are ready and prepared for revision.

Annotations tool

Use this to add your own notes, for example links to your wider reading, such as websites or other files. Or make a note to remind yourself about work that you need to do.

Activities strengthen your understanding

Apply your knowledge to new contexts in questions on the whole chapter

See your achievements throughout the chapter

Further investigations

Alongside your A level work, you might have the opportunity to carry out a major piece of investigative work for an Extended Project qualification (EPQ).

As you proceed through the course, keep a note of any areas you might like to pursue; we have included some suggestions under the heading Further investigations, but any unanswered question that intrigues you could form the basis of an EPQ.

FURTHER INVESTIGATION

Measure and compare the response times of LEDs and LCDs. Explore the factors that affect response time.

WORKING AS A PHYSICIST

Throughout this course, you will be developing your knowledge and understanding of what it means to work scientifically in the context of physics and its applications.

The examples, questions and activities in each chapter will help you to become increasingly competent in manipulating quantities and their units, in planning, carrying out and evaluating practical experiments, and in communicating your knowledge and understanding of physics. The Maths notes are written to help you to build up your skills in handling quantities and units, and the Practical skills references are designed to support your investigative and practical work.

You will also learn more about the ways in which the scientific community, and society as a whole, use physics and contribute to its progress. You will study many applications of physics in a wide variety of situations, and you will see how the risks and benefits of those applications are evaluated. You will also learn more about the ways in which the scientific community operates to validate new knowledge relating to physics, and to ensure that work on physics and its applications is carried out with integrity.

By the time you have completed your A level course, you will have developed your knowledge of Working as a Physicist, and these Achievements will be tested in the examinations. Table 1 lists the requirements for Working as a Physicist, copied from the examination specification, and refers to sections of this book where they are addressed. Note that these are not all the places that involve Working as a Physicist – that would make Table 1 very large. Rather, Table 1 lists the sections where aspects of Working as a Physicist are first introduced, where they are treated in most detail or where there are activities or questions specifically designed to help with these parts of your course. As you progress through the course, you will revisit these statements several times, helping you to build up your skills and to develop a broad and deep knowledge of what it means to work as a physicist. The lists of Achievements at the end of the chapters include some further examples of where particular aspects of Working as a Physicist are addressed.

Table 1 Achievements for Working as a Physicist

Statement from examination specification		Chapter	Section(s)
1	know and understand the distinction between base and derived quantities and their SI units	Maths notes	2
		TRA	2.2, 3.3, 4.2
		MDM	3.2
		PRO	3.2
		BLD	2.2, 3.3, 4.2
		STA	2.5, 4.1
2	demonstrate their knowledge of practical skills and techniques for both familiar and unfamiliar experiments	Practical skills	all
		TRA	4.2
		MDM	1.1, 1.3, 3.1, 3.2
		PRO	3.4
		BLD	2.1, 2.4, 4.2, 4.3
		STA	2.2, 2.4, 3.1, 4.2
3	be able to estimate values for physical quantities and use their estimate to solve problems	TRA	1.1, 3.2
		MDM	3.1
		PRO	3.5, 4.3
		BLD	1.2
		STA	2.7, 4.2
4	understand the limitations of physical measurement and apply these limitations to practical situations	Practical skills	4
		Maths notes	7
		TRA	4.2, 5.2, 5.3
		MDM	3.3
		PRO	3.4
		BLD	2.1, 2.4, 4.2, 4.3
		STA	2.2, 2.4, 3.1, 4.2

Statement from examination specification		Chapter	Section(s)
5	be able to communicate information and ideas in appropriate ways using appropriate terminology	TRA	3.4, 5.4, 7.1
		MDM	1.4, 3.3, 3.4
		PRO	2.2, 2.4
		BLD	2.1, 4.2, 4.4, 5.1
		STA	3.1, 4.1, 4.2, 5.3
6	*understand applications and implications of science and evaluate their associated benefits and risks	TRA	6.1
		MDM	3.3
		BLD	5.1
7	understand the role of the scientific community in validating new knowledge and ensuring integrity	PRO	4.1
		STA	5.1, 5.2, 5.4
8	understand the ways in which society uses science to inform decision making	TRA	6.1
		MDM	3.3
		BLD	1.2

* All the chapters discuss applications and implications of physics. These references indicate where risk and benefit are discussed.

CHAPTER 1

TRANSPORT ON TRACK

Why a chapter called *Transport on Track*?

In 1825 George Stephenson's steam-powered *Locomotion* (developed from his famous *Rocket*) drew the world's first passenger/goods train from Stockton to Darlington, a 25-mile journey, at an average speed of only 8 miles per hour. Trains can now travel from Britain to France in roughly 20 minutes, and many cities offer rapid commuter transport in the form of trains or trams (see Figure 1.1).

The desire to improve, extend and speed up the rail transport system has constantly tested scientists and engineers. The rails themselves need to be strong enough to cope with the weight of fully loaded trains, while incorporating designs to limit the effect of expansion. Engine and coach production has had to make use of new techniques of propulsion and new materials of construction to become faster and cleaner. Finally, and most importantly, increasing speeds, passenger numbers and service density (number of trains or trams in a given time) have necessitated continuous advances in the area of railway safety, using the foremost technologies of the day. Even though people see the running of a train system as being a very simple process, the scientific and technological challenges involved are considerable. Many people believe that this century will see the birth of a new railway age, as environmental worries and road congestion make trains and trams the sensible way forward. The extraordinary technical success of the Channel Tunnel and electrified East and West Coast main lines, along with the urban tram, light-railway and underground systems, may just be the start of an exciting era with less reliance on petroleum-based transport.

Figure 1.1 (a) A *Eurostar* train

Figure 1.1 (b) The *Locomotion* No 1

Figure 1.1 (c) A modern electrified urban transport system

OVERVIEW OF PHYSICS PRINCIPLES AND TECHNIQUES

This chapter builds on your understanding of basic electrical circuits and mechanics gained from earlier chapters. Ideas about magnetism will be extended and this will lead into electromagnetism, electric motors and the laws of electromagnetic induction. You will also investigate circuits containing capacitors and resistors used to time events. By considering what happens when objects start, stop and collide, you will revisit the concept of momentum and will apply energy conservation. Finally, you will consider how science works in relation to some of the issues involved in the safe running of a rail network.

In the course of this chapter you will be involved in a wide range of practical activities, many of which will have an electrical basis. You will also use spreadsheets to model situations involving the motion of trains, and you will explore a circuit-modelling application to test circuit designs.

In this chapter you will extend your knowledge of:

● DC circuits from the chapters *Technology in Space* and *Digging Up the Past*

● momentum from *Higher, Faster, Stronger*

● energy conservation from *Higher, Faster, Stronger, Spare Part Surgery* and *Technology in Space*.

In other chapters you will do more work on:

● momentum and magnetic fields in *Probing the Heart of Matter*

● exponential changes in *The Medium is the Message* and *Reach for the Stars*

● capacitance in *The Medium is the Message*

● forces and motion in *Build or Bust?*

● kinetic energy and work in *Reach for the Stars*.

1 Getting on track

The following introductory section provides some initial information and sets the scene for the chapter.

1.1 Tunnel trains

The Channel Tunnel was designed as a quick and reliable route for passengers and freight between Britain and continental Europe. *Le Shuttle* trains, specially designed and constructed for the tunnel, run between purpose-built terminals at each end. *Eurostar* trains provide a fast link to the continent – London to Paris in 2 hours 15 minutes. To meet the demand for cross-channel transport, the Eurotunnel company needs to send around 25 trains through the tunnel per hour. In the tunnel, these trains travel at up to 160 km per hour, forcing air pressure pulses in front of them that generate high winds, making it impossible to work in the tunnels. In the entry and exit sections of the tunnel, trains travel over gradients as steep as 1 in 90.

The *Eurostar* train in Figure 1.1(a) carries hundreds of people on board, all of whom hope to arrive safely at their destination. It is almost half a kilometre long and has roughly the mass of a thousand small cars; and yet it can travel at 300 km per hour ($83 \, \text{m s}^{-1}$). When moving at speed, it has enough energy to raise itself and its contents to the top of the Eiffel Tower. Each of its two electrically powered locomotives, or engines, needs to be equivalent to 20 Formula 1® racing cars.

Figure 1.2 shows a *Le Shuttle* train. It is three-quarters of a kilometre long and can travel at 140 km per hour ($39 \, \text{m s}^{-1}$). At this speed, it needs 1.5 km to stop. It contains hundreds of vehicles: cars, lorries, coaches and their passengers. To cope with passenger demand, there is generally a similar train not far behind it, and one in front.

Figure 1.2 *Le Shuttle* locomotives are the most powerful electric railway locomotives in the world; each train has one at each end

Both the *Eurostar* and *Le Shuttle* trains travel through a narrow tunnel 7.6 m wide, only slightly wider than the train itself (Figure 1.3). The Channel Tunnel actually consists of three separate tunnels, each 50 km long. There are two rail tunnels, one for traffic in each direction, and a central service tunnel. They were bored, largely through chalk, at an average depth of 45 m below the seabed; for most of its length the Channel Tunnel runs through the virtually impermeable stratum of chalk marl (Figure 1.4).

Figure 1.3 Channel Tunnel

Finally, the potentially dangerous environment of the Channel Tunnel, where access is difficult, creates its own set of problems in the event of a train breaking down. To address this (within the original design brief) was the requirement that each engine should have the power to push a second train out of the tunnel to a point of safety, even up its steepest gradient.

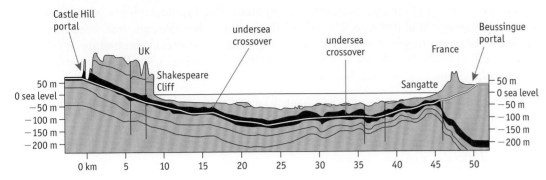

Figure 1.4 Geological cross-section showing the Channel Tunnel

QUESTIONS

Q 1   **(a)** Paragraph 2 identifies the speed of a *Eurostar* train in km h^{-1} and m s^{-1}. Rewriting the speed as $\dfrac{300\,\text{km}}{1\,\text{hour}}$, convert km to m and hours to seconds to show $300\,\text{km h}^{-1}$ is approximately $83\,\text{m s}^{-1}$.

     (b) Given that 1 mile is approximately 1.6 km, convert the speed to mph.

Q 2   Paragraph 2 also states that *Eurostar*, moving at $83\,\text{ms}^{-1}$, has enough energy to raise itself and its contents to the top of the Eiffel Tower. By assuming equality between kinetic energy at this speed and gravitational potential energy, find an estimate of the tower's height. (Hint: you should spot that you do not need the mass of *Eurostar*.)

Q 3   One of the steepest parts of the Channel Tunnel is the section at the English entrance, from 0 km to 8 km, as shown on Figure 1.4. Show that its gradient is approximately 1 in 90, as stated at the beginning of Section 1.1.

1.2 Tram systems

The Channel Tunnel train is a wonderful example of physics in action, albeit one that mostly lies out of sight underground. However, it shares many of its scientific and technical challenges with the more visible electrified tram and light-railway systems now found around many cities in Britain and Europe that began to be developed in the late twentieth century.

The priority given to the automobile had led to a loss in quality of urban life, particularly in large cities where smog, traffic congestion, sound pollution and parking became problematic. Acknowledging this, some authorities re-thought their transport policies and the advantages of the tram became obvious. Britain began replacing its local transport with light rail in the 1980s, starting with the Tyne and Wear Metro in Tyneside and followed by the Docklands Light Railway in London. The first UK modern street-operating light rail system opened in Manchester in 1992, with Italian-built vehicles. The trend to light rail was by then firmly established with the success of this and Sheffield Supertram (1994), followed by Midland Metro in Birmingham (1999), Tramlink in London (2000) and Edinburgh Trams (2014). Manchester Metrolink has 92 stops along 57 miles (92 km) of track, making it the largest light rail system in the UK and one of the largest modern tram networks in Europe. In 2013–14, 29.2 million passenger journeys were made on the system.

Tram vehicles are usually lighter and shorter than conventional trains, but the size of trams (particularly light-rail vehicles) is rapidly increasing. Some trams or light-railway systems run on ordinary rail tracks but, particularly in cities, sharing roads with other vehicles is common and requires careful design. Traditionally trams had high floors, requiring passengers to climb

Figure 1.5 Manchester Metrolink

several steps in order to board. However, since the 1990s this design has been largely replaced by low-floor trams, or occasionally by high-floor trams with level boarding platforms, as in Manchester's Metrolink, which allow passengers in wheelchairs or with pushchairs or shopping trolleys easier access (Figure 1.5).

Most trams today use electrical power, often fed by an overhead pantograph (see page 14) in contact with an overhead live wire. The rails guide the wheels but also complete the electric circuit for the power supply. Some details of the electrical infrastructure required around the track, as well as inside the vehicles, will be covered later in this chapter, as will the technical considerations in making the tram bodies safe.

All these train and tram systems, and the Channel Tunnel railway, share obvious safety problems including:

- How can the enormous electrical power required to operate these vehicles be delivered safely and used effectively to provide motion?

- How can the vast energies of these vehicles be dissipated safely as they brake?

- How can we ensure that the trains/trams will not collide with each other or anything else in their busy working environments?

To solve these problems, some of which affect all modern railway systems, sophisticated engineering techniques are used. For example, in the Channel Tunnel, an automated electronic signalling system is employed to allow very small gaps between trains. This involves a technically advanced sensing system to ensure the position and speed of each train is known at all times. Despite the complexity of modern rail safety systems, you will find their principles of operation are based in fundamental physics and are usually straightforward to understand.

This chapter looks at some of the physics behind making high-speed, high-momentum trains and trams acceptably safe. It focuses on three areas:

- *Stopping and starting*: bringing trains to a halt in an acceptably safe time and distance interval, and starting them moving again, if necessary on a gradient.

- *Sensing speed*: the technique of detecting a train and thereby providing signals to indicate its presence, as well as monitoring its speed to enable safe intervals to be computed and applied.

- *Structure and safety*: how to make trains 'crashworthy' – that is, able to undergo collisions, with as little damage and danger to passengers as possible.

PRACTICAL SKILLS REFERENCE

Scientific questions and information research

See Practical skills notes 1.2–1.3

⚙ ACTIVITY 1 GETTING ON TRACK

Having read the information above, think about which aspects of physics you might expect to find being used in each of the three areas mentioned.

Use the Internet to find more information on Channel Tunnel trains and about an electrified tram or light-railway system. Compare the two systems.

2 Stopping and starting

2.1 The technical challenge of *Eurostar*

Moving and stopping the massive *Eurostar* trains (Figure 1.6) has proved an enormous technical challenge. It has required the application of intriguing uses of electrical power and electromagnetism, plus an understanding of dynamics. The operation of city tram systems requires exactly the same physical ideas, but for now we will set the scene by concentrating on the trains.

Figure 1.6 *Eurostar* locomotive

Supplying the electricity

The *Eurostar* service running through three European countries is complicated because each has a different electrical power supply. The demands of imposed restrictions on axle loads means that the power developed in the new motors had to be 50 per cent higher than in any previous locomotive for less weight.

Motors

Eurostar has AC traction motors rated at 1400 kW, with two motors per locomotive. Each drive system includes a power conversion unit to change a DC or AC fixed-voltage supply into an AC supply of variable voltage and variable frequency. Domestically we run hairdryers, electric drills, etc., using motors of a type called commutator motors. They work very effectively on 50 Hz AC, but only need a few hundred watts. *Eurostar* needs more powerful AC motors, generally of a type called AC induction motors. An alternating current in the winding on a static part generates a changing magnetic flux that induces current in copper bars in a rotating part. These currents react with the changing flux to produce rotation, as we shall see shortly. The copper bars look like the sort of exercise wheel given to hamsters or pet mice, and – presumably for this reason – AC induction motors are often called 'squirrel-cage motors' (see Figure 1.7).

Figure 1.7 Induction motor

Moving off

A high force of friction between the wheels and the track enables a large driving force or tractive effort to be developed. For *Eurostar*, up to 5.6 MW of power can be provided using two locomotives. This gives around 400 kN of force to start the train off.

Slowing down

In the 19th century, mechanical braking was used for wagons and trains. When a train went downhill the brakemen had to turn a cranking wheel to set the brakes on each coach, and repeat the process to turn the brakes off. Single mechanical systems were found to be impossible for longer trains made of separate freight wagons or passenger coaches. The invention of the air brake eventually solved this problem for the railways and is a system still in use. A simple high-pressure air hose can easily be connected or disconnected between coaches, and the system can be supplied from a single high-pressure air pump in the locomotive.

Even when *Eurostar* is slowing down quite gradually, it needs a braking power of many megawatts. To achieve this braking power, the train uses multiple braking systems. Air-lines operate the conventional friction action of disc brakes (just as in a car) on the passenger coaches, while on the locomotive forms of electric braking are available (we will explore these later). Knowledge of the stopping distances required by these new trains is essential if the signals are to be put at suitable intervals along the line.

2.2 Moving theory

In this section we will consider the motion of trains as they start and stop. This will involve building on the basic theories and equations of motion that you met in the earlier chapter *Higher, Faster, Stronger*.

STUDY NOTE

For a summary of information about motion, see TRA Additional Sheet 1.

STUDY NOTE

1 tonne (1 t) is 1000 kg.

QUESTION

Q 4

(a) If the *Eurostar*, mass 750 t, is subject to a net tractive (pulling) force of 400 kN, what is its acceleration?

(b) What power is required to produce this acceleration when the train reaches a speed of 10 m s^{-1}?

Momentum

In *Higher, Faster, Stronger* you learned that the relationship between force, mass and acceleration can be expressed as:

$$F = ma$$

and that this can also be written as:

$$F = \frac{m\Delta v}{\Delta t} = \frac{m(v - u)}{\Delta t}$$

These equations are one way of expressing **Newton's second law of motion**, but in this chapter we look at another way.

If the velocity changes from u to v in time t, then:

$$\frac{\Delta v}{\Delta t} = \frac{(v - u)}{\Delta t}$$

so:

STUDY NOTE

Here we are using symbols F, u, v and p to represent the magnitudes of vector quantities.

$$F = \frac{(mv - mu)}{\Delta t}$$

$$F = \frac{\Delta(mv)}{\Delta t} \tag{1}$$

The quantity *mass × velocity* is the **momentum** of the object. Momentum is a vector quantity and is usually given the symbol p. It has SI base units $kg\,m\,s^{-1}$ (equivalent to Ns).

Equation 1 can be written as:

$$F = \frac{\Delta p}{\Delta t} \tag{1a}$$

This equation is more in line with Newton's original meaning. It shows that force on an object is equal to its rate of change of momentum.

If Δp and Δt are small, this can give an instantaneous value for force, written:

$$F = \frac{dp}{dt} \tag{1b}$$

When Newton's second law is expressed in terms of momentum change, we do not have to assume that the mass remains constant. If the mass changes from an initial value of m_1 to a final one of m_2:

$$\Delta mv = (m_2 v - m_1 u)$$

This widens the application of Newton's second law to cover, for example, a space rocket burning an enormous amount of fuel in a short period of time to achieve take-off.

Another rearrangement gives us:

$$F\Delta t = \Delta(mv) = \Delta p \tag{2}$$

This relationship is used so frequently that the product of force and time has gained its own name. It is called the **impulse** of a force and is particularly useful when analysing impacts, since the force and time of an impact must be the same for both objects involved in the collision. Equation 2 can be expressed in words as:

impulse = change in momentum

Note that impulse and momentum are both vector quantities, so the direction as well as magnitude should be included, and positive and negative values can occur.

We are now able to address such important questions as 'How much force is needed to stop a train?' Use Equations 1 and 2, and other equations of motion that you have met earlier, to answer Questions 5 to 10.

QUESTIONS

Q 5 What are two possible SI units for impulse and momentum?

Q 6 A railway coach of mass 50 t is in motion at $+20\,m\,s^{-1}$. Calculate the change in momentum needed to produce each of the following final velocities (1 t = 1000 kg):

(a) $v = +40\,m\,s^{-1}$

(b) $v = -20\,m\,s^{-1}$

(c) $v = 0\,m\,s^{-1}$

Q 7 A 1.2 t car accelerates uniformly from 0 to $25\,m\,s^{-1}$ in 10 s. Calculate:

(a) the change in momentum

(b) the tractive force applied (ignore any drag due to air resistance, etc.)

(c) the distance travelled

(d) the power of the engine.

Q 8 A *Le Shuttle* train (Figure 1.8) has mass of 2400 t, i.e. 2000 times greater than the car in Question 7. To achieve the same performance it would require a power 2000 times greater, i.e. 75 MW, but it is pulled by power units rated at only 5.6 MW. Why is this sufficient?

Figure 1.8 *Le Shuttle*

Q 9   A fully laden *Eurostar* of 750 t leaves the tunnel at $44\,\mathrm{m\,s^{-1}}$ and takes 14 minutes to get to its top speed of $83\,\mathrm{m\,s^{-1}}$.

(a) What tractive force was required?

(b) How many times smaller is the tractive force required to give the same acceleration to a train of mass 250 t?

(c) How does your value for the tractive force compare with the braking force needed to slow down the same train by the same amount over the same distance?

(d) For both *Eurostar* and the lighter train, drag at high speeds is about 40 kN. What difference does that fact make to your answer to (a) and (b)?

Q 10   Suppose a train must come to a halt in 60 s.

(a) Find the braking force required to do this inside the tunnel, where the maximum speed is $44\,\mathrm{m\,s^{-1}}$, when the train is (i) a fully loaded *Shuttle*, mass 2400 t, and (ii) a *Eurostar*, mass 750 t.

(b) Above ground the top speed of *Eurostar* is $83\,\mathrm{m\,s^{-1}}$. If it stops in 3 minutes, what net braking force must be applied?

(c) In Question 9 you calculated tractive forces. Compare the sizes of the forces required from the engine to those needed by the brakes.

Stopping distances

Once the brakes are applied to a fast-moving train it still takes some time to come to a stop, during which the train can move a considerable distance. City-centre trams are required to stop and start even more frequently on their journeys. Information about the probable stopping distance is therefore critical for normal service in order to ensure that the train or tram stops exactly alongside its platform. It is also important in deciding on the positions of signals. This type of question can be tackled using force and momentum and other equations of motion, but another approach, using energy, is more direct.

Imagine that a train of mass m is moving at an initial speed u, before it is brought to a standstill by a force F acting in the opposite direction to its motion. All the kinetic energy E_k of the train is transferred by the work of the resultant braking force. We can use:

> work done = energy transferred

and:

> work done = force × distance moved in direction of force

Let Δs be the distance covered while the brakes are applied. Since the force acts along the direction of motion, we have:

$$-F\Delta s = \Delta E_k = \frac{1}{2}mv^2 - \frac{1}{2}mu^2$$

Since the final kinetic energy of this train is zero:

$$-F\Delta s = -\frac{1}{2}mu^2 \tag{3}$$

Rearranging Equation 3, the stopping distance for a known brake system can be calculated from:

$$\Delta s = \frac{mu^2}{2F}$$

QUESTIONS

Q 11   A city-centre tram has mass 46.5 t when empty, and can carry a maximum of 250 passengers when full. The maximum emergency deceleration of the tram is $3.0\,\mathrm{m\,s^{-2}}$.

(a) Assuming an average mass for each passenger of 65 kg, find the maximum braking force the tram requires (i) empty and (ii) with its full passenger load.

(b) If the maximum tram speed is 50 mph ($22\,\mathrm{m\,s^{-1}}$), what would be the likely emergency stopping distance for a fully loaded tram?

Q 12 (a) Calculate the stopping distances for a *Shuttle* train that is partly loaded and has a mass 2100 t; it is stopped from a speed of $40\,\mathrm{m\,s^{-1}}$ by a braking force of $1.2\,\mathrm{MN}$.

(b) A *Eurostar*, mass 750 t, is travelling at $80\,\mathrm{m\,s^{-1}}$ and must stop within 1500 m. What force is required?

(c) If there is a drag force of 40 kN at a speed of $80\,\mathrm{m\,s^{-1}}$, suggest what difference that would make to your answers to **(a)** and **(b)**.

Q 13 (a) How long would it take the *Eurostar* in Question 12 to come to rest?

(b) What would be its acceleration?

Q 14 (a) The speed of a train is doubled. What is the change in **(i)** its momentum and **(ii)** its kinetic energy?

(b) Suppose two trains are travelling at the same speed, but one has twice the mass of the other. What is the ratio of **(i)** their momenta and **(ii)** their kinetic energies?

In Question 14 we could go on with all the possible combinations of initial speed, mass of train and distance to station, and calculate the braking force required to stop in time, but it would take ages using a calculator. In Activity 2 you will let a computer take the strain.

⚙ **ACTIVITY 2 STOP THE TRAIN**

Set up spreadsheets to calculate the braking forces required to stop a *Eurostar* train within a specified distance or time. The initial speed could be anything up to 300 km h⁻¹. Analyse any patterns that you find by plotting graphs from your spreadsheets.

MATHS REFERENCE

Testing mathematical relationships

See Maths note 5.5

These same mathematical approaches can be used to analyse the motion of any object – as is the case in Activity 3, which uses data from the *Highway Code*.

In the *Highway Code* there is a table of stopping distances for cars travelling at different speeds (Figure 1.9). It is there to inform drivers of the correct distance to allow between themselves and the vehicle in front, in case they have to make an emergency stop. The figures include an allowance for a delay between recognising danger and applying the brakes – the thinking distance.

Typical Stopping Distances

Figure 1.9 Stopping distances from the *Highway Code*

⚙ **ACTIVITY 3 HIGHWAY CODE**

Use the information from Figure 1.9 and set up a spreadsheet or use your calculator to find out the values assumed for thinking time, acceleration and braking force for a typical car.

Data: 30 mph ≈ $13\,\mathrm{m\,s^{-1}}$

Mass of typical car ≈ 1 t

An uphill struggle

It is important that Channel Tunnel trains do not get stuck – especially not underground. The engine specifications for Channel Tunnel trains particularly mention that the two locomotives must be able to restart the train from rest with an acceleration of $0.13\,m\,s^{-2}$ up the steepest slope on the track. This section has a gradient of 1 in 90. If one locomotive fails the other must still be able to restart the train, albeit at a reduced speed.

Whatever the situation, the forces do not all act along the same direction. In order to analyse what is going on, you need to resolve force vectors into components, as this worked example shows.

STUDY NOTE

To review your knowledge of vectors and components, see TRA Additional Sheet 2.

WORKED EXAMPLE

A tram of mass $6.0 \times 10^4\,kg$ stands on a track with a 1 in 10 gradient.

(i) What force must be provided, parallel to the slope, to hold the tram stationary? (Use $g = 9.81\,N\,kg^{-1}$.)

(ii) How much extra force will then be required to give the tram its maximum acceleration of $1.2\,m\,s^{-2}$?

(iii) How long will this tram take to reach its maximum speed of 50 mph ($22\,m\,s^{-1}$)?

(iv) How far will the tram have travelled in that time?

Figure 1.10 shows the arrangement; note the gradient markings 1 and 10.

Figure 1.10 Forces acting on a tram on a slope

Resolving forces parallel to the slope:

(i) F (upwards) = $mg\sin\theta$ (downwards) for holding the position (equilibrium)

So holding force $F_h = 6.0 \times 10^4\,kg \times 9.81\,N\,kg^{-1} \times \dfrac{1}{10}$

$$= 5.9 \times 10^4\,N \text{ (to 2 significant figures)}$$

(ii) Using accelerating force $F_a = ma$ we get

$F_a = 6.0 \times 10^4\,kg \times 1.2\,m\,s^{-2}$

$\quad = 7.2 \times 10^4\,N$

(iii) Rearranging Equation 1 to make Δt the subject of the equation, we get the time value as:

$$\Delta t = \frac{\Delta(mv)}{F}$$

where $\Delta(mv) = m(v - u)$ and F = acceleration force F_a

So $\Delta t = \dfrac{6.0 \times 10^4\,kg \times (22 - 0)\,m\,s^{-1}}{7.2 \times 10^4\,N} = 18.5\,s$ (3 significant figures)

(iv) We can use Equation 3 adapted from $F\,\Delta s = \left(\dfrac{1}{2}\right)mv^2$ to get:

$$\Delta s = \frac{mv^2}{2F} = \frac{6.0 \times 10^4\,kg \times (22^2\,m\,s^{-1})^2}{2 \times 7.2 \times 10^4\,N} = 205\,m \text{ (3 significant figures)}$$

QUESTIONS

Q 15 **(a)** How much tractive force is required to give the specified acceleration of $0.13\,\mathrm{m\,s^{-2}}$ to a fully loaded *Shuttle* on the flat against a constant drag force of $40\,\mathrm{kN}$? (*Shuttle* mass is $2400\,\mathrm{t}$.)

(b) If the train stops on a slope of 1 in 90, how much forward tractive force does it need to exert to stay still; i.e. to stop rolling back? (Use $g = 9.81\,\mathrm{N\,kg^{-1}}$.)

(c) Combine your answers to determine the total force needed to accelerate the train from rest (with the required acceleration) on this slope.

(d) How much time does it take for the train in **(c)** to get up to a speed of $44\,\mathrm{m\,s^{-1}}$?

(e) **(i)** If a single locomotive has a tractive force of $400\,\mathrm{kN}$, what acceleration can it produce on the slope?
(ii) How long is it before the *Shuttle* speed reaches $44\,\mathrm{m\,s^{-1}}$?
(iii) How far has it travelled?

Q 16 A *Shuttle* loses all engine power at the lowest point in the Channel Tunnel when it is travelling at $44\,\mathrm{m\,s^{-1}}$ and has $25\,\mathrm{km}$ to travel along a 1 in 90 slope to get up and out.

(a) Assuming no friction, would the train be able to free-wheel to the surface?

(b) If the drag forces average $30\,\mathrm{kN}$, how far will the free-wheeling train travel?

Figure 1.11 Hauling a train uphill

⚙ ACTIVITY 4 SUMMING UP PART 2

Part 2 has involved mainly revision of earlier work from *Higher, Faster, Stronger*. Before you move on to later parts of this chapter, make sure you have reminded yourself of the work you did in *Higher, Faster, Stronger* and that you know the meaning of all the words printed in **bold** in Part 2.

3 Electrical systems on track

3.1 Trams and trains

The electrical systems that support tram and train motion are very similar and relatively simple. As both run on rails, they do not require steering controls. However, they do require controls for the smooth application of power, for braking and to control forward and reverse running. Even these circuits have limited gearing and are really not much more complicated than fast-operating switches dictating the power supplied to electric motors and magnetic brakes.

Figure 1.12 Electical features of a tram viewed from the front

Figure 1.12 shows the key electrical features of a tram. The system has the folllowing key features.

The sub-station

The electrical power comes from a local sub-station containing a transformer that steps the high 11 kV alternating (AC) voltage down to the lower level required, and a rectifier (based on a thyristor circuit – page 22) that converts alternating voltage to a direct 750 V (DC) supply.

The overhead line

The 750 V supply is fed underground to each of the insulated masts alongside the tracks. The cables then travel up through the mast (with an isolation switch near ground level) then out through the top and down into the overhead line, which is held under tension between the masts.

The pantograph

This provides the contact between the tram and the powered overhead line. It is pushed up against the line with compressed air and has a conducting top surface that rubs against the conducting overhead line. The internal tram wiring then conducts electricity down through the pantograph struts into the tram, where another isolation switch is available if required. Some of the electrical power is deployed in lighting, heating and door-control circuits, but the primary requirement is to power the electric motor circuits.

The electric motor

Each pair of wheels on an axle is powered by its own motor, comprising a rotating coil inside an electromagnet. To ensure the smooth acceleration and deceleration to keep passengers safe (whether seated or standing), the motor output passes into the chopper unit (also thyristor controlled), which can pulse the output in adjustable time intervals. It is this finely adjusted pulsed signal that is fed to the axles. When braking is required, various options are possible. Electric current can be diverted from the motors into electromagnets situated close to the tracks to provide eddy-current braking. Also, the still-rotating motor coil can act like a dynamo, converting kinetic energy into electrical energy by regenerative braking.

The wheel-rail contact

As well as providing the traction force against the ground to ensure motion, the conducting wheel touching the conducting track surface provides the electrical contact to complete the electrical circuit, as the track itself is earthed with contacts made near the mast bases.

Figure 1.13 shows a simplified cross-sectional image of an electric train for comparison. The overhead line and its contents are virtually identical to that of the tram, except that the larger distances involved make it sensible to retain high voltage alternating (AC) supply in the line. This then requires the transformer and rectifier circuits needed to reduce the AC voltage and then convert it to a direct (DC) supply to be situated inside the train itself. Other than this, the circuitry involved is very similar to the tram, and so details have been omitted in the diagram.

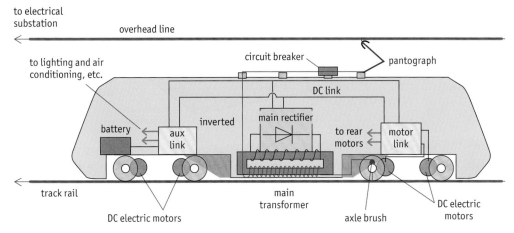

Figure 1.13 Electric train

STUDY NOTE

Our challenge now is to understand the physics principles behind the electrical components referred to above. To review your knowledge of DC electricity, see TRA Additional Sheets 1, 2 and 3.

3.2 Motors

Both *Eurostar* and trams depend on electrical motors and brakes to operate at high speeds. These motors and braking systems exploit electricity and magnetism to generate an electromagnetic force that drives motors and to recover some of the energy dissipated during braking.

STUDY NOTE

Brakes are the subject of Section 3.3.

> ⚙ **ACTIVITY 5 BRUSH UP ON MAGNETISM**
> Use some permanent magnets and a current-carrying coil to review your knowledge of the nature and behaviour of magnetic fields.

In an electric motor, the magnetic effect of a current is used to drive and maintain the rotation of the wheels. Activities 6 and 7 build up the basic principles behind the working of an electric motor.

ACTIVITY 6 AN ELECTROMAGNETIC FORCE

Use the arrangement shown in Figure 1.14 to explore the force on a current-carrying conductor when it is in a magnetic field. Use a top-pan balance to measure the size of the **electromagnetic force** between an electric current in a wire and a magnetic field (Figure 1.15). Vary the conditions to show how the size of the force varies with the length of wire, the strength of field and the size of the current.

Figure 1.14 Demonstrating the direction of an electromagnetic force

Figure 1.15 Measuring the size of an electromagnetic force

The results of the first part of Activity 6 are usually expressed in a rule known as **Fleming's left-hand rule**. The thumb and first two fingers of the left hand are held so that they are all at right angles to each other, like the corner of a box (see Figure 1.16). If the first finger points in the direction of the magnetic field and the second finger points in the direction of the current, then the thumb shows the direction of the motion and therefore of the force.

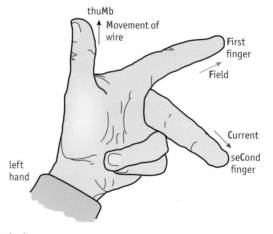

Figure 1.16 Fleming's left-hand rule

The size of the force

Several factors determine the size of the force: the strength of the magnet, the size of the current and the length of the wire. Experiments such as the second part of Activity 6 lead to the conclusion that the size of the force, F, on a wire is proportional to:

● the current, I, in the wire

● the length of the wire, l, that lies within the magnetic field

● the strength of the magnetic field.

Furthermore, F is a maximum when the current is at right angles to the magnetic field and zero when the current and magnetic field are parallel. The force still acts in a direction that is at right angles to both the current and the magnetic field, but its magnitude decreases as the angle between the magnetic field and current is reduced.

When the magnetic field and current are at right angles, the relationship could be written as:

$$F = kIlm$$

where k is a constant of proportionality and m represents the strength of the magnets. In practice, though, this relationship is used to define what we mean by the strength of the magnetic field and, rather than introducing an extra constant, we express it as:

$$F = BIl \tag{4}$$

where the symbol B represents an important quantity called the **magnetic flux density** (loosely speaking, the 'strength' of the field). The SI unit of magnetic flux density is the tesla, T (named after Nikola Tesla, 1856–1943, a Croatian–American physicist). $1\,T = 1\,N\,A^{-1}\,m^{-1}$. A field of $1\,T$ containing a wire of length $1\,m$ carrying $1\,A$ at right angles to the field will produce a force of $1\,N$ on the wire. How large is a tesla? – very large, as it happens. The magnetic field of the Earth is typically about $50\,\mu T$ and most laboratory solenoids produce no more than a few mT at their centres.

MATHS REFERENCE

Constant of proportionality

See Maths note 5.1

We use the term **magnetic field** to describe the three-dimensional region of space where the magnet has some influence. However, the term 'flux density' requires some explanation, as 'flux' is a word generally associated with flow – you met it in the chapter *Technology in Space*, where it was used to describe the 'flow' of energy associated with radiation. The same word is used to describe the flow of fluids. The formal language to describe electromagnetism was developed by Michael Faraday (1791–1867), and he adopted the term 'flux' because there are some similarities between the mathematical descriptions of magnetic field patterns and fluid flow patterns. If you choose to think of a magnetic field plotted out by field lines, then B represents the density of the lines – where the lines are closer together, the magnetic flux density is higher. Loosely speaking, the magnetic flux density is related to the number of lines crossing a given area.

Angles

To see how the size of the force varies with angle between field and current, we need to note that magnetic field and current are both vectors – they have direction as well as size and can be represented with arrows.

Figure 1.17 shows a three-dimensional space. Imagine a magnetic field acting in the negative z-direction. Now add an imaginary 'wire' running in the positive x-direction. If the current goes left to right, the wire will want to move up the page in the positive y-direction (as described by Fleming's left-hand rule). The length l of wire, carrying current I, takes up the full width of the magnetic field so the electromagnetic force will be:

$$F = BIl \qquad \text{(Equation 4)}$$

Figure 1.17 Wires carrying a current in a magnetic field: (a) 3D view (b) viewed from above. The arrows labelled B show the direction of the magnetic field.

STUDY NOTE

To review your knowledge of components, refer to the chapter *Higher, Faster, Stronger*.

Now consider the 'wire' rotating in the xz-plane. Only the component of the wire's length at right angles to the magnetic field, $L \sin \theta$, produces a force. The component $L \cos \theta$ is parallel to the field and ineffective. The full version of Equation 4 now becomes:

$$F = BIl \sin \theta \qquad \text{(4a)}$$

WORKED EXAMPLE

Using $F = BIL \sin \theta$, and taking $B = 1\,T$, $I = 1\,A$, and $L = 1\,m$, we get:

When $\theta = 90°$:

$F = BIL \sin 90° = 1\,T \times 1\,A \times 1\,m \times \sin 90° = 1.0\,N$

When $\theta = 60°$:

$F = BIL \sin 60° = 1\,T \times 1\,A \times 1\,m \times \sin 60° = 0.866\,N$

When $\theta = 30°$:

$F = BIL \sin 30° = 1\,T \times 1\,A \times 1\,m \times \sin 30° = 0.5\,N$

So, in rotating the wire from its 90° position to its 30° position, the force has halved in magnitude but still acts in the same positive y-direction, as described by Fleming's left-hand rule.

STUDY NOTE

The symbols for current (*I*) and length (*l*) can look very similar. To avoid confusion the symbols *i* (current) and/or *L* (length) are sometimes used so Equation 4 might be printed $F = Bil$, $F = BIL$ or $F = BiL$.

As we are dealing with a three-dimensional situation, we sometimes need a way to represent a vector at right-angles to the plane of the diagram. The symbol that represents the field (or any other vector) directed into the page is a small circle with a cross in its centre; think of an arrow travelling away from you – you see its feathers, which are represented by a cross. A field (or other vector) directed towards you out of the page is represented by a dot in a circle (the tip of the arrow). These symbols are used in Question 21.

QUESTIONS

 (a) Using the same method as described above, complete Table 1.1 using $B = 1\,T$, $I = 1\,A$, and $L = 1\,m$.

Angle θ	0°	10°	20°	30°	40°	45°	50°	60°	70°	80°	90°
$F = BIL \sin \theta$/N				0.50				0.87			1.00

Table 1.1 Table of values for Question 17

(b) Sketch a graph of F on the *y*-axis against angle θ on the *x*-axis.

(c) Sketch new graphs of force against angle for each of the following:
  (i) current direction is reversed, all other values remaining the same as originally
  (ii) magnetic field direction is reversed and the flux density halved, all other values reverting to original values
  (iii) both current and magnetic field directions are reversed and their values doubled compared with the original conditions.

Q 18 There is a current in each of the wires shown in Figure 1.18. State the direction of the force on the wire in each case.

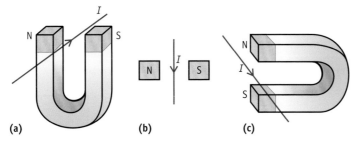

Figure 1.18 Diagram for Question 18

Q 19 The flux density between the poles of a strong electromagnet is 2 T. What is the force exerted on 10 mm of wire carrying 4 A when the wire is:

(a) at right angles to the field?

(b) parallel to the field?

(c) at an angle of 30° to the field?

Q 20 A horizontal wire of length 100 mm lies perpendicular to the field between the poles of a magnet. When a current of 5.0 A is passed through the wire, it is pushed vertically upwards. A rider (i.e. a small mass that rests on the wire) of mass 15 g placed on the wire just brings it back to its original position. Calculate the mean magnetic flux density between the poles of the magnet. ($g = 9.81\,\mathrm{N\,kg^{-1}}$)

Q 21 **(a)** Sketch the magnetic field patterns for the situations shown in Figure 1.19(a).

(b) Now consider the situation where the current-carrying conductor is placed between the magnetic poles as in Figure 1.19(b) so their magnetic fields combine.

Assume that, to a reasonable approximation, the two fields are the same size at each of points A, B, C and D. Think about the two fields adding as vectors. Draw arrows at A, B, C and D to represent the size and direction of the combined field.

(c) Explain how the size and direction of the combined magnetic field varies along the vertical line through points A and C.

(d) Taking the ideas from **(b)** and **(c)**, sketch the complete combined magnetic field.

(e) Going back to the previous work on electromagnetic force and Fleming's left-hand rule, identify the direction of the force. Can you spot a way of predicting this electromagnetic force direction from the shape of the combined magnetic field? (Hint: this situation is sometimes called the 'catapult effect'.)

(f) Finally, consider the two current-carrying conductors X and Y in Figure 1.19(c). If X and Y are rigidly fastened together and pivoted at the midpoint between them, what consequent motion do you predict?

(a) (i) the field between 2 magnetic poles

(ii) the field around a current-carrying conductor with current flowing vertically out of the page

(b)

(c)

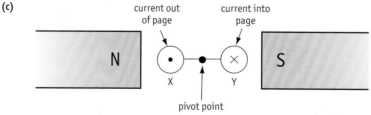

Figure 1.19 Diagrams for Question 21

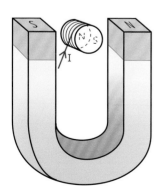

Figure 1.20 Current-carrying coil in a magnetic field

A force that just kicks a wire into or out of a horseshoe magnet is not sufficient to move a train from London to Brussels. The wheels have to keep turning for hours, driven by forces acting to produce a turning effect – a torque. Question 21, Figure 1.20 and Activity 7 show how a force on a wire can produce a torque that will maintain the rotation of an axle. This is the basis of a simple electric motor.

When a wire is wound into a coil and the current switched on, the two faces have magnetic poles. Freely suspend such a coil and place it at an angle to a magnetic field, and the pull of the attraction between opposite poles will turn the coil. That is the first stage of a continuously moving motor. Unfortunately, the turning stops once the coil faces its opposite poles. This problem goes away if the current then reverses, since the poles are then on the opposite faces of the coil and the coil has to move on.

In the following activity you will look at motors. The paper clip version maintains rotation by inertia – the coil turns past the place where it is facing the poles because it is moving rapidly and there is nothing to stop it.

 ACTIVITY 7 A SIMPLE ELECTRIC MOTOR

Construct a simple motor such as one of those shown in Figure 1.21. If you are working with other groups of students, compete to see whose motor can rotate for more than 30 s.

Motors have their own vocabulary: armature, split ring commutator, brushes, etc. Use this activity to make sure that you know and recognise this vocabulary.

As an optional extension, use the motor demonstration program available via www.pearsonhotlinks.co.uk.

PRACTICAL SKILLS REFERENCE

Carrying out practical work
See Practical skills notes 3.1–3.2

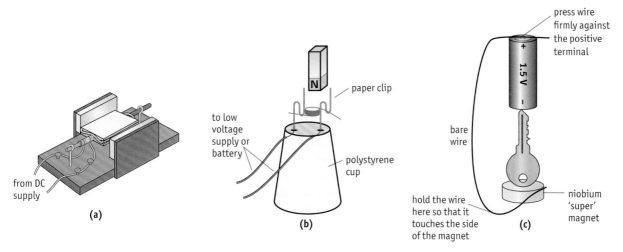

Figure 1.21 Simple electric motors: (a) Westminster model, (b) paper-clip model, (c) key model

QUESTION

Q 22 The *Eurostar* demands very large forces from its motors (around 400 kN). Assume the engineers have designed a modified motor of the sort you have met in Activity 7. Using Equation 4, make estimates for the necessary sizes of the quantities involved if the forces are to be anything approaching the right order of magnitude. (There is one extra quantity you need to include – think back to the construction of the coils.)

Real motors

In Activity 7 you made a very basic motor consisting of a single coil inside a field made of permanent magnets. You probably also found that you had to position the coil so its commutator contacts touched the brushes to start the motor, or maybe you flicked the coil to initiate the motor's rotation. This starting process is clearly not feasible for a train or tram motor.

Real motors consist of many coils orientated in a full 360° about the axis of rotation. This is known as an **armature**. Groups of coils have their own commutator so that whatever angle the armature finished at after the last rotation, there will be commutators in contact with brushes to start a new rotation (no flicking needed). Also, the magnetic field is not provided by permanent magnets but by electromagnet **field coils** powered by the same current used to supply the motor armature.

This real DC motor has low resistance field coils and armature circuit. So when voltage is applied across it, current will be large ($I = V/R$). High current in the field coils means magnetic flux density is large. Together with a large current in the armature this produces a large force, F ($F = BIl \sin \theta$), producing the large torque needed to start the train or tram. Disadvantages are that too large a current could overload and damage the wiring of the motor, or produce such a large torque that the wheels slip on the track rather than rolling. One solution is to put extra variable series resistance in the circuit. This can then be increased to limit current at start-up, and then decreased to zero once the motor stops accelerating. We will consider further uses of this series resistance when we look at braking in Section 3.3.

Using resistance to control DC motors wastes energy supplied by the overhead supply line as only a proportion of energy is used to accelerate the vehicle, the rest being lost to the atmosphere as heat. A more modern approach replaces some resistance with thyristors. A thyristor is a device that only allows current flow in one direction, and the flow only occurs after the thyristor has been 'gated' (meaning switched on at a chosen voltage). Gating is used to provide a pulsed voltage output with controllable time interval ΔT, as shown in Figure 1.22.

 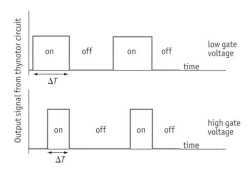

Figure 1.22 Voltage controlled by a thyristor

Variation of the time interval ΔT provides a fine adjustment of the average voltage applied to the motors. At start-up ΔT will be small, but these short time pulses will then be enlarged as speed increases. This segmenting or chopping of the supply voltage gives rise to the name 'chopper' for the circuitry involved that is found in most train and tram power systems (Figure 1.23).

3.3 Braking

Trams operate in a city environment, surrounded by pedestrians and other road users, where stopping quickly is as important as accelerating. We have seen the motor effect used to get vehicles accelerating, but it can also offer a means of stopping them, bearing in mind it takes just as much force to give a negative acceleration as a positive one. Activity 8 shows you how electromagnetism can be used to achieve braking using **eddy currents**, which are currents circulating within the body of a conductor.

Figure 1.23 Inside a motor

ACTIVITY 8 EDDY CURRENT BRAKING
Use the apparatus shown in Figure 1.24 to explore the strange effects of magnetic brakes.

Figure 1.24 Diagrams for Activity 8: (a) linear system, (b) disc system

In Activity 8 you saw examples of eddy-current braking. Trams have large coils fitted to the base of their carriages to position them just above the tram rails, known as track brakes (Figure 1.25). In sudden braking, current is diverted from the motor circuit into these coils, turning them into strong electromagnets moving just above the metal tracks. The resulting eddy currents in the tracks provide magnetic fields to generate forces that will oppose the motion of the track brake magnet attached to the tram. So here we have moving magnets above fixed conductors, rather than the moving conductors and fixed magnets in Activity 8, but the braking effect is just the same.

Figure 1.25 Track brakes

In a wire circuit the path of the electrons is clearly round the wire, but if the conducting material is a whole thick sheet of metal, the possible loops for a flow of electrons are not so clear. Indeed, in the changing situations we are talking about here, the eddies of current change path from one minute to the next (Figure 1.26), rather like the swirls of water (eddies) in a fast-moving river.

Each temporary current loop produces its own magnetic field and so gives rise to forces of attraction and repulsion between it and the external magnet. Yet how do the eddy currents arise in the first place? In order to produce a current, an emf is required. One way to provide an emf is to use a battery or a solar panel, but as you will see in Activity 9, an emf can also be produced using magnetism – this effect is called **electromagnetic induction**, and the emf and current thus produced are referred to as an **induced emf** and an **induced current**.

Figure 1.26 Eddy currents

ACTIVITY 9 ELECTROMAGNETIC INDUCTION

Use the apparatus shown in Figure 1.27 to show how a current can be induced in a wire using magnetism. Explore the factors that affect the size of the induced emf and current and then answer Question 23.

STUDY NOTE

You met the term 'emf' in the chapter *Technology in Space*.

Figure 1.27 Demonstrations of electromagnetic induction: (a) moving a wire in a magnetic field, (b) moving a magnet relative to a coil, (c) changing a magnetic field in a coil

PRACTICAL SKILLS REFERENCE

Carrying out practical work
See Practical skills notes 3.1–3.2

QUESTION

Q 23 Based on your observations from Activity 9:

    (a) What causes an induced emf?

    (b) What factors can be changed to make an induced emf bigger?

    (c) How can a positive induced emf be made negative?

We can now describe some of the properties of electromagnetic induction, i.e. when there is relative motion between a conductor and a magnetic field, an electromotive force is induced in the circuit. By relative motion we mean that the movement causes the conductor to 'cut across' the magnetic field. In each case there is an induced emf when there is a change in the magnetic field around the conductor. This might be caused by the conductor moving, the magnetic field moving or changing in size, or some combination of these. The size of the induced emf, $\mathscr{E}$, is proportional to:

- the strength of the magnetic field
- the rate at which the wire or coil is moved, or the rate at which the magnet is moved
- the number of wire loops or turns on the coil.

The emf changes direction if the magnet is reversed or the direction of movement is reversed.

In the third part of Activity 9 you saw that an emf can be induced in one solenoid when the current in another, nearby solenoid is changed. Again, the key point is that an emf is induced in the first solenoid because it is in a changing magnetic field – only now the change is made electrically rather than by moving a magnet or the coil.

Laws of electromagnetic induction

We can make the above description more formal using Faraday's way of describing magnetic fields. Think first of the situation when an emf is induced by moving a magnet into a single loop of wire (Figure 1.28) and then picture the magnet's field lines.

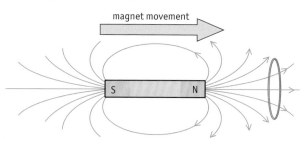

Figure 1.28 Moving a magnet into a loop of wire

When the magnet is some distance from the loop only a few lines thread through the loop, but as the magnet approaches the coil the field within the loop becomes stronger – more lines thread through the loop. Faraday expressed the 'amount of field' within a loop in terms of the **magnetic flux** enclosed by (or 'linking') the loop. Figure 1.29(a) shows a (flat) loop square-on to the field direction.

    magnetic flux = magnetic flux density × area of loop

Flux is usually given the symbol Φ, so in symbols:

$$\Phi = BA \tag{5}$$

where A is the area of the loop. If you think of B as the number of magnetic field lines per square metre, then the flux Φ is the total number of lines enclosed by the loop. The SI unit of magnetic flux is the weber, Wb. $1\,\text{Wb} = 1\,\text{T m}^2 = 1\,\text{N A}^{-1}\,\text{m}$. Magnetic flux density is sometimes expressed in units of Wb m^{-2} rather than tesla – yet the two are exactly equivalent.

If the normal to the loop makes an angle of θ to the field, as in Figure 1.29(b), then Equation 5 becomes:

$$\Phi = BA \cos\theta \tag{5a}$$

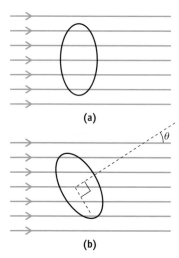

Figure 1.29 Single loop of wire in a magnetic field (a) square-on (b) at an angle

STUDY NOTE

The normal is a line at 90° to the plane of the loop. You met the term in the chapter *Good Enough to Eat* in the context of reflection and refraction at a surface.

QUESTIONS

Q 24 Express the derived SI units Wb and T in terms of SI base units.

Q 25 **(a)** If the coil in Figure 1.29 has radius 0.15 m and lies in a field of magnetic flux density 5.0 mT, what will be the magnetic flux linking the loop when $\theta = 0°$, 30°, 60° and 90°?

(b) Sketch a graph of how magnetic flux Φ varies with angle θ for a full 360° rotation using the results from part **(a)**.

The size of the induced emf

Faraday noted that the size of the emf, $\mathcal{E}$, induced in a loop is proportional to the rate at which the flux within the loop is changed; this relationship is known as **Faraday's law of electromagnetic induction** and is written mathematically as:

$$\mathcal{E} \propto \frac{\Delta\Phi}{\Delta t} \qquad (6)$$

If a coil contains N turns (N loops) of wire, then an emf is induced in each loop as the loops are connected in series. The net emf from the whole coil is then multiplied by N, and Equation 6 becomes:

$$\mathcal{E} \propto \frac{\Delta(N\Phi)}{\Delta t} \qquad (6a)$$

$N\Phi$ is called the **magnetic flux linkage**. For a coil with more than one loop, Faraday's law states that the size of the induced emf is proportional to the rate of change of flux linkage. Flux linkage is sometimes given the unit weber-turns to distinguish it from the flux through a single loop – but N is just a number so 1 weber-turn has the same dimensions as 1 weber.

In SI units, i.e. with Φ in webers, t in seconds and $\mathcal{E}$ in volts, the size of the emf is *equal* to the rate of change of flux linkage – not merely directly proportional. However, there is one important point that we have not yet included, and that is the direction of the induced emf (and hence of the induced current). Here, an energy argument is useful.

> **MATHS REFERENCE**
>
> Dimensions
>
> See Maths note 2.5

The direction of the induced emf

People have long searched for a perpetual motion machine, i.e. one that needs no input of energy to keep on turning for all eternity. It is a pipe dream, of course. At first sight the demonstration of induction in Activity 8 appears to offer untold riches. For example, in a bicycle dynamo or generator, movement of a coil inside a magnet produces an emf that drives a current through the coil, which – from Activity 6 – you know will make it rotate inside the magnet. You might think as the coil moves faster it will generate a larger emf, which in turn will drive a larger current making the coil rotate faster still, and so on. A small initial movement would produce runaway motion, apparently generating kinetic energy out of nothing. So what is the catch? It is that emf is, in practice, always induced in a direction such that any forces will oppose the motion that produces it. As illustrated in Question 26, a magnet moving into a wire is slowed down by repelling action, and a coil rotating clockwise is opposed by a torque that is trying to rotate it anti-clockwise.

QUESTION

Q 26 When a magnet moves in or out of a coil, an emf is induced across the coil and currents flow in the coil so it becomes an induced electromagnet in one of two polarities (labelled A and B in Figure 1.30).

Figure 1.30 Polarities of an electromagnet

According to the ideas expressed above that emfs and currents are induced to produce forces opposing motion, which of the two options (A or B) will apply in each of the situations shown in Figure 1.31?

Figure 1.31 Diagrams for Question 26

Lenz's law sums up the situation: 'The direction of the current induced in a conductor by moving it relative to a magnetic field is such that its own field opposes the motion.' Mathematically, Lenz's law is expressed by inserting a minus sign into Equation 6:

$$\mathcal{E} = -\frac{\Delta(N\varPhi)}{\Delta t} \tag{6b}$$

thus producing an expression that encapsulates both Faraday's and Lenz's laws. In calculus notation:

$$\mathcal{E} = -\frac{d(N\varPhi)}{dt} \tag{6c}$$

The change in flux linkage can be brought about by a change in any or all of N, B or A. If only one of these is changed, then we can rewrite Equation 6 in yet more ways. Changing B only:

$$\mathcal{E} = -NA\frac{dB}{dt} \tag{6d}$$

Changing A only:

$$\mathcal{E} = -NB\frac{dA}{dt} \tag{6e}$$

Changing N only:

$$\mathcal{E} = -\varPhi\frac{dN}{dt} \tag{6f}$$

QUESTIONS

Q 27 Show that the SI units on the left-hand side of Equation 6 (any version) are the same as those on the right.

Q 28 A magnetic flux of 10 Wb passes through a coil of 50 turns. The flux decreased to zero over 2.0 s. What would be the size of the emf induced in the coil?

Q 29 A 10-turn coil has a cross-sectional area of 0.03 m². It is placed so that the plane of the coil is perpendicular to a field of magnetic flux density 2 T. It is then rotated through 90° in 0.2 s so that it lies parallel to the field. What size emf would be induced in the coil?

Q 30 If large currents are turned on or off near your radio you may hear a crack or a click. Suggest an explanation for this.

Another look at electromagnetic induction

We started by considering a single wire loop that can be extended to make a coil with many turns, but Equation 6 also applies to all other ways of inducing an emf. In Figure 1.27(c), changing the current on one solenoid changes the magnetic flux linking the other solenoid; in Figure 1.27(a), the movement of the wire between the poles of magnet changes the flux linking the circuit made by the wire and the meter. It is sometimes useful, too, to consider an isolated section of wire, as shown in Figure 1.32, moving in a magnetic field. The wire is straight, has a length l, and moves at a velocity v in a direction that is at right angles to its length and to the direction of a uniform field B.

Imagine that the wire is connected in a circuit. In a time Δt, the wire moves through a distance $v\Delta t$ and so the area of the circuit changes by an amount:

$$\Delta A = lv\Delta t$$

The motion will thus give rise to an emf which, from Equation 6e with $N = 1$, is given by:

$$\mathcal{E} = -\frac{Blv\Delta t}{\Delta t} = -Blv \tag{7}$$

The emf could be detected as a potential difference by a voltmeter connected between the ends of the wire. If the wire is connected into a complete circuit, the induced emf will give rise to an induced current whose size will depend on the emf and on the resistance of the circuit.

Inspection of Figure 1.30 and consideration of Lenz's law provides us with another useful way of predicting the direction of the induced emf and current. We know that the electromagnetic force due to the induced current must oppose the motion, i.e. it must act to the left of Figure 1.30. Using Fleming's left-hand rule with the first finger pointing into the page and the thumb pointing to the left shows that the current must be flowing towards the top of the page.

The direction of the induced current can be deduced directly using **Fleming's right-hand rule** (see Figure 1.33). Now the thumb points in the direction of the motion giving rise to the induced current, and the first and second fingers have the same meaning as in Fleming's left-hand rule. But it is easy to confuse left and right, and forget which hand to use in which situation, so think of rIght hand for Induced current.'

Uniform magnetic field directed into page

Figure 1.32 Moving an isolated wire in a magnetic field

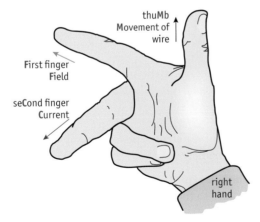

Figure 1.33 Fleming's right-hand rule

QUESTIONS

Q 31 Imagine an aeroplane as two metal conductors (the fuselage and wing tip to wing tip) connected together in the shape of a cross. What are the electromagnetic consequences as it flies towards the Earth's magnetic north pole?

Q 32 At a certain location, the Earth's magnetic flux density has a vertical component of 4×10^{-5} T. The metal body of a train is 4 m wide and 5 m high, and the train is moving to the right at $60 \, \text{m s}^{-1}$ (see Figure 1.34).

(a) State the direction of any induced emf and current resulting from moving across the vertical component of the Earth's magnetic field.

(b) Calculate the size of the induced emf that is likely to develop.

Figure 1.34 Moving train in Question 32

Back to brakes

We are now in a position to explain eddy-current braking such as you saw in Activity 8, when a metal disc rotated in a magnetic field. The motion of the metal in the field induces emfs in the disc that vary according to the local flux density and the speed of the motion. The net effect is to drive eddy currents in the disc. The currents, plus the external magnetic field, give rise to electromagnetic forces that, by Lenz's law, must oppose the motion, i.e. they retard the disc.

QUESTIONS

Q 33 (a) A bicycle brake operates by a block of rubber being pressed against the rim of the wheel. It is a contact brake. Is an eddy-current brake a contact brake?

(b) Is an eddy-current brake any use as a parking brake? Explain.

Q 34 In designing an eddy-current brake, how would you design the disc so as to ensure that the induced currents were large?

Q 35 From a given speed, any braking system must allow strong or gentle braking. Suggest *one* way to get this kind of control from eddy-current braking.

Types of braking

Question 35 illustrates the need for different types of braking for different situations. Most trains and trams have up to five types, namely:

● *Air braking*: The most common type of brake, which uses compressed air from a motor-driven compressor to apply brake pads to the wheels, slowing them by friction.

● *Regenerative braking*: When the brake switch is selected the traction motors are made to act as dynamos and the induced current produced is fed back into the electrical supply line and used to power other vehicles. So a tram going downhill and braking converts kinetic energy to electricity, which can then help a tram going uphill gain gravitational potential energy. However, this regenerated energy cannot always be absorbed by the electrical supply, so the logical step is to dispose of generated current within the tram/train using:

● *Rheostatic braking*: Here the power circuit connections are switched so regenerated current is fed into banks of resistors that disperse the electricity as heat (similar to electric fire elements). In trams these resistors are mounted on the roofs of carriages, enabling air cooling. The resistance is variable to control deceleration, making it gradual for passenger safety.

- *Track braking*: We have already seen that in an emergency current can be switched into special large coils hung beneath a tram directly over the rails. The moving electromagnets this creates induce eddy currents in the rails, which create braking forces.

- Finally, there is a *parking brake* used to hold an unattended vehicle in place when the electrical system is switched off and other braking systems are shut down. It acts like a car handbrake.

Regenerative braking uses a dynamo to 'recover lost energy' by generating a current while braking. Figure 1.35 shows a circuit in which a motor, when connected to a power supply, turns a flywheel. If no power is connected to it, the motor runs as a dynamo as long as it is rotated by something. Activity 10 demonstrates this.

> ### ⚙ ACTIVITY 10 ELECTRICAL BRAKING
>
> Use the apparatus shown in Figure 1.35 to demonstrate regenerative braking. First, allow the flywheel to reach a steady speed, then switch off the power and allow it to freewheel to a stop. Repeat, but this time connect the circuit through the light bulb when the power is switched off. Note any differences and explain them in terms of energy transfer.

PRACTICAL SKILLS REFERENCE

Carrying out practical work

See Practical skills notes 3.1–3.2

to battery

Figure 1.35 Circuit for demonstrating regenerative braking

Points of contact

The electric railway or tramway expects to have a safe, economical power supply that can be accessed immediately. To achieve this, the overhead wires need to be able to carry currents of thousands of amps and remain in line with the route whatever the weather conditions (heat, cold, winds, etc.). Lines have to be held in tension (tension forces of over 2 MN are likely for wire lengths of about 1 km) both in the straight sections and round curves. Pantographs on top of the train or tram collect the electricity from the overhead line. Contact is maintained by compressed air pushing the pantograph upwards. 'Horns' at the end of the upper contact surface reduce the risk of the pantograph hooking itself above the overhead line, which would probably pull the line down (obviously an exceedingly dangerous event).

The running rails are held at earth potential to provide the return path for electricity from the vehicle to the substation. Good electrical contact is needed between the wheels and the track to ensure this current flow. Problems can occur, such as leaves on the track, whose waxy surface coatings are good insulators. The return electrical path also relies on the rails being better conductors than the neighbouring ground so current does not leave the chosen path as if a parallel resistor had been added. A wet and salty environment alongside the track is particularly likely to cause this to happen, as well as corroding the rails. So keeping the Channel Tunnel dry is crucial.

3.4 Summing up Part 3

This has been a long part of the chapter. You have met two important areas of the physics of electricity and magnetism: you saw how a current-carrying wire can experience an electromagnetic force in a magnetic field, and how a changing magnetic field can give rise to electromagnetic induction.

Activities 11 and 12 are designed to take you back through some of the ideas you met earlier. At the same time they give you an opportunity to extend your knowledge of electromagnetism.

PRACTICAL SKILLS REFERENCE

Carrying out practical work

See Practical skills notes 3.1–3.2

ACTIVITY 11 MORE ELECTROMAGNETISM

Explore some more examples of electromagnetism and electromagnetic induction. In each case, describe and explain your observations using *at least three* of the terms printed in bold in Part 3 of this chapter.

ACTIVITY 12 INDUCTION MOTOR

Construct and investigate an induction motor. Compare its operation with that of the motor that you used in Activity 7.

QUESTIONS

Q 36 Imagine a situation where engineers were ignorant of the effects of electromagnetic induction and the Channel Tunnel was lined with hoops of metal connected together. A *Eurostar* motor, as you have seen, has magnets as a vital element in its motor construction and is going to rush towards the tunnel at $44\,\text{m s}^{-1}$. Why would this cause problems?

Q 37 An aeroplane is travelling horizontally with a velocity of $268\,\text{m s}^{-1}$ in a region where the vertical component of the Earth's field is $4.1 \times 10^{-5}\,\text{T}$ and the horizontal component is $1.8 \times 10^{-5}\,\text{T}$. If the wing-span is $47\,\text{m}$, find the size of the emf induced between the wing tips.

Q 38 A rectangular coil with 100 turns, $5.0\,\text{cm}$ by $8.0\,\text{cm}$, hangs in a vertical plane with a north-west to south-east orientation. A horizontal magnetic field of size $3.5\,\mu\text{T}$ is aligned due north.

(a) What flux passes through the coil's area?

(b) What is the flux linkage through the coil?

(c) If the magnetic field through the coil collapses to zero in $0.10\,\text{ms}$, find the size of the emf induced in the coil.

4 Sensing speed

Drivers of trains and trams rely on information from instruments and signals telling them about their own vehicle's performance and about conditions on the track ahead. The problem is how to get information to a fast-moving object reliably under all conditions. The key elements in the solution are the design of sensors and the electronics to decipher the outputs.

The driver of a train has access to computer-generated information about the journey. This is shown on a screen and includes a three-digit number, which is either a maximum allowable speed (if all is clear) or a target speed for the end of the block section (if slowing down is required). If a signal is flashing, it means that there will be a new signal at the beginning of the next block section, thus giving the driver some forewarning of a required change. For each block a new target speed is set. A graph modelling the projected journey ahead is shown in Figure 1.36 above the sequence of signals. This shows a *Shuttle* train that needs to be brought to rest from $140\,\mathrm{km\,h^{-1}}$.

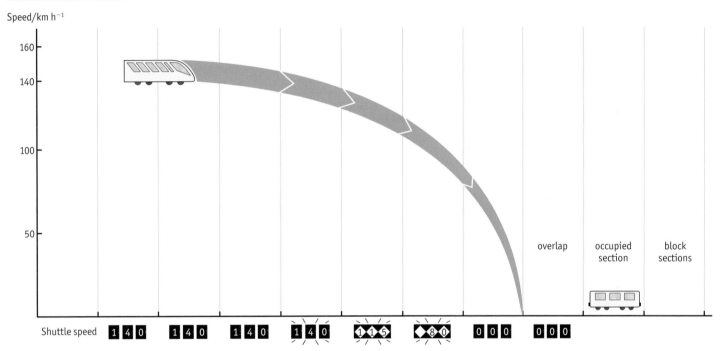

Figure 1.36 *Slowing down Le Shuttle*

The most important task of a driver is to control the train's speed; for example, in anticipating a sharp bend, the approach to a station or the speed limits of the Channel Tunnel. Speed is continuously monitored directly on the train, and if the driver exceeds a maximum speed or ignores a target speed, automatic braking comes into effect operated by the on-board computer. Measuring speed is the subject of this part of the chapter.

4.1 Inductive speed sensors

You will probably have had some experience at GCSE using light gates to measure the linear speed of vehicles. Activity 13 introduces a similar method using magnetism.

⚙ ACTIVITY 13 SENSING TRAIN SPEED

Use a system of magnets and coils, such as the one shown in Figure 1.37, to show how electromagnetic induction can be used to sense the speed and acceleration of a model train. Figure 1.38 shows some typical results for you to analyse if you are unable to obtain your own.

Figure 1.37 Arrangement for sensing the speed of a model train

18Sep2000 10:40

Figure 1.38 Output from coils in Activity 13 displayed on Pico Technology Picoscope

The arrangement in Activity 13 illustrates some basic principles of the inductive speed sensors used on trains, although a practical speed sensor must have all its parts on the train.

4.2 Timing

In measuring speed electronically, one very important requirement is the ability to generate a voltage pulse that is exactly one second long. To do this, we can use devices called **capacitors**, which can store electric charge and release it at a precisely known and predictable rate. Such devices are the subjects of this section.

Introducing capacitors

Capacitors are components that can store electrical charge. The simplest capacitor design consists of two parallel metal plates separated by an air gap. For this reason, capacitors are symbolised as two parallel lines, one of which may be shown as an open plate or with a positive sign to indicate which terminal of the capacitor should be connected nearest to the positive terminal of a power supply (see Figure 1.39).

Connect this side to positive supply

Figure 1.39 Capacitor symbols

Figure 1.40 Diagram for Activity 14

⚙ ACTIVITY 14 CAPACITORS

Explore the way the current in a circuit changes as various capacitors are charged from a battery and then discharged using the circuit shown in Figure 1.40.

For a circuit as shown in Figure 1.41, when the switch is closed, charge flows on to one side of the capacitor from the battery; an equal amount flows away from the other side of the capacitor into the other battery terminal. Charge flows until the potential difference across the plates becomes the same as that across the battery. The capacity then has a charge $+Q$ on one side and $-Q$ on the other. In a sense it 'stores' a charge Q – though 'separates' would be a better word – and it is actually more accurate to say that a capacitor stores energy by separating charge. The charge that a capacitor can store is directly proportional to the battery voltage:

$$Q \propto V$$

The value of $\dfrac{Q}{V}$ is constant, and is defined to be the **capacitance**, C, of the capacitor:

$$Q = CV \qquad (8)$$

Capacitance is the amount of charge stored per unit potential difference across the plates. The SI unit of capacitance is the farad, F (named after Michael Faraday).

$1\,\text{F} = 1\,\text{C\,V}^{-1}$. One farad is a very large capacitance. Capacitors used in electronic circuits generally range between 1 picofarad (pF) and 10 000 microfarads (µF).

If the potential difference across the capacitor plates is too high, the insulation (called dielectric) between the plates may start to conduct. For this reason capacitors are marked with a maximum working voltage above which the capacitor should not be operated.

Charging and discharging

In order to see how a capacitor can be used in a timing circuit, it is important to know what happens to voltage and current in a capacitor–resistor circuit as the capacitor accumulates charge and releases it. In Activity 15 you will investigate charging and discharging experimentally using a large-value capacitor and resistor, which will slow down the changes in the circuit so that they can be easily observed.

> **ACTIVITY 15 SLOW CHARGE AND DISCHARGE**
> Observe the way voltages and currents change as a capacitor charges and discharges.

> **ACTIVITY 16 MODELLING CHARGE AND DISCHARGE**
> Use *Yenka* or other modelling software to investigate more closely how the charging and discharging of a 100 µF capacitor through a 100 kΩ resistor affects the current and voltage in the circuit, and to see how capacitance and resistance values determine the time of charging and discharging.

In Activities 15 and 16 you have seen graphs representing the charge and discharge of a capacitor. They belong to a family of graphs known as **exponentials**. Many naturally occurring changes are exponential, as you will see when you study the chapters *The Medium is the Message* and *Reach for the Stars*. One important characteristic of an exponential discharge (or decay) graph is that *the changing quantity changes by equal fractions in equal times*. You should have seen this when you found the times for the voltage to halve, halve and halve again.

We will look at various ways in which the shape of a capacitor discharge graph can be described mathematically; they are all equivalent to one another.

Figure 1.42 shows a capacitor, C, discharging through a resistor, R. The capacitor behaves rather like a cell with a changing terminal potential difference. This pd, V, is related to the charge, Q, stored on the capacitor:

$$V = \frac{Q}{C} \qquad \text{(from Equation 8)}$$

As the capacitor discharges, Q decreases and hence V also decreases. Since we also know that the current, I, in the resistor is related to the pd, we can write:

$$I = \frac{V}{R} = \frac{Q}{RC} \qquad (9)$$

Figure 1.41 Capacitor circuit

STUDY NOTE

You will learn about capacitors as energy storage devices in the chapter *The Medium is the Message*.

MATHS REFERENCE

SI prefixes

See Maths note 2.4

PRACTICAL SKILLS REFERENCE

Carrying out practical work

See Practical skills notes 3.1–3.2

Figure 1.42 Capacitor discharging through a resistor

Since Q is decreasing, I must also be decreasing. Remembering that current is the rate of flow of charge:

$$I = \frac{\Delta Q}{\Delta t}$$

We can write:

$$\frac{\Delta Q}{\Delta t} = -\frac{Q}{RC} \tag{10}$$

The negative sign indicates that the charge is decreasing with time.

If we are dealing with very small time intervals, we can use calculus notation:

$$\frac{dQ}{dt} = -\frac{Q}{RC} \tag{10a}$$

Equation 10 says that *the rate of flow of charge is proportional to the charge itself.* This is another important characteristic of exponential discharge. Equation 10 is actually equivalent to the 'equal fractions in equal times' pattern, as you can see by rearranging it:

$$\frac{\Delta Q}{Q} = -\frac{\Delta t}{RC} \tag{10b}$$

This version tells us that the fraction $\frac{\Delta Q}{Q}$ is directly proportional to the time interval Δt.

MATHS REFERENCE

Exponential changes
See Maths note 9.1
Exponential functions
See Maths note 9.2

Equation 10 describes a capacitor discharge curve by telling us how the gradient at any one time is related to the stored charge at that time. However, it does not directly tell us how much charge is stored at any one time or answer the most commonly asked questions such as: 'How much charge is left after 0.5 s?' For this, we need another equation, which can be derived from Equation 10 using calculus:

$$Q = Q_0 e^{-t/RC} \tag{11}$$

Here Q_0 is the charge stored when $t = 0$, and e is a number that arises from the maths: $e \approx 2.718$.

We can use Equation 11 to tell us how the voltage and current change as well. Substituting $Q = CV$ (from Equation 8) and cancelling C, we get:

$$V = V_0 e^{-t/RC} \tag{12}$$

STUDY NOTE

If you are familiar with calculus, you can show that the 'equal fractions in equal times' pattern will always be described by an expression such as Equation 11. See TRA Additional Sheet 6.

Then similarly, using $I = \frac{V}{R}$ and cancelling R, we get:

$$I = I_0 e^{-t/RC} \tag{13}$$

In other words, the voltage and current also decrease exponentially.

WORKED EXAMPLE

Q A 100 μF capacitor is discharged through a 250 kΩ resistor for 20 s. What fraction of its original charge will remain stored at this time?

A From Equation 11:

fraction of charge left $= \dfrac{Q}{Q_0} = e^{-t/RC}$

$RC = 250 \times 10^3\,\Omega \times 100 \times 10^{-6}\,F = 25\,\Omega F$

$\dfrac{t}{RC} = \dfrac{20}{25} = 0.80$

$e^{-t/RC} = 0.45$ (to 2 significant figures)

QUESTIONS

Q 39 Initially the pd across the plates in the worked example above was 10 V. How much charge remained stored after 20 s?

Q 40 **(a)** By inspecting Equation 10, argue convincingly that RC must have dimensions of time (and hence SI units of seconds).

(b) A capacitor, C, discharges through a resistor, R. What fraction of the starting voltage (or charge) will remain across the capacitor after a time equal to RC?

If you answered Question 40 correctly, you should have seen that approximately 0.37 (or 37%) of the initial voltage or charge will remain on a capacitor after a period equal to RC. The time RC is called the **time constant** of the circuit and is used as a measure of how fast the resistor–capacitor combination discharges. It is sometimes represented by the Greek letter τ 'tau':

$$\tau = RC \tag{14}$$

As the charge and voltage drop by 63% in each interval of τ, we can argue that they never actually reach zero. In practice, provided the time interval is several times τ, we can say the capacitor is fully discharged. Question 40 and Activity 17 illustrate this.

⚙ ACTIVITY 17 EXPONENTIAL DISCHARGE

Analyse the discharge curve shown in Figure 1.43 and verify that it is exponential. Determine the time constant for the discharge. Use a spreadsheet to examine the discharge of a capacitor over a succession of small time intervals.

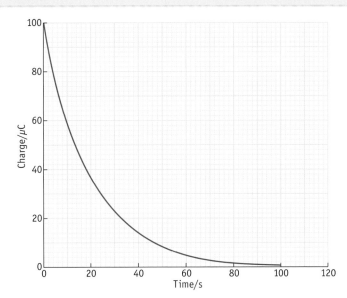

Figure 1.43 Discharge curve for Activity 17

More on charging and discharging

You will have seen in Activities 15 and 16 that when a capacitor *charges* through a resistor (Figure 1.44), the curves for Q and V (Figure 1.45) are the same shape as the discharge curves, but upside down. The charge and voltage approach their *final* values Q_0 and V_0 exponentially:

$$Q_0 - Q = Q_0\, e^{-t/RC} \tag{15}$$

This can be rearranged to give:

$$Q = Q_0 \left(1 - e^{-t/RC}\right) \tag{15a}$$

Similarly:

$$V = V_0 \left(1 - e^{-t/RC}\right) \tag{16}$$

Figure 1.44 Circuit for charging a capacitor through a resistor

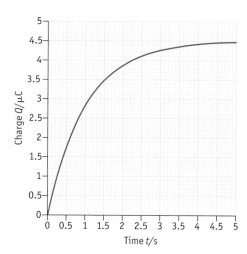

Figure 1.45 Charging curve for a capacitor

Equation 15 and 16 show that when $t = RC$:

$$Q = Q_0\left(1 - \frac{1}{e}\right)$$

$$\approx Q_0 \times (1 - 0.37)$$

$$\approx 0.63\, Q_0$$

Similarly, using Equation 16:

$$V \approx 0.63\, V_0$$

As current is the rate of flow of charge, the size of the current, I, is given by the gradient of the graph of Q against time. The current is large initally and deceases exponentially to zero, just as it does when the capacitor discharges:

$$I = I_0\, e^{-t/RC} \tag{Equation 13}$$

QUESTION

Q 41   By considering the potential differences across the capacitor and the power supply when a capacitor is charged through a resistor, explain why the current is initially large but decreases exponentially to zero.

⚙ ACTIVITY 18   CAPACITOR CHARGE AND DISCHARGE

Display and analyse the potential difference (pd) across a capacitor as it charges (and discharges) through a resistor. Use this to determine the time constant of the circuit.

Equations 11, 12 and 13 enable us to calculate the charge, current and voltage as a capacitor discharges. Sometimes we need to do the reverse calculation and find how long it takes for the charge, current or voltage to reach a given value, or to change by a given amount. In other words, we need to make time, t, the subject of the equations. To do this, we need to use the inverse of the exponential function, which is the **natural log**, ln, and we need to manipulate the logs of numbers that are multiplied or divided by one another.

Taking the natural logs of both sides of Equation 11:

$$\ln Q = \ln Q_0 - \frac{t}{RC} \tag{17}$$

PRACTICAL SKILLS REFERENCE

Scientific questions and information research

See Practical skills notes 1.1–1.4

Planning and experimental design

See Practical skills notes 2.1–2.2

Carrying out practical work

See Practical skills notes 3.1–3.2

Analysis and interpretation of data

See Practical skills notes 4.1–4.2

Conclusion and evaluation

See Practical skills notes 5.1–5.2

MATHS REFERENCE

Logs: multiplication and division

See Maths note 8.3

Natural logs

See Maths note 8.5

Exponentials and logs

See Maths note 9.3

Equation 17 can be rearranged to make t the subject:

$$\frac{t}{RC} = \ln Q_0 - \ln Q \tag{17a}$$

$$= \ln\left(\frac{Q_0}{Q}\right)$$

$$t = RC \ln\left(\frac{Q_0}{Q}\right) \tag{17b}$$

In exactly the same way, we can produce equations involving charge and current by taking natural logs of Equations 12 and 13.

$$\ln V = \ln V_0 - \frac{t}{RC} \tag{18}$$

$$\ln I = \ln I_0 - \frac{t}{RC} \tag{19}$$

WORKED EXAMPLE

Q A 10 µF capacitor is charged by connecting it to a pd of 4.5 V. It then discharges through a 500-kΩ resistor. How long does it take for the pd across the capacitor to fall to 4.0 V?

A Substituting values into Equation 17b:

$$t = 500 \times 10^3\,\Omega \times 10 \times 10^{-6}\,F \ln\left(\frac{4.5}{4.0}\right)$$

$$= 0.59\,s \text{ (to 2 significant figures)}$$

QUESTIONS

Q 42 A capacitor, C, discharges through a resistor, R. In terms of the time constant τ, how long does it take for the charge, voltage and current to reach a fraction, f, of their initial values?

Q 43 (a) What percentage of a capacitor's initial charge will remain after three time constants?

(b) A student claims that a discharging capacitor is 'effectively fully discharged' after five time constants. Do a calculation and comment on this viewpoint.

Q 44 Figure 1.46 shows a timing circuit. When the push switch is pressed, the capacitor charges almost instantly. When the switch is released, it discharges more slowly through the resistor. As it does so, the voltage at the output falls and pd across the capacitor falls.

(a) What is the voltage at the output after the push switch is pressed?

(b) What is the time constant for the resistor–capacitor combination?

(c) What will be the output voltage of the discharge circuit after 5 s?

(d) How long does it take for the voltage of the discharge circuit to reach 0.6 V?

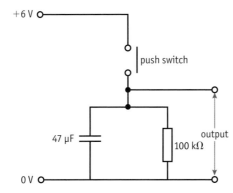

Figure 1.46 Timing circuit for Question 44

Figure 1.47 An electronic timer

An electronic timer

In Question 44 you saw how a simple capacitor circuit might be used to produce a set time delay.

Through the use of a small, electronic integrated circuit (chip) we can produce a timing device that uses capacitor charge and discharge. The 555 chip (Figure 1.47) can be used to create a voltage pulse of any duration – and one second is as good as any.

Near the start of this part of the chapter, we considered a method of counting pulses to determine the speed of a train and said that it was necessary to create a voltage pulse exactly one second long. Now that we have the means to do this accurately, we are in a position to go back to our model with greater understanding.

4.3 Summing up Part 4

In this part of the chapter you have learned about capacitors and about some properties of exponential graphs. You have also seen how the discharge of a capacitor can be used to control a timing circuit.

> ⚙ **ACTIVITY 19 SUMMING UP PART 4**
>
> Check through your work from this part of the chapter and make sure you know the meaning of all the terms highlighted in bold.

QUESTION

 A model-rail enthusiast wants to create a block signalling system that gives a red light when a train is occupying one section of track and a green light 10 s after the train has left the block. The circuit is shown in Figure 1.48.

Figure 1.48 Model railway circuit for Question 45

X and Y are connected to each of the metal tracks in the 'block' to be monitored.

The NOT gate turns an analogue (continuously varying) input voltage into a digital signal. If the input voltage to the NOT gate is above 1.5 V the output is low (close to 0 V), and if the input is below 1.5 V the output is high (around 6 V).

The LEDs (light-emitting diodes) emit light only when the potential decreases in the direction of the arrow that forms part of their symbol.

(a) Explain how the circuit works.

(b) Show that, for a time delay of 10 s, $RC \approx 7.2$ s.

(c) Suppose the circuit was made with $R = 1000 \, \text{k}\Omega$ and $C = 10 \, \mu\text{F}$. Comment on whether this would produce a time delay that was too short or too long.

5 Structure and safety

5.1 How safe is the Channel Tunnel?

Train passengers want to get to their destination on time and in safety. As trains move ever faster and travel in more extreme conditions the potential for danger increases, but so does the skill of the engineers applying physics in the modern world.

Despite what some people think, flooding is not a major concern as the amount of sea water actually seeping into the tunnel is very slight; some of it evaporates and pumps remove the rest. Fire is the biggest worry, and there is a complex system of fire detection, suppression and smoke control in place. In the case of a fire, the passengers would be sealed off and would breathe recycled air until the fire was under control.

General wear and tear is also a concern. High speeds and heavy loads deform and move the track under a train. Tiny kinks and dips develop and become exaggerated with time – like potholes in roads. Eventually they can lead to the train swaying about (which disturbs the passengers) and, if left for a very long time, they could cause a derailment. *Eurostar* reduces the problem of wear by being lighter than traditional trains. Axle loads are limited to 17 tonnes and the line is maintained every single night so that the high-speed service can continue.

The lighter weight may be ideal for track maintenance, but seems to offer poor protection in the event of hitting something or being hit. In fact, you have already seen one advantage of reducing the mass of a train in Part 3: it makes it easier to stop a train before it collides with anything dangerous, as less force is required.

The manufacturers use materials that may be light but are also strong and so resist deformation. Most important of all, the locomotives and coaches incorporate sophisticated impact-absorbing structures. These designs were the result of Rail Safety and Standards Board (RSSB) formerly British Rail Research's Crashworthy Development Programme. To protect passengers in a collision, this team constructed trains with collapsible zones in the ends of the coaches. Hydraulic couplers will now sheer off at high-speed impact and slide back into the undercarriage. Each cab and coach end has special anti-climber bars fitted to prevent riding up in a crash. Winston Rasaiah, project engineer, commented, 'With crash damage restricted to collapsible zones, the passenger compartment remains intact'.

We need to be able to predict the likely damage caused by moving objects, so we need to know the size of any forces involved and also the energy transferred. This part of the chapter deals with both these aspects of collisions.

5.2 Forces in collisions

A train crash can be caused by many different events. Here we focus on collisions between two trains that are unfortunately on the same track. A theory of collisions will enable you to calculate the likely outcome of such a meeting or to deduce the speeds just prior to impact.

> ### ⚙ ACTIVITY 20 OBSERVING COLLISIONS
>
> Observe some collisions between trolleys on a runway or between air-track vehicles. Include situations where one vehicle is initially at rest and when both are initially moving towards one another. Include some collisions in which the two vehicles become coupled together and some in which they bounce off one another. Notice what happens to the velocities after impact. Try altering the mass of one or both vehicles.

In Activity 20 you probably noticed that, if the moving mass is increased (as when a moving vehicle couples to a stationary one and they both move off together), the velocity decreases. Also, if two vehicles of equal mass approach one another with equal speeds, they come to rest when they couple together. Mass and velocity are both important when it comes to determining the outcome of a collision, and so are the material properties of the colliding vehicles (do they bounce, stick together or crumple?). To gain more insight into what happens, we need to use Newton's laws of motion and the concept of momentum that you met in *Higher, Faster, Stronger* and in Part 2 of this chapter. If one train meets another train we may have situations similar to the one shown in Figure 1.49.

STUDY NOTE

You may find it helpful to refer back to your work on momentum and collisions in the chapter *Higher, Faster, Stronger*.

PRACTICAL SKILLS REFERENCE

Carrying out practical work

See Practical skills notes 3.1–3.2

Figure 1.49 Two trains about to collide – the rear train is travelling faster

For all collisions, the impact forces on the two bodies during the collision are equal and opposite and last for equal times. This is a direct application of Newton's third law and is always true regardless of whether the two bodies bounce apart or stick together. In Part 2 of this chapter you met the impulse of a force ($F\Delta t$). Both objects in a collision experience the same impulse during the impact. In Figure 1.50, the force of the collision may cause unequal amounts of damage but the force will be of equal size on each train. The *Eurostar* will experience a force of magnitude F to the left and the steam train a force of the same magnitude to the right; both forces will act for exactly the same time interval Δt.

As you saw in Part 2, the change in momentum of a moving object (mass × velocity) is related to the impulse:

$$F\Delta t = \Delta(mv) \qquad \text{(Equation 2)}$$

Since the forces at impact act in opposite directions, the changes in momentum produced are also in opposite directions (remember that both force and momentum are vector quantities). Let the initial velocities be u_1 and u_2, and the final velocities v_1 and v_2, as shown in Figure 1.51 for trains 1 and 2 with masses m_1 and m_2.

For train 1:

$$-F\Delta t = m_1v_1 - m_1u_1$$

For train 2:

$$F\Delta t = m_2v_2 - m_2u_2$$

So:

$$-(m_1v_1 - m_1u_1) = m_2v_2 - m_2u_2$$

This can be rearranged to give:

$$m_1u_1 + m_2u_2 = m_1v_1 + m_2v_2 \qquad (20)$$

Equation 20 can also be written using p to represent momentum:

$$p_1 + p_2 \text{ before collision} = p_1 + p_2 \text{ after collision} \qquad (20a)$$

It can also be generalised to a system consisting of more than two interacting objects:

$$\Sigma p = \text{constant} \qquad (20b)$$

(In words: the sum of all the momenta remains constant.)

The terms on the left of Equation 20 represent the total momentum of the system before the impact, and those on the right represent the total momentum of the system afterwards. This is very important! Expressed in words, *the total momentum of a system remains constant provided no external forces act on the system*; this is the **principle of conservation of linear momentum**. ('Linear' means it applies to objects moving in a straight line.) This applies to *all* interactions without exception. The idea of a system meaning 'a collection of interacting objects' is important. For the duration of the impact the only forces that count must be between the two trains. In Activity 20, you should have found that the velocities before and after impact were (approximately) as described by Equation 20.

Explosions

At first sight an explosion is very different from a collision, but this is another event where the principle of conservation of momentum can be applied. When a gun fires a shell forward, the gun recoils backward. The two forces produced by the explosion have to be equal but opposite in direction. This means that impulses are equal for gun and shell, and so the total momentum of the gun-shell system is unchanged (see Figure 1.51).

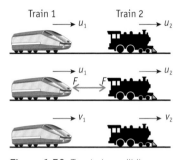

Train 1   Train 2

Figure 1.50 Two trains colliding

MATHS REFERENCE

The summation sign
See Maths note 0.3

STUDY NOTE

You used the conservation of linear momentum in the earlier chapter *Higher, Faster, Stronger*.

before firing

recoil

after firing

Figure 1.51 An explosion

QUESTIONS

Q 46 A 5 t wagon runs into the back of a stationary 41 t *Eurostar* locomotive and gets jammed there. They move off together at $5\,\text{m}\,\text{s}^{-1}$. How fast was the wagon travelling before the collision?

Q 47 As shown in Figure 1.52, two railway coaches are about to collide head on. After the collision the 90 t mass moves on in the same direction, at a velocity of $30\,\text{m}\,\text{s}^{-1}$. What is the velocity of the 20 t mass after the interaction?

$50\,\text{m}\,\text{s}^{-1}$ $30\,\text{m}\,\text{s}^{-1}$

90 t   20 t

Figure 1.52 Diagram for Question 47

Q 48 A party trick is to let the air out of a balloon and watch it fly all over the room. How is this an example of an 'explosive' interaction?

Impulse and momentum change

The principle of conservation of momentum is useful for analysing the situation immediately before and after a collision. However, if we want to know about possible safety implications, we need to know something about the forces that act during the collision. You have already met useful expressions for this:

impulse = $F\Delta t$

and:

$$F = \frac{\Delta p}{\Delta t}$$ (Equation 2)

If the impact time is known, then the (average) force acting can be found. Activity 21 and Questions 49 to 52 illustrate this.

⚙ ACTIVITY 21 FORCES IN COLLISIONS

Use *Tracker*, *Logger Pro* or other motion analysis software that you met in the chapter *Higher, Faster, Stronger* to explore the forces acting in various impacts.

PRACTICAL SKILLS REFERENCE

Carrying out practical work

See Practical skills notes 3.1–3.2

QUESTIONS

Q 49 A snooker ball is struck by a cue. Figure 1.53 shows the force–time graph for the shot from the point of view of the ball.

(a) Draw the force–time graph for the same event for the cue.

(b) What information can you obtain from the area under such a graph?

(c) What shape graph would you get for uniform acceleration?

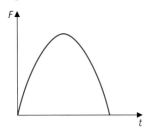

Figure 1.53 Diagram for Question 49

Q 50 When you catch a hard cricket ball it hurts less if you pull your cupped hands back at impact. Why does this work?

Q 51 A hole in a water pipe leading from a tanker in the Channel Tunnel allows 70 kg of water to escape every second in a horizontal jet moving at $30 \, \text{m s}^{-1}$. The jet strikes the tunnel wall and then falls to the ground.

(a) What horizontal force does the wall exert on the water?

(b) What is the horizontal force on the wall?

Q 52 Suppose the arrival of a hurricane is predicted for the Channel area with wind speeds of 120 mph. The forces on the side of a *Eurostar* train are known for a 60 mph gale. Will the hurricane forces on the same train be:

A   the same?

B   twice as large?

C   three times as large?

D   four times as large?

⚙ ACTIVITY 22 IMPACT FORCES AND BUFFERS

In this activity, a model train is allowed to run, out of control, down an inclined track and is brought to rest by an energy-absorbing buffer. Use a force sensor to investigate the relationship between the force exerted on the train and its change of momentum. Hold a competition to see who can construct the safest buffer. Use your knowledge of dynamics to analyse your results and to in this way test your structures as fully and quantitatively as possible.

5.3 Energy in collisions

There is another useful way to look at what happens in a collision. Figure 1.54 shows test coaches that have been damaged in a collision. Each coach experiences a force, and the force does work, producing some plastic deformation of structure. So in an **inelastic collision** there is a net loss of kinetic energy (although energy overall is conserved). On the other hand, an **elastic collision** is one in which there is no net loss of kinetic energy – any energy that is temporarily stored due to elastic deformation is recovered as kinetic energy as the colliding objects spring apart.

STUDY NOTE

This might remind you of some ideas you met in the chapters *Good Enough to Eat* and *Spare Part Surgery*, where you considered elastic and plastic deformation of materials.

Figure 1.54 Result of collision between a test coach and a stationary object

Mathematically, an elastic collision can be described as follows:

$$\frac{1}{2}m_1u_1^2 + \frac{1}{2}m_1u_2^2 = \frac{1}{2}m_1v_1^2 + \frac{1}{2}m_1v_1^2 \qquad (21)$$

or:

$$E_{k1} + E_{k2} \text{ before} = E_{k1} + E_{k2} \text{ after} \qquad (21a)$$

Here, E_k represents kinetic energy and the other symbols have the same meanings as in Equation 20. For a system consisting of more than two interacting objects, Equation 21 can be written in a more general form:

$$\Sigma E_k = \text{constant} \qquad (21b)$$

From the viewpoint of a safety engineer, which type of collision is more desirable – elastic or inelastic? How can trains or cars be designed to be 'crashworthy'?

From the perspective of train crashes, the chances of an elastic collision are remote and positively undesirable. Bits rebounding all over the place would be a nightmare in an accident. The whole point of crashworthiness is that kinetic energy must be absorbed in ways that do as little damage as possible. One way to achieve this is to make the outer parts of a vehicle from a material that readily undergoes plastic deformation, such as is done in the 'crumple zones' of cars, while surrounding the passengers with a strong and fairly rigid structure.

In some situations, though, elastic collisions are desirable, as in some items of sports equipment; for example, a tennis racket that can return a demon serve with little loss of kinetic energy is very desirable (Figure 1.55). You may have seen a desk toy called Newton's cradle (Figure 1.56), which also relies on collisions being almost elastic in order that the ball bearings can continue to bound back and forth for a long time.

Figure 1.55 Tennis racket distorted as a result of impact with a ball

Figure 1.56 Newton's cradle

QUESTIONS

Q 53   The new collapsible front on trains mentioned in Section 5.1 provides for a total of 1 MJ energy absorption in a deformation length of 1 m at each end of the locomotive, and further energy absorption at the interfaces between coaches. A multiple unit of 24 t is built to this specification.

(a) What is the approximate acceleration as the front of the locomotive deforms? Comment on the size of your answer and the likely effect of this acceleration on people within the vehicle.

(b) If a 750 tonne *Eurostar* has the same features, what is its acceleration on deforming?

Q 54   In Part 1 of this chapter it was stated that the kinetic energy of *Eurostar* at full speed ($83\,\mathrm{m\,s^{-1}}$) is almost exactly that required to raise its 750 t from the ground level to the top of the TV aerial at the top of the Eiffel Tower, which is 318 m high.

(a) Calculate the energies involved to show how close 'almost' is to having enough energy to do this. ($g = 9.81\,\mathrm{N\,kg^{-1}}$)

(b) In practice, could the *Eurostar* be made to rear straight up in the air to such a great height? (Imagine this taking place outside the Channel Tunnel!)

Q 55   Examine some of the collisions from Questions 46 and 47 and Activities 20 and 21 to see whether they were elastic.

5.4 Coming together

In this part of the chapter, you have seen how collisions can be analysed first in terms of force, momentum and impulse and then in terms of energy transfer. It is important to be clear that momentum and energy are *different* quantities. Momentum is always conserved in any event where no external forces act. Energy, too, is always conserved, but *kinetic* energy is conserved only in *elastic* collisions. So, when analysing a collision or explosion, it is usually best to deal with momentum first and then see what happens to the kinetic energy.

The questions and activities in this section involve both momentum and energy and therefore provide you with an opportunity to review your knowledge and understanding.

ACTIVITY 23 SUMMING UP PART 5

Check through your work on this part of the chapter and make sure you know the meaning of all the terms printed in **bold**.

ACTIVITY 24 CRASHWORTHY?

At a press conference, after a nasty railway accident, a railway engineer was asked why his trains weren't built to be as strong as it is physically possible to build. He replied, 'Our trains are as strong as is necessary to withstand normal service loads but we use clever engineering at the ends of the vehicles to disperse energy by structural collapse while maintaining the integrity of the occupied zone'.

Explain in your own words what he meant, using *at least three* of the terms printed in **bold** in this part of the chapter.

QUESTION

Q 56 A man of mass 70 kg stands on a trolley of mass 330 kg that rolls smoothly on frictionless rails. The man fires a rivet gun in a direction parallel to the rails. The gun contains 50 g rivets that leave the gun at a speed of $20\,\mathrm{m\,s^{-1}}$ relative to the trolley.

(a) How does firing a single rivet affect the speed of the man-plus-trolley?

(b) If the gun fires continuously at a rate of two rivets per second, calculate the speed of the trolley after the first minute.

(c) Which acquires more kinetic energy per minute, rivets or man-plus-trolley?

(d) The man gets a warning of an oncoming train, 5 km away, approaching at a constant speed of $10\,\mathrm{m\,s^{-1}}$. Can he get up enough speed to escape?

6 Running a railway

In Victorian times tycoons including Isambard Kingdom Brunel (Figure 1.57) set up railway companies all over the UK. Since then there have been a succession of entrepreneurs such as Richard Branson (Figure 1.58) who want to run their own railway. Trams and light railways have enjoyed a modern-day comeback too (Figure 1.59). So what does it take to run a railway?

Figure 1.57 Two early rail tycoons: Isambard Kingdom Brunel (top right, facing right) and George Stephenson (left, light-coloured coat)

Figure 1.59 A Manchester Metrolink tram

Figure 1.58 Richard Branson and a Pendolino train

6.1 How safe is safe?

The majority of modern trains are very expensive, with most costing millions of pounds. What sort should you use? What are the limitations? In Europe, fast trains run on dedicated high-speed track, but most UK companies operate on standard track. HS1 is the UK's first and (in 2015) only, high-speed rail link. It takes *Eurostar* trains from St Pancras to Kent.

Automatic train protection safety systems have been installed on trains to apply the brakes automatically to prevent signals being passed at danger (SPAD). Many UK fast passenger trains, such as *Eurostar* and West Coast Main Line trains, have it. However, the colossal cost (around £6 billion in 2003) prevents it being fitted everywhere.

In Parts 2 and 5 of this chapter, the emphasis has been on rail safety. Yet making railways safe comes at a cost, and this must also be considered when designing a rail system.

Acceptable risk?

The Rail Safety and Standards Board (RSSB) is in charge of ensuring the safety of UK railways. The railway companies have a legal duty to make the risk of accidents and injury on the railways 'as low as reasonably practicable' (ALARP).

The more money that is spent on reducing a risk, the lower the chance of an accident. If this principle is followed to a logical conclusion, a company could spend millions of pounds preventing an accident that is extremely unlikely to happen. However, the cost of travel on such a railway would be so high that no one could afford it, and so the company would go bankrupt. To decide whether to spend money on a safety improvement, over and above the statutory legal requirements to ensure safety, companies use several ways of analysing risk and benefits.

Figure 1.60 shows the tolerability of risk (TOR) framework for reducing risks. The Health and Safety Executive (HSE) suggests guidelines of a 1:1 000 000 fatality risk per year for the boundary between broadly acceptable and tolerable, and 1:10 000 per year (for the public) and 1:1000 per year (for the workforce) fatality risk for the boundary between tolerable and unacceptable.

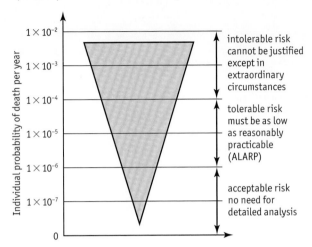

Figure 1.60 Tolerability of risk (TOR) framework

Cost–benefit analysis

Another method that companies use to consider risk is to do a cost–benefit analysis (CBA). This involves calculating the value of preventing a fatality (VPF) and the costs of an accident, and comparing them to the cost of implementing the safety measure. Safety measures are then put in place if the value of the benefits is greater than the cost. All the costs and benefits to everyone, not just the company, must be included in the calculations. You can see immediately that it is important to fix the VPF at the right level.

Costs such as medical costs, damage to trains, track and buildings, and the service disruption can be calculated directly. The cost of loss of life and permanent injury is more difficult, and the method used to value the prevention of a fatality is to ask members of the public how much they would be willing to pay for safety measures that gave a small reduction in risk. These surveys are then used to estimate a mean value. The official 2015 value of preventing a fatality was set at £1.889 million. It is important to understand that this fatality is a statistical fatality – the actual number may be higher or lower in any year. The value put on a life does not represent the life of a real person, as is shown by the fact that if an accident occurs every effort is made to save everyone; CBA would be unethical once real people were involved.

Other indirect costs of accidents include the fact that some people may stop using the train because they think it is not safe. These are very difficult to judge in advance, so the results of a CBA are used with caution.

The main objection to CBA is that it balances the cost of safety against other financial considerations and that this is unethical. However, when you think about it, this always happens – when people consider buying a car, they do not usually find out which is the safest model on the market and then buy it regardless of cost. What CBA does is to make the choice clear and objective, and make it more likely that the interests of the general public will be given a high priority. Another objection is that railways can put 'profit before safety', but provided firstly, that all the costs and benefits are taken into account, not just those of the railway companies, and secondly, that the value of the VPF is not set by the railway company but by the general public, a proper CBA should prevent this.

Figure 1.61 shows how the CBA and TOR are used together to decide what safety measures should be implemented.

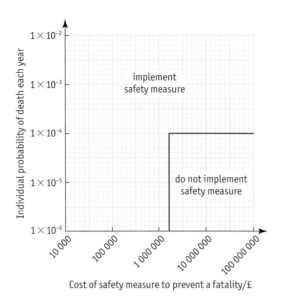

Figure 1.61 How tolerability of risk (TOR) and cost–benefit analysis (CBA) are used to decide whether to implement safety procedures

QUESTION

Q 57 Make a rough sketch of the CBA diagram shown in Figure 1.61.

(a) Divide your diagram into four regions and label them A, B, C, D so that:

A: low risk, high cost of implementing safety measure (the area labelled 'Do not implement …')

B: low risk, low cost of implementing safety measure

C: high risk, low cost of implementing safety measure

D: high risk, high cost of implementing safety measure.

(b) Mark on your diagram a point representing a risk of 1 in 100 000 of a fatality, which could be prevented by a safety measure costing £10 million. Say whether this would be implemented.

(c) Use your diagram to give an example of a risk and safety measure cost that:
(i) would be implemented
(ii) would not be implemented.

LINK THE LEARNING

7 Journey's end

7.1 On track

In this chapter, you have encountered several areas of physics: mechanics, electromagnetism and DC electricity. You have met some important fundamental physical laws and principles: momentum conservation, and Faraday's and Lenz's laws. And you have used a mathematical description of a common type of naturally occurring change: exponential decay. The following activities, and the questions in the next section, are designed to help you look back over your work from this chapter.

⚙ ACTIVITY 25 TRACKING THE PHYSICS

You learned something about each of the following in this chapter. Look back through your work and make brief notes under each heading, listing the examples that were used in the chapter and noting any ways in which the ideas were developed or refined. Add cross-references to earlier chapters where you met the same areas of physics. You could add references to later chapters when you have studied them.

- **Capacitors:** behaviour in DC circuits, charge and pd, exponential discharge, time constant
- **Electromagnetism:** magnetic flux and flux density, electromagnetic force, electromagnetic induction, transformers
- **Mechanics:** momentum and impulse, kinetic energy, elastic and inelastic collisions

---

⚙ ACTIVITY 26 HOLD THE FRONT PAGE

Watch a video of a train crash test. Either act as reporter for a non-specialist publication and write an article on train safety or the nature of the collision you have observed, or act as a member of a company that makes and sells devices that help make trains safe, and produce a technical advertisement for one of your products.

⚙ ACTIVITY 27 DOWN THE TUBES

Observe a small cylindrical magnet being dropped down three different vertically mounted tubes: a plastic tube much wider than the magnet; a plastic tube just wider than the magnet; and a copper or aluminium tube just wider than the magnet. Explain your observations.

FURTHER INVESTIGATIONS

Explore ways in which cracks in rails might be detected and located.

Investigate the effectiveness of seatbelts or airbags in cars.

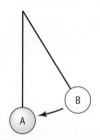

Figure 1.62 Diagram for Question 58

7.2 Questions on the whole chapter

Q 58 Figure 1.62 is taken from the notes of an accident investigator.

The driver of car A said he stopped at the junction and looked. When he continued he was struck by car B. He thinks that car B must have been speeding and he had not seen him because of the bend in the road.

The driver of car B says she was driving within the speed limit and car A did not stop, but drove out in front of her.

Car A has a mass of 1450 kg and travelled 26.7 m after the collision in the direction shown on the map.

Car B has a mass of 2020 kg and travelled 22.8 m after the collision in the direction shown on the map.

Investigators measured a value for the deceleration produced by the road surface after the collision of $3.10\,\mathrm{m\,s^{-2}}$.

(a) Show that:
  (i) the speed of car A immediately after the collision was about $13\,\mathrm{m\,s^{-1}}$
  (ii) the speed of car B was about $12\,\mathrm{m\,s^{-1}}$.

(b) Calculate the momentum of **(i)** car A and **(ii)** car B immediately after the collision.

(c) Calculate the component of momentum to the east, immediately after the collision of **(i)** car A and **(ii)** car B.

(d) Use your answer to **(c)** to calculate the speed of car B immediately before the collision.

(e) The speed limit was 40 mph ($17.8\,\mathrm{m\,s^{-1}}$). Was car B speeding before the collision?

(f) Use a similar analysis to determine the speed of car A immediately before the collision.

(g) Comment on your answer to **(f)**.

Q 59 A student makes some measurements using the plastic toy shown in Figure 1.63. Table 1.2 shows the data for one collision.

Figure 1.63 Diagram of toy for Question 59

Mass of sphere A, m_A	63 g
Mass of sphere B, m_B	31 g
Speed of sphere A just before collision, u_A	0
Speed of sphere B just before collision, u_B	$2.45\,\mathrm{m\,s^{-1}}$
Speed of sphere A just after collision, v_A	$0.90\,\mathrm{m\,s^{-1}}$

Table 1.2 Data for Question 59

(a) Show that the speed, v_B, of sphere B just after the collision is about $0.6\,\mathrm{m\,s^{-1}}$.

(b) Determine whether this is an elastic or inelastic collision.

(c) When experimenting with another toy in which both spheres have $m = 30\,\mathrm{g}$, the student finds that sphere B is always stationary after the collision.

Show that for this toy:
  (i) the speed v_A of sphere A after the collision is always the same as the speed, u_B, of B immediately before
  (ii) the collision is always elastic.

Q 60 Figure 1.64 shows a so-called 'everlasting torch' that operates without batteries. When the torch is shaken, a strong permanent magnet moves through a copper coil and generates an electric current.

(a) By referring to Faraday's law, explain:
  (i) how the current is generated
  (ii) how the speed of shaking affects the size of the current.

The current is used to charge a capacitor, which provides a power source for an LED. Figure 1.65 shows how the capacitor voltage varies with charge.

Figure 1.64 Everlasting torch

Figure 1.65 Capacitor voltage when the torch is shaken

(b) Calculate the capacitance of the capacitor.

Figure 1.66 shows how the capacitor voltage changes with time when the torch is in use after the shaking has stopped.

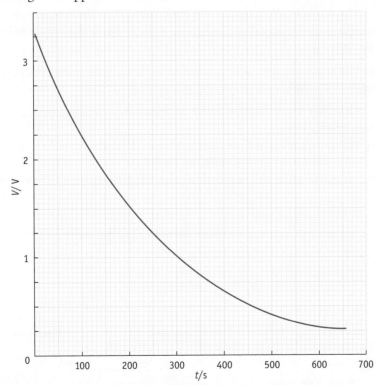

Figure 1.66 Capacitor voltage when the torch is used

(c) Explain why the brightness of the LED fades over a few minutes.

(d) Use the graph in Figure 1.66 to calculate:
  (i) the time constant for the capacitor–LED circuit
  (ii) the circuit's resistance.

Q 61 Tram and train carriages often have a system of flashing warning lights and beeps to warn passengers that the carriage doors are about to close automatically (Figure 1.67). The duration of the warning period can be determined by a resistor-capacitor (RC) combination.

Figure 1.67 'Stand clear of the doors'

Suppose the warning period ends when the voltage across the output falls to $V_0/4$, where V_0 is the initial pd.

(a) Write an equation for the time taken to fall to $V_0/4$ in terms of the circuit's time constant $\tau\,(= RC)$.

(b) Explain what would happen to the duration of the warning period if the following changes were made:
   (i) the same capacitor C discharges instead through two *series* resistors each of resistance R, as shown in Figure 1.68(a)
   (ii) the capacitor C discharges through two *parallel* resistors, each of resistance R.

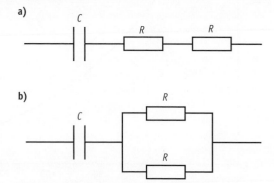

Figure 1.68 Diagrams for Question 61(b)

(c) Suggest a reasonable duration, in seconds, for the warning period. In this way determine a value for the time constant, τ, for the original RC combination.

7.3 Achievements

Now you have studied this chapter you should be able to achieve the outcomes listed in Table 1.3.

TABLE 1.3 ACHIEVEMENTS FOR THE CHAPTER *TRANSPORT ON TRACK*		
Statement from examination specification		Section(s) in this chapter
1	know and understand the distinction between base and derived quantities and their SI units	2.2, 3.3, 4.2
2	demonstrate their knowledge of practical skills and techniques for both familiar and unfamiliar experiments	4.2
3	be able to estimate values for physical quantities and use their estimate to solve problems	1.1, 3.2
4	understand the limitations of physical measurement and apply these limitations to practical situations.	4.2, 5.2, 5.3
5	be able to communicate information and ideas in appropriate ways using appropriate terminology	3.4, 5.4, 7.1
6	understand applications and implications of science and evaluate their associated benefits and risks	6.1
8	understand the ways in which society uses science to inform decision making	6.1
97	understand how to use the equation impulse = $F\Delta t = \Delta p$ (Newton's second law of motion)	3.2, 5.2
98	CORE PRACTICAL 9: Investigate the relationship between the force exerted on an object and its change of momentum	5.2
101	understand how to determine whether a collision is elastic or inelastic	5.3
116	understand that capacitance is defined as $C = \dfrac{Q}{V}$ and be able to use this equation	4.2
118	be able to draw and interpret charge and discharge curves for resistor–capacitor circuits and understand the significance of the time constant RC	4.2
119	CORE PRACTICAL 11: Use an oscilloscope or data logger to display and analyse the potential difference (pd) across a capacitor as it charges and discharges through a resistor	4.2
120	be able to use the equation $Q = Q_0 e^{-t/RC}$ and derive and use related equations for exponential discharge in a resistor-capacitor circuit, $I = I_0 e^{-t/RC}$, and $V = V_0 e^{-t/RC}$ and the corresponding log equations $\ln Q = \ln Q_0 - \dfrac{t}{RC}$, $\ln I = \ln I_0 - \dfrac{t}{RC}$, $\ln V = \ln V_0 - \dfrac{t}{RC}$	4.2,
121	understand and use the terms *magnetic flux density B*, *flux Φ* and *flux linkage NΦ*	3.2, 3.3
123	be able to use the equation $F = BIL \sin \theta$ and apply Fleming's left-hand rule to current-carrying conductors in a magnetic field	3.2
124	understand the factors affecting the emf induced in a coil	3.3
125	understand the factors affecting the emf induced in a coil when there is a change of current in another coil linked with this coil	3.3
126	understand how to use Lenz's law to predict the direction of an induced emf and how the prediction relates to energy conservation	3.3
127	understand how to use Faraday's law to determine the magnitude of an induced emf and be able to use the equation that combines Faraday's and Lenz's laws $\mathscr{E} = \dfrac{-\mathrm{d}(N\Phi)}{\mathrm{d}t}$	3.3

Answers

Q 1 **(a)** Speed $= \dfrac{300\,\text{km}}{1\,\text{hour}}$

$= \dfrac{300 \times 1000\,\text{m}}{3600\,\text{s}} = 83.3\,\text{m s}^{-1}$

(b) Speed $= \dfrac{300\,\text{km}}{1\,\text{hour}}$

$= \dfrac{300\,\text{km}/(1.6\,\text{km mile}^{-1})}{1\,\text{hour}} = \dfrac{187.5\,\text{mile}}{1\,\text{hour}}$

Q 2 Assuming all the kinetic energy is converted into gravitational potential energy, then:

$\dfrac{1}{2}mv^2 = mgh$

Cancelling m and rearranging gives:

$h = \dfrac{v^2}{2g} = \dfrac{(83\,\text{m s}^{-1})^2}{(2 \times 9.81\,\text{m s}^{-2})} \approx 350\,\text{m}$

Q 3 The small scale of Figure 1.4 makes this difficult, but from 0 km to 8 km on the English side the vertical drop is approximately 50 m above sea level to 40 m below sea level.

So gradient $\approx \dfrac{90\,\text{m}}{8000\,\text{m}} = \dfrac{1}{89}$

Strictly, gradient $= \dfrac{\text{vertical drop}}{\substack{\text{hypotenuse of the} \\ \text{right-angled triangle}}}$

$= \dfrac{90}{(90^2 + 8000^2)^{1/2}}$

$= \dfrac{90}{8000.5} \approx \dfrac{1}{89}$

Q 4 **(a)** Using $F = ma$, $a = \dfrac{F}{m}$:

$a = \dfrac{400 \times 10^3\,\text{N}}{750 \times 10^3\,\text{kg}} = 0.53\,\text{m s}^{-2}$

(b) Power = work done per second, $P = Fv$:

$P = 400 \times 10^3\,\text{N} \times 10\,\text{m s}^{-1}$

$= 4 \times 10^6\,\text{W} = 4\,\text{MW}$

Q 5 N s or kg m s^{-1}. The units are interchangeable (see Equation 2).

Q 6 **(a)** $\Delta p = 50\,000\,\text{kg} \times (+40\,\text{m s}^{-1}) - 50\,000\,\text{kg} \times (+20\,\text{m s}^{-1})$

$= 1.00 \times 10^6\,\text{kg m s}^{-1}$

(b) $\Delta p = 50\,000\,\text{kg} \times (-20\,\text{m s}^{-1}) - 50\,000\,\text{kg} \times (+20\,\text{m s}^{-1})$

$= -2.0 \times 10^6\,\text{kg m s}^{-1}$

(c) $\Delta p = 50\,000\,\text{kg} \times 0\,\text{m s}^{-1} - 50\,000\,\text{kg} \times (+20\,\text{m s}^{-1})$

$= -1.00 \times 10^6\,\text{kg m s}^{-1}$

Q 7 **(a)** $\Delta p = (1200\,\text{kg} \times 25\,\text{m s}^{-1} - 1200 \times 0\,\text{m s}^{-1})$

$= 30 \times 10^3\,\text{kg m s}^{-1}$

(b) $F = \dfrac{\Delta p}{\Delta t} = \dfrac{30 \times 10^3\,\text{kg m s}^{-1}}{10\,\text{s}} = 3 \times 10^3\,\text{N}$

$= 3\,\text{kN}$ in direction of the motion

(c) $\Delta s = \dfrac{1}{2}(u + v)\,\Delta t$

$s = 0.5(0\,\text{m s}^{-1} + 25\,\text{m s}^{-1}) \times 10\,\text{s} = 125\,\text{m}$

(d) $P = \dfrac{\text{work}}{\text{time}} = \dfrac{\text{force} \times \text{distance}}{\text{time}}$

$P = \dfrac{(3000\,\text{N} \times 125\,\text{m})}{10\,\text{s}} = 3.75 \times 10^4\,\text{W}$

$= 37.5\,\text{kW}$

Q 8 Although the mass for the power unit to pull is so much greater, it would be very uncomfortable to leave a station at the acceleration quoted for a car. Trains contain passengers who expect to be able to walk about.

Q 9 **(a)** $F = \dfrac{\Delta(mv)}{\Delta t}$

$F = \dfrac{750 \times 10^3\,\text{kg} \times 83\,\text{m s}^{-1} - 750 \times 10^3\,\text{kg} \times 44\,\text{m s}^{-1}}{14 \times 60\,\text{s}}$

$= 35\,\text{kN}$

(b) To produce the same acceleration with only one-third the mass requires a tractive force one-third of that found in **(a)**.

(c) It takes the same size force to change speed from $83\,\text{m s}^{-1}$ to $44\,\text{m s}^{-1}$ over the same length of time, but it now acts in the opposite direction to the direction of travel.

(d) To produce the same resultant force, the tractive force provided by the engines must increase by 40 kN to 75 kN in the case of *Eurostar* and 52 kN for the lighter train.

Q 10 **(a)** Using $F = \dfrac{\Delta p}{\Delta t}$:

(i) $F = \dfrac{2400 \times 10^3\,\text{kg} \times 44\,\text{m s}^{-1}}{60\,\text{s}}$

$= 1.76 \times 10^6\,\text{N}$

$= 1.8\,\text{MN}$ opposed to the direction of travel

(ii) The same method gives $F = 550\,\text{kN}$ for *Eurostar*.

(b) $F = (750\,000\,\text{kg} \times 83\,\text{m s}^{-1} - 750\,000 \times 0\,\text{m s}^{-1}) \div 180\,\text{s}$

$= 345\,833\,\text{N}$

$= 350\,\text{kN}$ opposed to the direction of travel.

(c) Far larger forces are needed to stop these trains quickly than it does to get them going, because the time interval is shorter.

Q 11 **(a) (i)** For an empty tram:
$m = 46.5 \times 10^3$ kg and $a = -3.0$ m s^{-2}
(negative sign indicates direction opposing motion)
So braking force is:
$F = ma = 46.5 \times 10^3$ kg $\times -3.0$ m s^{-2}
$= -1.40 \times 10^5$ N

(ii) For a full tram:
$m = 46.5 \times 10^3$ kg $+ (250 \times 65$ kg$) = 62\,750$ kg
$F = -1.89 \times 10^5$ N

(b) From Equation 3 rearranged:

$$\Delta s = \frac{mu^2}{2F}$$

$$= \frac{62\,750 \text{ kg} \times (22 \text{ m s}^{-1})^2}{2 \times 1.89 \times 10^5 \text{ N}} = 80 \text{ m}$$

Q 12 **(a)** Equation 3: $\Delta s = \dfrac{mu^2}{2F}$

$\Delta s = 2100 \times 10^3$ kg $\times (40$ m s$^{-1})^2 \div (2 \times 1.2 \times 10^6$ N$)$

$= 1400$ m

(b) $F = \dfrac{mu^2}{2\Delta s}$

$= 750 \times 10^3$ kg $\times (80$ m s$^{-1})^2 \div (2 \times 1500$ m$)$

$= 1.6$ MN

(c) At high speeds the drag will help to slow down the trains, but 40 kN is small compared to the mega-newton braking forces. As speed decreases, so will the drag force. Both answers will thus be smaller than calculated in **(a)** and **(b)**, but only slightly.

Q 13 **(a)** Equation 1 is again the best one to use:

$\Delta t = \dfrac{\Delta p}{F} = \dfrac{750 \times 10^3 \text{ kg} \times 80 \text{ m s}^{-1}}{1.6 \times 10^6 \text{ N}}$

$= 37.5$ s ≈ 38 s

(b) $a = (0 - 80)$ m s$^{-1} \div 37.5$ s $= -2.1$ m s^{-2}

Q 14 **(a) (i)** Momentum doubles (mv to $2mv$).
(ii) Kinetic energy is four times greater:
$\frac{1}{2}mv^2 \rightarrow \frac{1}{2}m(2v)^2$

(b) Both quantities are in the ratio 2:1. For the more massive train, **(i)** momentum doubles (mv to $2mv$) and **(ii)** kinetic energy is also doubled
$\frac{1}{2}m\,v^2 \rightarrow \frac{1}{2}\,2\,m\,(v)^2$

Q 15 **(a)** *Resultant* force is found using:

$F = ma = 2.40 \times 10^6$ kg $\times 0.13$ m s$^{-2} = 312$ kN

Drag force = 40 kN, so to produce the required resultant, we need a tractive force $T = 352$ kN.

(b) Resolving along the slope, with T representing tractive force and W the weight of the train:

$T = W \sin \theta = mg \sin \theta$

$= 2.40 \times 10^6$ kg $\times 9.81$ N kg$^{-1} \times \left(\dfrac{1}{90}\right) = 262$ kN

(c) Total tractive force required = 352 kN + 262 kN
$= 614$ kN

(d) $\Delta t = \dfrac{\Delta p}{F}$

where F is the *net* accelerating force; i.e.:

$F = 312$ kN (from **(a)**)

$\Delta t = \dfrac{2.40 \times 10^6 \text{ kg} \times 44 \text{ m s}^{-1}}{312\,000 \text{ N}} = 338$ s = 5 min 38 s

(e) (i) From **(a)** and **(b)**, net force along slope is:

$F = 400$ kN $- 262$ kN $- 40$ kN $= 98$ kN

$a = \dfrac{F}{m} = \dfrac{98 \times 10^3 \text{ N}}{2.40 \times 10^6 \text{ kg}} = 0.041$ m s^{-2}

(ii) $\Delta t = \dfrac{\Delta v}{a} = \dfrac{(v-u)}{a}$

$= \dfrac{44 \text{ m s}^{-1}}{0.041 \text{ m s}^{-2}} = 1073$ s = 17 min 53 s

(iii) $\Delta s = \dfrac{\Delta E_k}{F}$

$= \dfrac{0.5 \times 2.40 \times 10^6 \text{ kg} \times (44 \text{ m s}^{-1})^2}{98 \times 10^3 \text{ N}}$

$= 2.37 \times 10^4$ m = 23.7 km

Q 16 (a) One possible approach is to note that acceleration due to gravity is a vector, and so can be resolved as shown in Figure 1.69.

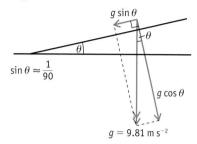

Figure 1.69 Diagram for the answer to Question 16

Along the slope, $u = 44\,\mathrm{m\,s^{-1}}$, $v = 0\,\mathrm{m\,s^{-1}}$,

$a = 9.81 \times \dfrac{1}{90}\,\mathrm{m\,s^{-2}} = 0.109\,\mathrm{m\,s^{-2}}$:

$v^2 = u^2 + 2as$

$s = \dfrac{(v^2 - u^2)}{2a}$

$= \dfrac{(-44\,\mathrm{m\,s^{-1}})^2}{2 \times 0.109\,\mathrm{m\,s^{-2}}}$

$= 8.89 \times 10^3\,\mathrm{m} \approx 9.0\,\mathrm{km}$;

i.e. not far enough to get out.

(We have chosen the positive direction to be the direction of travel, i.e. up the slope.)

(b) The net force along the slope = drag + component of weight acting along slope. This has magnitude:

$F = D + W \sin\theta$

which is related to the acceleration by $F = ma$.

So $a = \dfrac{F}{m} = \dfrac{D}{m} + g\sin\theta$

$= \dfrac{30 \times 10^3\,\mathrm{N}}{2400 \times 10^3\,\mathrm{kg}} + \dfrac{9.81\,\mathrm{m\,s^{-2}}}{90} = 0.122\,\mathrm{m\,s^{-2}}$

Taking the positive direction to be up the slope:

$a = -0.122\,\mathrm{m\,s^{-2}}$

Using the same method as in Question 19:

$s = \dfrac{(v^2 - u^2)}{2a} = \dfrac{(-44\,\mathrm{m\,s^{-1}})^2}{2 \times -0.122\,\mathrm{m\,s^{-2}}} = 7.9\,\mathrm{km}$

Q 17 (a) See Table 1.4.

Angle θ	0°	10°	20°	30°	40°	45°	50°	60°	70°	80°	90°
$F = BIL\,\sin\dfrac{\theta}{N}$	0.00	0.17	0.34	0.50	0.64	0.71	0.77	0.87	0.94	0.98	1.00

Table 1.4 The answers to Question 17(a)

(b) and (c) See Figure 1.70. All these graphs have the shape of the first quarter cycle of a sine curve.

(b)

(c)(i)

(c)(ii)

(c)(iii)

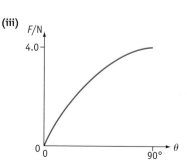

Figure 1.70 The answers to Question 17(b) and (c)

Q 18 (a) downwards

(b) out of the page

(c) to the right.

 Q 19 (a) $F = 2\,\text{T} \times 4\,\text{A} \times 10 \times 10^{-3}\,\text{m} = 0.08\,\text{N}$

(b) $F = 0\,\text{N}$

(c) $F = 2\,\text{T} \times 4\,\text{A} \times 10 \times 10^{-3}\,\text{m} \times \sin 30° = 0.04\,\text{N}$

Q 20 The electromagnetic force must be equal in size (but opposite in direction) to the weight of the rider:

$BIl = mg$

$$B = \frac{mg}{Il} = \frac{15 \times 10^{-3}\,\text{kg} \times 9.81\,\text{N kg}^{-1}}{5.0\,\text{A} \times 100 \times 10^{-3}\,\text{m}}$$

$= 0.3\,\text{N A}^{-1}\,\text{m}^{-1} = 0.3\,\text{T}$

Q 21 (a) and (b) See Figure 1.71.

(c) At A the fields of the magnetic pole pieces and of the current-carrying conductor are in the same direction, so add to produce a large field acting from left to right. The combined field will be largest near the conductor (where the conductor's field is greatest). At C we have already seen the two fields cancel because they are equal in size but opposite in direction. In the region between C and the current-carrying conductor, the combined field will point right to left, as the conductor's field is stronger than that of the magnetic pole pieces. The opposite is true on the other side of C, where the combined field will point left to right.

(d) See Figure 1.71.

Figure 1.71 The answers to Question 21(a), (b) and (d)

(e) From Fleming's left-hand rule, **F**irst finger (**F**ield) acts left to right, and se**C**ond finger (**C**urrent) is up out of the paper. This gives thu**M**b (**M**otion) directed from the bottom of the page to the top. Thinking of the field's lines acting like elastic, the current-carrying conductor would be catapulted towards the top of the page.

(f) From (d), using either Fleming's left-hand rule or the catapult effect, X is propelled from the bottom to the top of the page and Y is propelled from top to bottom. If they are joined rigidly the only possible motion is clockwise rotation until Y lies directly below X; the two equal forces have the same line of action and so no turning effect.

 Q 22 The missing ingredient is the number of turns, n. The force is increased for every extra turn in the coil. This gives $F = nBIl$ for the force on each side of the coil.

These are some plausible estimates: suppose we have current between 10 and 100 A, B between 0.1 and 1 T, length of coil is limited by space, say 2 m, n depends on thickness of wire (too thin and the resistance is too high for required high currents we need but thicker means fewer turns fit on coil) – let's say 1000 turns.

These values give:

$F \approx 1000 \times 1\,\text{T} \times 100\,\text{A} \times 2\,\text{m} = 2 \times 10^5\,\text{N}$

This is half our target, which is close. We can increase each quantity slightly, or we can always use two motors per locomotive and have two or more locomotives per train.

Q 23 (a) An emf is induced whenever there is relative motion between the magnetic field and the conductor such that the conductor appears to 'cut through' the imaginary field lines. In experiment (c) switching the power on and off has this effect.

(b) Either a stronger magnetic field or more conductor (e.g. turns of wire) 'cutting' the magnetic field, or the rate of cutting the field is greater (the conductor or magnet moves faster). In experiment (c) increasing the power supplied would have this effect.

(c) Any *one* of: turning the magnet round so N becomes S and vice versa; reversing the motion of the magnet; reversing the motion of the conductor. In experiment (c) reversing the power connections would work. Reversing the meter connections in all the experiments *appears* to make the emf negative but it has *not* altered the induced emf in the coil itself.

 Q 24 $1\,\text{Wb} = 1\,\text{N A}^{-1}\,\text{m}$. $1\,\text{N} = 1\,\text{kg m s}^{-2}$.
So in SI base units $1\,\text{Wb} = 1\,\text{kg m}^2\,\text{s}^{-2}\,\text{A}^{-1}$.

$1\,\text{T} = 1\,\text{Wb m}^{-2}$.
So in SI base units $1\,\text{T} = 1\,\text{kg m}^2\,\text{s}^{-2}\,\text{A}^{-1}\,\text{m}^{-2} = 1\,\text{kg s}^{-2}\,\text{A}^{-1}$

Q 25 (a) $\Phi = BA \cos \theta$

$$= 5.0 \times 10^{-3}\,\text{T} \times \pi\,(0.15\,\text{m})^2 \times \cos \theta$$

$$\approx 3.53 \times 10^{-4}\,\text{Wb} \times \cos \theta$$

See Table 1.5:

Angle θ	0°	30°	60°	90°
Flux $\Phi/10^{-4}$ Wb	3.53	3.06	1.77	0.00

Table 1.5 The answers to Question 25(a)

(b) See Figure 1.72

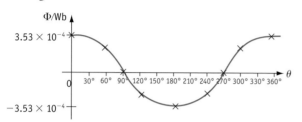

Figure 1.72 The answer to Question 25(b)

Q 26 (a) A

(b) B

(c) B

(d) A

(e) A as the N pole enters the coil, then B as the S pole enters. Then A as the N pole leaves the coil, then B as the S pole leaves.

(f) No induced emf or current.

Q 27 The SI units of the right-hand side are Wb s^{-1}:

$$1\,\text{Wb s}^{-1} = 1\,\text{N A}^{-1}\text{m s}^{-1}$$

But $1\,\text{N m} = 1\,\text{J}$, and $1\,\text{A s} = 1\,\text{C}$ (because $1\,\text{A} = 1\,\text{C s}^{-1}$), so:

$1\,\text{Wb s}^{-1} = 1\,\text{J C}^{-1}$ and $1\,\text{J C}^{-1} = 1\,\text{V}$ – which is the SI unit for emf as required for the left-hand side.

Q 28 Using Equation 6b, with $\Delta\Phi = 50 \times 10$ Wb, $\Delta t = 2.0$ s:

$$\mathcal{E} = \frac{5.0 \times 10^2\,\text{Wb}}{2.0\,\text{s}} = 250\,\text{V}$$

Q 29 Using Equation 6e, with $\Delta A = 0.03\,\text{m}^2$, $\Delta t = 0.2$ s, $B = 2$ T and $N = 10$:

$$\mathcal{E} = \frac{10 \times 2\,\text{T} \times 0.03\,\text{m}^2}{0.2\,\text{s}} = 3\,\text{V}$$

Q 30 The changing current produces a changing magnetic field. A radio receiver is sensitive to the changing fields that make up electromagnetic waves, so will also respond to the changing field produced by a current and therefore may produce an audible signal.

Q 31 As the field has a vertical component, any horizontal conducting path on the aeroplane will have an induced emf across it during flight. If flying due north, this will be from right to left (east to west) across wings and fuselage. (Check this with the right-hand rule looking down on the plane.) There will be no emf along the length of the fuselage as this is not 'cutting' across the field.

The induced emf is a potential danger as it could interfere with control systems within the plane.

Q 32 (a) As the train moves forward through the vertical field, an induced current will flow from right to left across the train. This current is produced by the induced emf, which also acts across the train from right to left.

(b) Use Equation 7 with $B = 4 \times 10^{-5}$ T, $v = 60\,\text{m s}^{-1}$, $l = 4$ m (the height is irrelevant):

$$\mathcal{E} = 4 \times 10^{-5}\,\text{T} \times 4\,\text{m} \times 60\,\text{m s}^{-1} = 9.6\,\text{mV}$$

Q 33 (a) Magnetic fields extend across space so no contact is required.

(b) No, it depends on motion for its effect so would not work when at rest.

Q 34 The disc should have low electrical resistance, i.e. it should be thick and made of good conducting material.

Q 35 Vary the distance of the magnet from the disc. Alter the strength of the magnet (only possible if the system uses an electromagnet).

Q 36 The tunnel lining would act like a giant solenoid. Any electric train, with very strong motor magnets, would set up an induced current in it, which (as described by Lenz's law) would be directed to push back on the train like a brake. More energy is then required to drive the train into the tunnel, i.e. increased fuel costs – a potentially expensive mistake.

Q 37 Only the vertical component of B contributes to the emf. Using Equation 7:

$$\mathcal{E} = 4.1 \times 10^{-5}\,\text{T} \times 47\,\text{m} \times 268\,\text{m s}^{-1} = 0.516\,\text{V}$$

Q 38 (a) $\Phi = BA \cos \theta$

$$B = 3.5 \times 10^{-6}\,\text{T}, \theta = 45°$$

$$A = 5.0 \times 10^{-2}\,\text{m} \times 8.0 \times 10^{-2}\,\text{m}$$

$$= 4.0 \times 10^{-3}\,\text{m}^2$$

$$\Phi = 3.5 \times 10^{-6}\,\text{T} \times 4.0 \times 10^{-3}\,\text{m}^2 \times \cos 45°$$

$$= 9.9 \times 10^{-9}\,\text{Wb}$$

(b) Flux linkage = $N\Phi$

$$= 100 \times 9.9 \times 10^{-9}\,\text{Wb} = 9.9 \times 10^{-7}\,\text{Wb}$$

(c) Use Equation 6 with $\Delta\Phi = 9.9 \times 10^{-7}$ Wb, $\Delta t = 1.0 \times 10^{-4}$ s

$$\mathscr{E} = \frac{9.9 \times 10^7\,\text{Wb}}{1.0 \times 10^{-4}\,\text{s}} = 9.9 \times 10^{-3}\,\text{V} = 9.9\,\text{mV}$$

Q 39 From Equation 8, initial charge:

$Q_0 = CV_0 = 100 \times 10^{-6}\,\text{F} \times 10\,\text{V} = 10^{-3}\,\text{C}$

After 20 s, $Q = 0.45 \times Q_0 = 4.5 \times 10^{-4}\,\text{C}$

Q 40 (a) In Equation 10b, the left-hand side has units of charge divided by charge, i.e. it is dimensionless.

The right-hand side, $\dfrac{\Delta t}{RC}$, must also be dimensionless.

As Δt has dimensions of time (SI units of second) then so must RC.

In terms of SI units:

RC has units $\Omega\,\text{F}$

$1\,\Omega = 1\,\text{V}\,\text{A}^{-1}$ (from $V = IR$)

$1\,\text{F} = 1\,\text{C}\,\text{V}^{-1}$ (from $Q = CV$)

$1\,\Omega\,\text{F} = 1\,\text{V}\,\text{A}^{-1} \times 1\,\text{C}\,\text{V}^{-1} = 1\,\text{C}\,\text{A}^{-1}$

$1\,\text{C} = 1\,\text{A}\,\text{s}$

so

$1\,\text{C}\,\text{A}^{-1} = 1\,\text{A}\,\text{s} \times 1\,\text{A}^{-1} = 1\,\text{s}$

(b) If $t = RC$, then $\dfrac{V}{V_0} = e^{-RC/RC} = e^{-1}$

$= \dfrac{1}{e} = \dfrac{1}{2.718} = 0.37$ to 2 significant figures

Q 41 Initially the capacitor stores no charge and there is no voltage across it. The circuit behaves as if the supply voltage V_0 is connected across the resistor R, and the current is $I_0 = \dfrac{V_0}{R}$. As the capacitor becomes charged, a voltage builds up acrross the capacitor, which acts in the reverse direction to the power supply voltage. The resultant voltage across the resistor decreases and so the current decreases, which reduces the rate of charging. Eventually the pd across the capacitor is equal to the supply voltage and no more charge flows.

Q 42 Writing $Q = fQ_0$, $V = fV_0$ and $I = fI_0$ and using any of Equations 17, 18 or 19:

$t = -RC \ln f = -\tau \ln f$

Q 43 (a) From Equation 10, $\dfrac{Q}{Q_0} = e^{-3} = 0.05 = 5\%$

(b) Similarly, $\dfrac{Q}{Q_0} = e^{-5} = 0.0067 = 0.7\%$

Strictly mathematically speaking, the charge *never* actually reaches zero, but more than 99% of the initial charge will have been released from a discharging capacitor after five time constants. Depending on how precisely one can detect this small amount of remaining charge, one can say that the capacitor is 'fully discharged' for all practical purposes.

Q 44 (a) 6 V

(b) $\tau = RC = 1 \times 10^5\,\Omega \times 47 \times 10^{-6}\,\text{F} = 4.7\,\text{s}$

(c) From Equation 12, $V = V_0 e^{-t/RC}$:

$\dfrac{t}{RC} = \dfrac{5\,\text{s}}{4.7\,\text{s}} = 1.06$, $V_0 = 6\,\text{V}$

$V = 6 \times e^{-1.06} = 2.08\,\text{V}\,(\approx 2\,\text{V})$

(d) Substituting in Equation 17b and using the answer to Question 44(b):

$t = RC \ln\left(\dfrac{V_0}{V}\right) = RC \ln\left(\dfrac{6.0}{0.6}\right) = 4.7\,\text{s} \times \ln 10$

$= 11\,\text{s}$

Q 45 (a) When a train is occupying the block, there is a connection across XY. This instantly charges the capacitor and also makes the NOT gate input high. The NOT gate output is therefore low, or close to zero volts, allowing the red LED to light. When the train leaves the block, the capacitor discharges and the pd across it gradually falls. Therefore, the input voltage to the NOT gate also falls. When this voltage drops below 1.5 V, it is read as low and the NOT gate output goes high, close to 6 V. This allows the green LED to light, thus indicating all clear.

(b) We need to show that the capacitor voltage falls to 1.5 V when $t = 10$ s. At this time:

$\dfrac{t}{RC} = \dfrac{10\,\text{s}}{7.2\,\text{s}} = 1.39$

From Equation 12, $V = 6\,\text{V} \times e^{-1.39} = 1.5\,\text{V}$ as required.

(c) $RC = 1000 \times 10^3\,\Omega \times 10 \times 10^{-6}\,\text{F} = 10\,\text{s}$. This is a little longer than the required τ of 7.2 s, so discharge will be slower, which gives a delay that is longer than the required 10 s.

Q 46 Using Equation 20 with m_1 = 5 t (wagon), m_2 = 41 t (*Eurostar* locomotive), u_2 = 0 m s^{-1} and v_1 = v_2 = v = 5 m s^{-1}, we have:

m_2u_2 = 0, and so:

$m_1u_1 = (m_1 + m_2)v$

$$u_1 = \frac{(m_1 + m_2)v}{m_1}$$

$$= \frac{46\,\text{t} \times 5\,\text{m s}^{-1}}{5\,\text{t}} = 46\,\text{m s}^{-1} \text{ (this is over 100 mph!)}$$

(Notice that we have kept all masses in tonnes. You could convert to kg, but the factors of 1000 would cancel at the final stage.)

Q 47 Using Equation 20 and taking left-to-right as positive:

m_1 = 90 t, m_2 = 20 t (as in Question 46, we have kept masses in t)

u_1 = +50 m s^{-1}, u_2 = −30 m s^{-1}, v_1 = +30 m s^{-1}

Momentum beforehand = $m_1u_1 + m_2u_2$

$$= 90\,\text{t} \times 50\,\text{m s}^{-1} - 20\,\text{t} \times 30\,\text{m s}^{-1}$$

$$= 3900\,\text{t m s}^{-1}$$

Momentum afterwards = $m_1v_1 + m_2v_2$

$$= 90\,\text{t} \times 30\,\text{m s}^{-1} + m_2v_2$$

$$= 2700\,\text{t m s}^{-1} + m_2v_2$$

Since momentum must be conserved:

$m_2v_2 = 3900\,\text{t m s}^{-1} - 2700\,\text{t m s}^{-1} = 1200\,\text{t m s}^{-1}$

$$v_2 = \frac{1200\,\text{t m s}^{-1}}{20\,\text{t}} = 60\,\text{m s}^{-1}$$

Notice that the final answer gives v_2 in the correct units (the tonnes cancel) and because the number is positive, we know the coach must be moving from left to right.

Q 48 Air plus balloon initially have no velocity and so no momentum. The air is held inside at high pressure and so once released the air moves out at high speed. To conserve momentum the balloon must move in the opposite direction with a momentum equal to that of the expelled air.

Q 49 (a) See Figure 1.73.

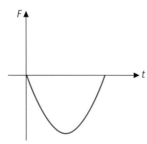

Figure 1.73 Answer to Question 49

(b) The area gives the impulse and therefore the change in momentum.

(c) The graph would be a straight line parallel to the time axis (since the force has to be constant).

Q 50 It lengthens the time of the impact. The required change in momentum to stop the ball can be achieved by either a large force for a short time (painful) or by a smaller force for a longer time (comfortable).

Q 51 (a) Using Equation 2: $F = \dfrac{\Delta(mv)}{\Delta t}$. In a time interval Δt = 1 s, 70 kg strikes the wall and its horizontal velocity decreases from 30 m s^{-1} to zero so:

$\Delta(mv)$ = 70 kg × 30 m s^{-1} = 2100 kg m s^{-1}

and so F = 2100 N in the opposite direction to the water's motion.

(b) The force exerted on the water is also 2100 N, but in the opposite direction, i.e. outwards.

Q 52 D

Force is equal to rate of momentum change (see Question 51). The air mass is moving twice as quickly so twice the mass hits the train per second. Twice the mass moving at twice the velocity has four times as much momentum, so the force will increase four-fold.

Q 53 (a) Using $\Delta W = F\Delta s$ and $F = ma$:

$$F = \frac{\Delta W}{\Delta s}, \; a = \frac{F}{m} = \frac{\Delta W}{m\Delta s}$$

$$a = \frac{1 \times 10^6\,\text{J}}{24 \times 10^3\,\text{kg} \times 1\,\text{m}} = 41.7\,\text{m s}^{-2} \approx 4 \times g$$

This is a very large acceleration, so would be very uncomfortable to experience.

(b) Using the same method as in (a), a = 1.3 m s^{-2} (which is much more comfortable but still a real jolt).

Q 54 (a) $E_k = \dfrac{1}{2}mv^2$

$$= 0.5 \times 750 \times 10^3\,\text{kg} \times (83\,\text{m s}^{-1})^2 = 2.58 \times 10^9\,\text{J}$$

$\Delta E_{grav} = mg\Delta h$

$$= 750 \times 10^3\,\text{kg} \times 9.81\,\text{N kg}^{-1} \times 318\,\text{m}$$

$$= 2.34 \times 10^9\,\text{J}$$

So there is *more* than enough energy to reach the top of the Eiffel Tower.

(b) The track would need to curve upwards, enabling the train to 'coast' to a great height – and it would need to be engineered so as to reduce frictional energy losses almost to zero.

Q 55 From Question 46, with $m_1 = 5\,\text{t}$ (wagon), $m_2 = 41\,\text{t}$ (*Eurostar* loco), $u_1 = 46\,\text{m s}^{-1}$, $u_2 = 0\,\text{m s}^{-1}$ and $v_1 = v_2 = v = 5\,\text{m s}^{-1}$:

before collision,

$$\text{total } E_k = \frac{1}{2}m_1u_1^{\,2} + \frac{1}{2}m_2u_2^{\,2}$$

$$= 0.5 \times 5 \times 10^3\,\text{kg} \times (46\,\text{m s}^{-1})^2 = 5.3 \times 10^6\,\text{J}$$

after collision,

$$\text{total } E_k = \frac{1}{2}(m_1 + m_2)v^2$$

$$= 0.5 \times 46 \times 10^3\,\text{kg} \times (5\,\text{m s}^{-1})^2 = 5.8 \times 10^5\,\text{J}$$

This collision is *not* elastic, since kinetic energy is not conserved.

From Question 47, with $m_1 = 90\,\text{t}$, $m_2 = 20\,\text{t}$:

$u_1 = +50\,\text{m s}^{-1}$, $u_2 = -30\,\text{m s}^{-1}$,

$v_1 = +30\,\text{m s}^{-1}$, $v_2 = 60\,\text{m s}^{-1}$

before, $E_k = 0.5 \times 90 \times 10^3\,\text{kg} \times (50\,\text{m s}^{-1})^2$
$\qquad\qquad + 0.5 \times 20 \times 10^3\,\text{kg} \times (30\,\text{m s}^{-1})^2$

$$= 1.215 \times 10^8\,\text{J}$$

after, $E_k = 0.5 \times 90 \times 10^3\,\text{kg} \times (30\,\text{m s}^{-1})^2$
$\qquad\qquad + 0.5 \times 20 \times 10^3\,\text{kg} \times (60\,\text{m s}^{-1})^2$

$$= 7.65 \times 10^7\,\text{J}$$

Again, this is not an elastic collision since there is a net loss of kinetic energy.

Q 56 **(a)** Suppose the system is initially at rest; i.e. momentum is zero. If a rivet, mass $m_1 = 50\,\text{g}$, is fired with a velocity $v_1 = 20\,\text{m s}^{-1}$, then the trolley-plus-man, total mass $m_2 = 400\,\text{kg}$, must acquire a velocity v_2 where $m_1v_1 + m_2v_2 = 0$ (because the total momentum must still be zero):

$$v_2 = \frac{-m_1v_1}{m_2} = \frac{0.005\,\text{kg} \times 20\,\text{m s}^{-1}}{400\,\text{kg}}$$

$$= -2.5 \times 10^{-3}\,\text{m s}^{-1}$$

That is, the trolley-plus-man moves at $2.5\,\text{mm s}^{-1}$ in the opposite direction to the rivet.

(b) One way to approach this problem is to find the force on the trolley-plus-man, using $F = \dfrac{\Delta p}{\Delta t}$

With $\Delta t = 1\,\text{s}$,

$$\frac{\Delta p}{\Delta t} = \frac{2 \times 400\,\text{kg} \times 2.5\ 10^{-3}\,\text{m s}^{-1}}{1\,\text{s}}$$

$$= 2\,\text{kg m s}^{-2} = 2\,\text{N}$$

(Alternatively, you could use the momentum change of two rivets to get the same result.)

Acceleration of trolley, $a = \dfrac{2\,\text{N}}{400\,\text{kg}}$

$$= 0.0050\,\text{m s}^{-2}$$

$v = u + at$, with $t = 60\,\text{s}$

$u = 0\,\text{m s}^{-1}$

$v = 0.0050\,\text{m s}^{-2} \times 60\,\text{s}$

$\quad = 0.30\,\text{m s}^{-1}$

(Or use the same method as in **(a)**, with $m_1 = 2 \times 60 \times 0.050\,\text{kg}$.)

(c) For rivets, in 1 minute:

$$\Delta E_k = 0.5 \times 2 \times 60 \times 0.050\,\text{kg} \times (20\,\text{m s}^{-1})^2$$

$$= 1200\,\text{J}$$

For trolley-plus-man, in 1 minute:

$$\Delta E_k = 0.5 \times 400\,\text{kg} \times (0.30\,\text{m s}^{-1})^2 = 18\,\text{J}$$

(d) No chance! The train takes only $500\,\text{s}$ to reach the man's original position. By this time, his speed is only $2.5\,\text{m s}^{-1}$ (using the same method as in **(b)**) and he will still be very close to the on-rushing train.

(By plotting s–t graphs for the train and for the trolley, perhaps using a spreadsheet, you could find the time for the train to catch up with the trolley.)

 Q 57 **(a)**

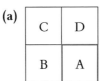

C	D
B	A

(b) risk = 1×10^{-5}

cost = £10 000 000

This point lies within region A of the graph so the safety measure would not be implemented.

(c) (Answers will vary.)

THE MEDIUM IS THE MESSAGE

Why a chapter called *The Medium is the Message?*

The world today depends on communications, whether speaking to someone, sending a letter, e-mailing, texting, hearing the news on the radio, television or Internet, or transferring data from and to the stock exchange or bank (Figure 2.1a). Recently there have been major advances in display technology (Figure 2.1b), providing bigger, clearer and brighter displays with lower power demands.

Electricity and magnetism first played their part in communication when the French physicist André Ampère (1775–1836) put forward the idea of telegraphy in 1820. Early versions involved switching electromagnets so that needles pointed to particular characters. By 1843 Samuel Morse (1791–1872) had invented his famous code, enabling the much faster transmission of messages. In 1876 Alexander Bell (1847–1922), the Scottish-born American scientist, invented the telephone. Radio followed a little later, with Guglielmo Marconi (1874–1937) sending a Morse-code signal across the Atlantic Ocean in 1901 and the first speech transmission by Reginald Fessenden in 1906. Television was established in 1925.

Figure 2.1 (a) Communications

Since then, electronic developments have come thick and fast, with the sending of signals through fibre-optic cables, faster broadband speeds, greater storage capacity of computers, digital television, plasma screens, satellite communications and our increasing reliance on mobile phones, where a wide range of information can be displayed on liquid-crystal displays. The world seems to want to communicate faster and more often, and progress in physics has helped this come about; the telecommunications industry is one of the major employers of physicists.

Figure 2.1 (b) Display

While we tend to think of communications in terms of messages and images sent between people, communication can involve any transmission, reception and display of information. In this chapter we will look mainly at how information is sent and displayed.

OVERVIEW OF PHYSICS PRINCIPLES AND TECHNIQUES

In this chapter you will learn how sensors based on potentiometers are used to gather information electronically. The output from sensors is often sent along a fibre-optic cable, and you will also explore how a signal is attenuated as it travels along such a cable.

Next you will explore the operation of a charge-coupled device (CCD), which uses the physics of charge, voltage and capacitance to produce and display two-dimensional images. Finally, you will look at the physics behind other displays — cathode-ray tubes, light-emitting diodes, liquid crystal and plasma displays — where electric and magnetic fields play a key part.

In this chapter you will extend your knowledge of:

- using graphs from *Higher, Faster, Stronger* and *Digging Up the Past*
- charge and potential difference from *Technology in Space* and *Digging Up the Past*
- the behaviour of light from *The Sound of Music, Good Enough to Eat* and *Technology in Space*
- electric current and voltage from *Technology in Space* and *Digging Up the Past*
- capacitors and exponential change from *Transport on Track*.

In other chapters you will do more work on:

- electric and magnetic fields in *Probing the Heart of Matter*
- electromagnetic radiation in *Reach for the Stars*.

1 Sensing and sending information

1.1 Sensing the situation

A lot of electronic equipment uses sensors to respond to the environment: for example, a doorbell responds to the pressure of a finger on the button on the door, and street lamps have sensors to measure the brightness of the sky so they turn on when it gets dark. There are many different sensors used in electronic equipment and we often want to transmit the measurement over some communication link, for example measuring the weather conditions at various sites around the world. One of the simplest types of sensor is a position sensor that measures how far something has moved from a reference position. Position sensors are used in factory machines and medical equipment, and in cars, trains and aircraft, where they convey information about the orientation of wheels, wing flaps, nose cones and so on.

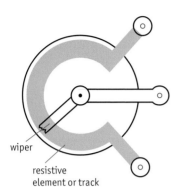

wiper

resistive
element or track

Figure 2.2 Inside a rotary potentiometer

The potentiometer

A simple but commonly used position sensor depends on the rotary **potentiometer** or **potential divider**. Figure 2.2 shows a rotary potentiometer with a resistive element of uniform thickness, width and resistivity. As you saw in the chapter *Digging Up the Past*, the potential difference between the sliding contact and a fixed end depends on the position of the sliding contact.

PRACTICAL SKILLS REFERENCE

Carrying out practical work

See Practical skills notes 3.1–3.2

ACTIVITY 1 SENSING POSITION

Discuss how a potentiometer, attached to a model aircraft part (e.g. a wing flap), could be arranged to give a voltage output that depends on position.

Then use a potentiometer attached to a model aircraft part to show how the voltage output varies with angular position (see Figure 2.3).

STUDY NOTE

For further discussion of why a meter's resistance can affect its measurement of voltage, see MDM Additional Sheet 1.

PRACTICAL SKILLS REFERENCE

Carrying out practical work

See Practical skills notes 3.1–3.2

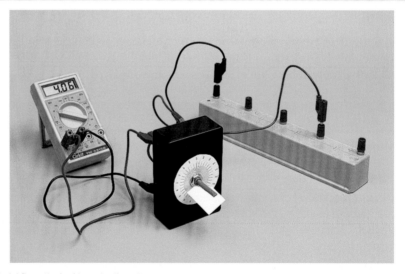

Figure 2.3 Model flap attached to potentiometer

In Activity 1 you should have found that the variation of a voltage output with angle was fairly linear. In practice the resistance of the device that is measuring the voltage output will affect this variation. You will explore this further in Activity 2.

ACTIVITY 2 WHAT IF THE VOLTMETER'S RESISTANCE IS LOW?

Investigate the effect of taking the potentiometer output to a low-resistance voltmeter. See how a voltage follower or buffer amplifier can help deal with the problems that arise.

The voltage output from the position sensor needs to be transmitted to the computers that are managing the aircraft's flight. This could be achieved by simply having long wires connected, via an appropriate interface or link, to the computers. However, long wires produce large voltage drops and so what started at 5 V might end up as only 2 V, with a consequent faulty interpretation of position. So other means of communicating information are needed that do not get degraded en route – though these lie outside the scope of this chapter.

This section has involved some revision of work from the chapters *Technology in Space* and *Digging Up the Past*. Use Questions 1 and 2 to check your progress and understanding.

QUESTIONS

1 Why is a large voltage drop associated with a current in a long cable?

2 Assume that the voltmeter in Figure 2.4 has infinite resistance. State the reading in each case. (Note that in (c), the contact is at the midpoint on the potentiometer.)

Figure 2.4 Diagrams for Question 2

1.2 Fibre-optic cables

Much communication requires signals being sent along a cable. Electrical signals can be sent along cables made of copper or other good electrical conductors. Alternatively, visible or infrared radiation can be sent along fibre-optic cables. Fibre-optic cables are used in most long distance communication systems, such as telephone systems and computer networks.

Figures 2.5 and 2.6 show the structure and relative sizes of the cables. We will concentrate on fibre-optic cables.

Figure 2.5 Coaxial cable

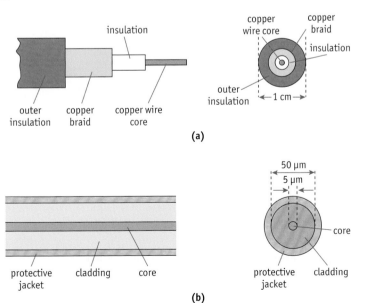

Figure 2.6 Structure of (a) coaxial cable and (b) fibre-optic cable

In fibre-optic cables the signal travels inside a thin transparent fibre, meeting the surface at an angle of incidence greater than the critical angle. It therefore undergoes total internal reflection and is confined to the interior of the fibre. Figure 2.7 shows this for a fibre made of a single material. The radiation is reflected at the surface, i.e. $i = r$ in Figure 2.7, and if the angle of incidence, i, is greater than the critical angle, C, then total internal reflection takes place. C depends on the refractive index, n, between the fibre and its surroundings:

$$n = \frac{1}{\sin C} \qquad (1)$$

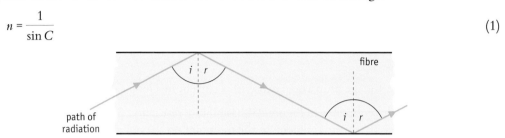

Figure 2.7 Radiation undergoes total internal reflection within an optical fibre

Unfortunately a pulse of radiation (light or infrared) entering a simple fibre is not just directed one way; it spreads sideways and results in many rays travelling by different routes down the fibre (Figure 2.8). The **dispersion** (spreading) means that a signal that starts as a sharp pulse is smeared out after travelling along the fibre. Because the refractive index is the same throughout the fibre, the speed of propagation of the light through it is the same regardless of route. Therefore ray C in Figure 2.8 takes the least time and ray A takes the most time. The result is that a sharp pulse gets spread out. The longer the fibre, the worse things get. This effect is known as **multipath dispersion**. The net effect is shown in Figure 2.9.

Figure 2.8 Rays travelling different routes in a fibre

Multipath dispersion can be reduced somewhat by limiting the range of angles of incidence that result in total internal reflection. This can be achieved by surrounding the core fibre with a cladding material whose refractive index is slightly lower than that of the core to produce a stepped-index fibre. Activity 3 illustrates how this works.

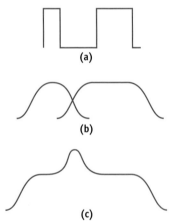

Figure 2.9 (a) Original pulses (b) after dispersion and (c) combined effect

⚙ ACTIVITY 3 MULTIPATH DISPERSION

Draw accurate diagrams to show how rays of light travel along (a) an optical fibre of refractive index $n_1 = 1.6$ surrounded by air and (b) $n_2 = 1.4$. Compare the ranges of angles of incidence that result in total internal reflection in each case.

Single or **monomode** fibre in which the rays can pass down only the centre of the fibre were invented in the 1980s and are now used in all long distance fibre-optic communications. You might then think that no dispersion would occur, as all the rays would have travelled the same path in the same time. This would be so if the light was of a single wavelength, but even laser light has a small spread of wavelengths: 1 to 2 nm or less. The refractive index of a material varies with wavelength, which causes the small spread of wavelengths to arrive at different times due to their different speeds. The result is again a slight spreading of the pulse.

QUESTION

Q 3 On aircraft or in a factory, cable lengths are typically less than 100 m. If the dispersion is 1 ns km^{-1}, and the pulse rate is 10^9 s^{-1}, explain whether dispersion is likely to be a major problem.

1.3 Attenuation

Both coaxial and fibre-optic systems have their advantages and drawbacks, and **attenuation** (loss of intensity) occurs in both. Attenuation in metal cables occurs where there is a transfer of energy from the signal to the cable due in part to its resistance and leakage through the insulation. Optical fibres cannot in practice be completely transparent, and some of the signal is absorbed by the material through which it passes. The attenuation in an optical fibre is modelled in Activity 4 using a jelly 'fibre' that absorbs strongly in the infrared region (see Figure 2.10). This activity also demonstrates a mathematical way of describing the attenuation.

Figure 2.10 Investigating attenuation in jelly fibre

⚙ ACTIVITY 4 THE JELLY FIBRE

Using an infrared-emitting diode as the source and an infrared-sensitive phototransistor as the detector, investigate attenuation in a glass fibre modelled with jelly.

Figure 2.11 shows some typical results from Activity 4. As you should have found, the intensity (as measured by the voltage) always changes by the same fraction when the length is changed in equal steps. For example, in Figure 2.11, a length of $x = 7$ cm reduces the voltage to half its initial value (V falls from 1.0 V to 0.5 V); after a further 7 cm the voltage has halved again ($x = 14$ cm, $V = 0.25$ V); after another 7 cm it has halved yet again ($x = 21$ cm, $V = 0.125$ V).

PRACTICAL SKILLS REFERENCE

Carrying out practical work
See Practical skills notes 3.1–3.2

Analysis and interpretation of data
See Practical skills notes 4.1–4.2

Conclusion and evaluation
See Practical skills notes 5.1–5.2

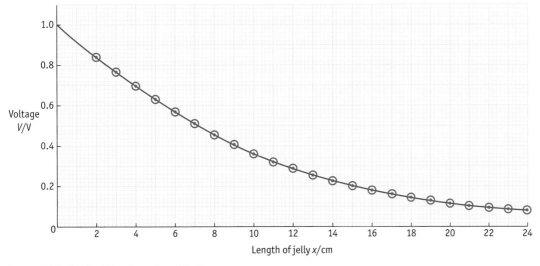

Figure 2.11 Graph of V against x for a jelly fibre

STUDY NOTE

In the chapter *Transport on Track*, you will have met another example of such a change when looking at the way a capacitor discharges.

Exponential change

The attenuation of radiation intensity in a fibre is an example of an **exponential** change. Exponential changes are characterised by the 'equal steps resulting in equal fractional changes' pattern. In the case of capacitor discharge, the steps are time intervals, whereas here they are equal changes in length.

Mathematically the exponential change in intensity with distance can be described by the equation:

$$I = I_0 e^{-\mu x} \tag{2}$$

where I_0 is the intensity of radiation that enters the jelly, I the intensity after it has travelled a distance x, and e is the exponential number (e ≈ 2.718). μ is the **absorption coefficient**, which describes how rapidly the intensity is attenuated with distance – if μ is large then $e^{-\mu x}$ rapidly becomes small as x is increased; in other words, the radiation is strongly attenuated.

After a distance $x = \dfrac{1}{\mu}$, the intensity will fall to $\dfrac{1}{e}$ of its initial value ($\mu x = 1$, and so $I = I_0 e^{-1}$).

The value of μ can therefore be found from a graph such as that in Figure 2.11 by finding the distance over which I falls to $\dfrac{1}{e}$ of its initial value, and then finding the reciprocal of that distance. This reasoning also tells us the units of μ: if x is expressed in m then μ has units m^{-1}, and if x is in cm then μ is in cm^{-1}.

MATHS REFERENCE

Reciprocals
See Maths note 3.3
Exponential changes
See Maths note 9.1
Exponential functions
See Maths note 9.2

STUDY NOTE

If you are familiar with calculus, you can show that the pattern of equal steps resulting in equal fractional changes will always be described by an equation such as Equation 2.

Many, but by no means all, naturally occurring changes are exponential. How can you tell whether a change is exponential? One way is to see whether it fits the 'equal steps result in equal fractional changes' pattern, but this can be time-consuming, and if there are large experimental uncertainties it is hard to decide where to draw a best-fit curve. A neater way is to plot a **log–linear graph**. Taking the log of both sides of Equation 2 gives:

$$\log(I) = \log(I_0) - \mu x \log(e) \tag{3}$$

Equation 3 describes a relationship of the form $y = mx + c$; $\log(I_0)$, μ and $\log(e)$ are all constants. If $\log(I)$ is plotted on the y-axis and x on the x-axis, then the graph is a straight line as shown in Figure 2.12. This gives another way to test whether a change is exponential, as a log–linear graph produces a straight line. This is usually quicker than looking for the equal-steps equal-fractions pattern, and it is often easier to tell whether points lie close to a straight line than it is to draw a best-fit curved line through points with error bars.

STUDY NOTE

A log–linear graph is so called because the y-axis has a logarithmic scale whereas the x-axis is linear.

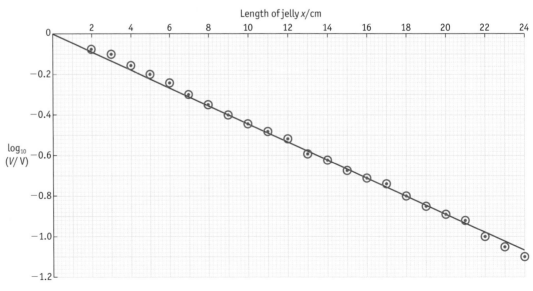

Figure 2.12 Log–linear graph of the data from Figure 2.11, using base 10 logs

If you use so-called natural logarithms (ln, logs to base e), then Equation 3 becomes:

$$\ln(I) = \ln(I_0) - \mu x \qquad (4)$$

because $\ln(e) = 1$. A graph of $\ln(I)$ against x is a straight line with gradient $-\mu$. Figure 2.13 shows a plot of the same data in Figures 2.11 and 2.12, using logs to base e. From the triangle drawn, the gradient of this graph is approximately -0.1, so the absorption coefficient in this case is approximately $0.1\ \text{cm}^{-1}$.

STUDY NOTE

The equations you used in *Transport on Track* for capacitor discharge have the same form as Equations 2 and 4.

MATHS REFERENCE

Logs to base e

See Maths note 8.5

MATHS REFERENCE

Using log graphs

See Maths note 8.7

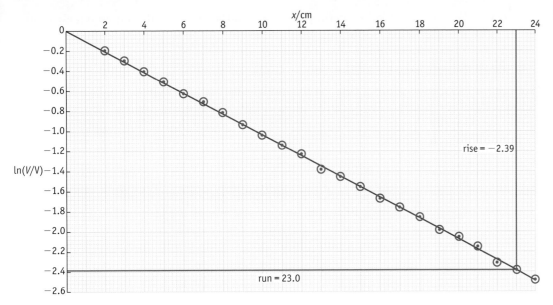

Figure 2.13 Log–linear graph of the data from Figure 2.11, using natural logs (logs to base e)

The fibre-optic cables used in communications have *much* smaller absorption coefficients than a jelly fibre – absorption coefficients are typically about $10^{-5}\ \text{m}^{-1}$ or less. Attenuation is negligible for the short distance connections between offices or homes and the telephone exchange. Optic-fibres are also used for long distance cables, such as those under the seas and oceans connecting countries together, so attenuation can still be a problem. For this reason it is important to reduce attenuation as much as possible. One way in which this can be done is by careful choice of fibre material and wavelength of radiation. As shown in Figure 2.14, attenuation can depend markedly on wavelength.

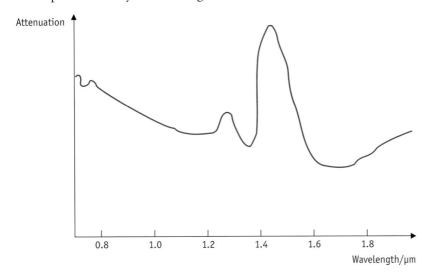

Figure 2.14 Attenuation against wavelength for a glass optical fibre

QUESTION

Q 4 An optical fibre has an absorption coefficient $\mu = 2.0 \times 10^{-6}\,\mathrm{m}^{-1}$.

(a) Over what distance will the intensity of a signal fall to $\dfrac{1}{e}$ of its initial value?

(b) What will be the intensity of the signal, expressed as a fraction of its initial value, after it has travelled 100 km along the cable?

(c) If the intensity of light entering the fibre is $200\,\mathrm{W\,cm}^{-2}$, and the output intensity must be at least $2\,\mathrm{W\,cm}^{-2}$, how long can the fibre be?

1.4 Summing up Part 1

Part 1 began by explaining how a potentiometer can be used as a position sensor. You then looked in detail at the degradation or attenuation of signals due to dispersion and absorption. Attenuation in a jelly fibre illustrated an exponential change, and showed how logarithmic graphs can help analyse such changes.

Use Activity 5 and Questions 5 and 6 to help you check your progress.

> ⚙ **ACTIVITY 5 SUMMING UP PART 1**
>
> Look back through your work and check that you know the meaning of all the terms highlighted in **bold**. Using some or all of these terms, write a paragraph describing how information about the angle of a flap in a piece of factory machinery could be sensed and then sent to a control centre via fibre optic communication.

QUESTIONS

Q 5 In a certain optical fibre, the intensity of the signal falls to half its initial value after travelling 50 km. What is the absorption coefficient of the material expressed in units of m^{-1}?

Q 6 To minimise attenuation, what would be the best wavelength to use for sending signals along the fibre in Figure 2.14?

2 Making an image

In Part 1 of this chapter, we were concerned with measurements (for example, of position) that could each be expressed as a single number or as a sequence of numbers that changed with time. However, much communication is concerned with encoding and transmitting two-dimensional pictures. Several examples of communicating two-dimensional images will be familiar to you – TV pictures, for example. A less familiar example might be the device shown in Figure 2.15, which is used to give a pilot or a soldier 'night vision'. It works by first producing an infrared image of the surroundings and then displaying the image in a helmet-mounted display. The image is projected into the user's field of vision, to give a realistic picture of the view ahead.

2.1 Charge-coupled devices

This section of the chapter is about one particular device: the CCD (charge-coupled device), which is used to obtain two-dimensional images. CCDs have replaced photographic plates previously used for recording astronomical images obtained by telescopes.

CCDs are usually rectangular, with length and widths ranging from a few millimetres to a few centimetres. Each is divided up into millions of little picture elements (pixels), which each have a tiny light sensor (see Figure 2.16). When a photon is absorbed by a pixel, a photoelectron is released from the semiconductor – this is rather similar to what happens in the photovoltaic cells that you used in the chapter *Technology in Space*. However, in this case, the released electron remains trapped within the pixel. As more photons are absorbed, more electrons are released. If an image is projected onto a CCD (for example, through a system of lenses) then charge builds up in each pixel according to the number of photons that have reached it, in this way recording the image.

In order to 'read' a CCD image, the electrons are electrically shunted along the CCD step by step, producing a small pulse of current as each 'package' reaches the edge. The size of each pulse can be detected and encoded to produce a digital signal, and the signals can then be processed and transmitted in just the same way as those that represent single measurements.

If the image is bright, then only a very short exposure time is needed, as the CCD images that give a soldier 'night vision' are produced so rapidly that there is no noticeable time delay. However, if a CCD is being used in an astronomical telescope to build up an image of a very faint galaxy, then this takes much longer. Figure 2.17 shows the image of a galaxy recorded from a CCD attached to a telescope. A CCD sensor can be made more sensitive to light by increasing the surface area of each pixel so that they each collect more light; this involves increasing the total size of the sensor and/or decreasing the number of pixels on the sensor.

Figure 2.15 Helmet-mounted display

Figure 2.16 CCD 'chip'

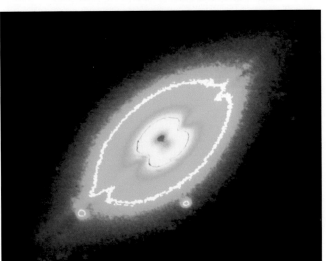

Figure 2.17 CCD image taken through a telescope

STUDY NOTE

You met capacitors in the chapter *Transport on Track*.

Each pixel in a CCD behaves rather like a tiny solar cell attached to a **capacitor** – a device that stores energy by separating electric charge. The simplest type of capacitor consists of two parallel conducting plates separated by an air gap. When a capacitor becomes charged, electrons simply redistribute themselves between the plates, making one positive and one negative. There is no net charge added to the device (see Figure 2.18). Most practical capacitors (Figure 2.19) consist of thin layers of conducting material separated by thin layers of electrical insulation and fitted into a compact space, so concealing the layered structure.

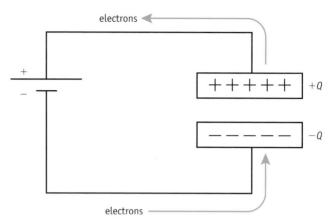

Figure 2.18 Charging a capacitor

Figure 2.19 Various types of capacitor

PRACTICAL SKILLS REFERENCE

Carrying out practical work

See Practical skills notes 3.1–3.2

⚙ ACTIVITY 6 MODEL CCD IMAGER

Use an array of capacitors and solar cells, each connected as in Figure 2.20, to model the behaviour of a CCD imager.

In Activity 6 you probably took it on trust that the voltage across a capacitor is proportional to the charge it stores. You can explore this in more detail in Activity 7.

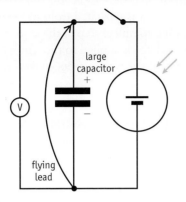

Figure 2.20 Circuit of model CCD pixel

In Activity 7 you will have found that equal additions of charge raised the voltage across the capacitor by equal amounts. In other words: $Q \propto V$. The constant of proportionality, C, is the **capacitance** of the device.

$$Q = CV \quad \text{or} \quad C = \frac{Q}{V} \tag{5}$$

The SI unit of capacitance is the farad, F, where $1\,\text{F} = 1\,\text{C}\,\text{V}^{-1}$. Most practical capacitors are rated in microfarads, μF ($1\,\mu\text{F} = 1 \times 10^{-6}\,\text{F}$) or picofarads, pF ($1\,\text{pF} = 1 \times 10^{-12}\,\text{F}$); the ones you used in Activity 6 were unusual in being rated at about 1 F – they were 'memory back-up' capacitors for use in computers.

QUESTIONS

Q 7 A typical MOS (metal oxide semiconductor) capacitor in a CCD imager has a capacitance of 1 pF and can store a charge of $Q \approx 0.16\,\text{pC}$.

    (a) How many electrons would have to be collected by the MOS capacitor to give it a charge of 0.160 pC? (Electron charge $e = 1.60 \times 10^{-19}\,\text{C}$)

    (b) What will be the voltage across such a capacitor when it is charged to this extent?

Q 8 The capacitors used in the model CCD imager were rated at about 1 F. If the highest voltage the solar cells can provide is 0.45 V, what is the maximum charge that one of these capacitors could have stored?

Q 9 If a solar cell provides an average current of 50 mA at 0.45 V to a 1 F capacitor, calculate how long it is before it reaches its maximum charge for that voltage. (The current actually varies while the capacitor charges, as you may recall from the chapter *Transport on Track*.)

STUDY NOTE

To find out how capacitors can be combined in series and in parallel to produce different values of capacitance, see MDM Additional Sheet 2.

CCD read-out

In order to extract information from the array, each packet of charge has to be moved to an output system (like the voltmeter in the model in Activity 6) and measured. Each package of charge is moved to a separate read-out register. To enable this to be done, the CCD is designed so that each tiny light-sensitive capacitor is separated from its neighbour by another capacitor that remains unaffected by light. One complete step in the read-out for one pixel is shown in Figure 2.21. First, a temporary voltage is applied to the 'empty' capacitor so that charge spreads into it. The first capacitor's voltage is then reduced to zero, leaving all the charge on the adjacent one.

This process is repeated until each capacitor's initial charge gets to an output sensor where the voltage across the capacitor is finally measured. All of this is done with the aid of a computer, which controls the application of the voltages, and a device (an analogue to digital converter), which encodes each measured voltage in order to determine how brightly to display an image at each point on a screen.

Figure 2.21 CCD read-out

2.2 Energy and capacitors

A capacitor is a device that stores energy by separating charge. In Activity 8 you will see that the stored energy is related to the capacitor voltage.

PRACTICAL SKILLS REFERENCE

Carrying out practical work

See Practical skills notes 3.1–3.2

⚙ ACTIVITY 8 HOW MUCH ENERGY IS STORED?

Charge a 10 000 µF capacitor to different voltages and discharge it into various numbers of small light bulbs. Use a charged 10 000 µF capacitor to drive a motor that lifts a small load. Measure the height through which the load is lifted as the capacitor discharges. Change the voltage used to charge the capacitor and see how this affects the height.

Suggest a relationship between capacitor voltage and energy stored.

In Activity 8 you should have found that 3 V provided enough energy to light one bulb with a moderately bright flash; doubling the voltage to 6 V provided enough energy to light four bulbs, and tripling the voltage enabled nine times as many bulbs to be lit. This suggests that the stored energy is proportional to the square of the capacitor voltage. The following argument shows why this indeed the case.

We already know that charge stored is proportional to the voltage to which the capacitor has been charged. A graph of voltage, V, against charge, Q, will therefore be a straight line through the origin, as in Figure 2.22.

Now think of a very small amount of extra charge (δQ), which has been placed on the capacitor – it is so small that the voltage across the capacitor remains at V. The small amount of work done (δW) is given by:

$$\delta W = V \times \delta Q \tag{6}$$

which is the shaded area in Figure 2.22. (Remember in the chapter *Technology in Space* you saw that $W = QV$.)

To charge the capacitor from zero to V would require the addition of lots of extra tiny amounts of charge δQ, each of which would need work done equal to the area of another tiny strip on the graph. The areas of all the tiny strips added together would give the total energy transferred; that is, the area under the graph. Since the graph is a straight line, we can write:

$$W = \frac{1}{2}QV \tag{7}$$

where W is the total work done when the capacitor is charged to a potential difference V. Since energy is conserved, this must also be the energy stored in the capacitor and which is recovered when it discharges.

Combining Equations 5 and 7 we get:

$$W = \frac{1}{2}CV^2 \tag{8}$$

and:

$$W = \frac{1}{2}\frac{Q^2}{C} \tag{9}$$

Figure 2.22 Graph of voltage, V, against charge, Q, for a capacitor

QUESTION

Q 10 How much energy is stored in a 10 000 µF capacitor when it is charged to:

(a) 3 V?   (b) 6 V?   (c) 9 V?

2.3 Summing up Part 2

Part 2 has shown you how a CCD responds to incident radiation by building up a pattern of charge that represents the brightness of an image. It has also involved some aspects of the physics of capacitors.

⚙ ACTIVITY 9 SUMMING UP PART 2
Look back through your work and make sure you know the meaning of each term printed in **bold**.

3 On display

Over the past decade or so, there have been many developments in display technology that have involved cutting-edge research and major investment from industry. These provide much of the material for this chapter. However, we will also look a bit further back and examine some devices that were developed during the 20th century. One area that has made great use of advances in display technology is the aircraft industry, so we will use the example of aircraft flight deck displays to illustrate the rapid and fundamental changes that have taken place.

A pilot needs to be able to see clearly how an aircraft is functioning, and so the display of information is a crucial aspect of avionics. Various techniques have been developed to improve the quality of such displays so that they can be easily seen and interpreted. Many of these techniques are not specific to avionics and can also be found in many other types of communication system.

Modern aircraft, such as the A380, have arrays of instruments giving details of speed, heading, altitude, position, as well as a myriad of other information (see Figure 2.23). Some displays appear on a screen only when required.

Figure 2.23 Instrument displays in the A380

(a)

(b)

Figure 2.24 Needle dial instrument displays: (a) electrical meter and (b) pressure gauge

Originally, virtually all aircraft instruments used needle dial displays, much like the analogue electrical meters and pressure gauges (see Figure 2.24) you probably met earlier in your science and technology education.

Displays in modern aircraft are of three types:

- In head-down displays (HDDs) the viewer has to look directly at the instrument. This might be a case of looking down but could also be looking upwards or sideways.

- Head-up displays (HUDs) project an image directly in front of the viewer to combine with the view of the environment (see Figure 2.25).

- Helmet-mounted displays (HMDs) are essentially HUDs provided direct to an image-forming system attached to the pilot's helmet (see Figure 2.25). In Part 2 of this chapter, you will already have come across one of the key elements of a helmet-mounted display – the CCD imager. The display moves with the head.

Here we will concentrate on head-down displays: making these into head-up or helmet-mounted displays involves optical projection of the image, which lies outside the scope of this chapter. We will deal in turn with three types of display – the light-emitting diode (LED), the liquid-crystal display (LCD) and the cathode-ray tube (CRT).

3.1 Light-emitting diode displays

LEDs have been used for many years as indicators on all kinds of electrical and electronic equipment. Large LED displays are made using millions of small LEDs. With the invention of the blue LED it became possible to make full colour LED devices. By grouping the LEDs in sets of red, green and blue mounted close together, a full-colour image can be produced by adjusting the brightness of each LED appropriately. To make a moving image, each LED is switched on or off by a computer at the correct time to produce the picture. Switching times can be quite short – in the nanosecond (10^{-9} s) region.

Figure 2.25 Head-up display

LEDs are **semiconductor** devices. Semiconductors are materials that conduct less well than metals do (they have a much lower density of mobile charges) but they are not insulators. The first known semiconductors were crystals of elements whose atoms have four electrons in their outer shell, such as silicon or germanium. Silicon is still the most common semiconductor used in electronics, but LEDs and many other modern semiconductors use a mixture of elements to form a semiconductor crystal. For example, gallium nitride (GaN) is used to make LEDs; the gallium has three electrons in its outer shell and the nitrogen has five electrons in its outer shell, giving an average of four electrons per atom. GaN is an example of a III–V semiconductor (the Roman numerals refer to the numbers of outer electrons). When a semiconductor crystal forms, the sharing of electrons by neighbouring atoms in the crystal gives each atom eight electrons in its outer shell, producing the lowest energy for these materials. Figure 2.26 shows how the two elements can share electrons in this way. Note that all electrons are the same – the colours show how 3 and 5 can be arranged to make 8.

atom with 3 electrons in the outer shell
(e.g. aluminium, gallium or indium)

atom with 5 electrons in the outer shell
(e.g. nitrogen, phosphor or arsenic)

Figure 2.26 Schematic diagram of outer electrons in III–V semiconductor crystal

Doping

Semiconductors conduct much better (their **resistivity** is lowered) if a small number of atoms of another element are introduced to the crystal to make the average number of electrons per atom slightly more than four or slightly less than four. This is called **doping**.

In a **p-type semiconductor**, the average number of outer electrons per atom is less than four. Some of the atoms have only seven electrons around them with space for another one – we call this space a **hole** (Figure 2.27). An electron from a neighbouring atom can easily move into the hole, allowing good electrical conduction and leaving a hole next to another atom. If an electron moves from the left into a hole, then the hole appears to move to the right. Many texts talk about holes moving as conduction of positive charge, but this is just negatively charged electrons moving in the opposite direction. Although the p in p-type stands for 'positive', the material is electrically neutral (the electrons balance the positive charges on the nuclei) and does not have an overall positive charge.

If the average number of electrons per atom is more than four, then the extra electrons cannot fit into the shell containing eight electrons so they have to occupy a higher energy level shell. These higher-energy electrons can move more easily, reducing the resistivity. This is an **n-type semiconductor**.

STUDY NOTE

You met resisitivity in the chapter *Digging Up the Past*.

hole

Figure 2.27 Schematic diagram of outer electrons in p-type III–V semiconductor crystal

STUDY NOTE

You met electron energy levels and photons in *The Sound of Music* and the eV unit of energy in *Technology in Space*.

Semiconductor junction

An LED uses a piece of n-type semiconductor joined to a piece of p-type semiconductor; the place where the two types of semiconductor meet is called a **junction**. Electrons near the junction move from the high energy level in the n-type material to fill the holes in the lower energy level in the p-type material. As it loses energy, each electron emits a photon. In the example shown in Figure 2.28, the energy difference is 3.4 eV, so that is the photon energy.

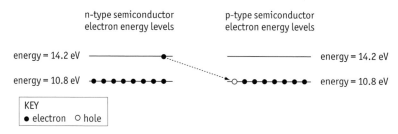

Figure 2.28 Electron crossing a p–n junction

When the electrons move from the n-type material, this leaves the n-type material near the junction positively charged and makes the p-type material near the junction negatively charged. The negative charge near the junction in the p-type material repels further electrons. The region near the junction where the free electrons move across the junction to fill the holes on the other side is called the **depletion zone** because the n-type material is depleted of free electrons and the p-type material is depleted of holes.

The potential difference produced across the depletion zone is equal to the change in energy when a free electron moves into a hole. To make electrons move continuously across the junction so that photons are continuously produced, a potential difference needs to be applied across the junction to push more electrons from the n-type material into the p-type material; so the n-type material needs to be made negative and the p-type material needs to be made positive. The pd applied using an external power supply must be at least as great as the pd produced across the depletion zone by the migration of electrons.

Activity 10 involves an exploration of LEDs. It leads into some revision of ideas about charge and voltage that you met in the chapter *Technology in Space*, and about photons and energy levels that you met in *The Sound of Music* and *Technology in Space*.

PRACTICAL SKILLS REFERENCE

Planning and experimental design
See Practical skills notes 2.1–2.4

Carrying out practical work
See Practical skills notes 3.1–3.2

Analysis and interpretation of data
See Practical skills notes 4.1–4.2

Conclusion and evaluation
See Practical skills notes 5.1–5.2

⚙ ACTIVITY 10 ILLUMINATING LEDS

Using LEDs of at least two different colours, find the voltage that will just light each one. Comment on whether this apparatus can be used to determine a value for the Planck constant.

In Activity 10 you should have found that different coloured LEDs required different voltages to make them light. As described above, in a semiconductor LED, charge carriers are accelerated across a gap between energy bands, which results in a photon being emitted whose energy is equal to the size of the energy gap. The voltage that is just big enough to light the diode corresponds to charge carriers just crossing the gap. If a particle with charge q is accelerated through a potential difference ΔV, then it acquires energy ΔE where:

$$\Delta E = q\Delta V \tag{10}$$

If this energy is then radiated as a photon, the frequency f and wavelength λ of the radiation are related to the energy via the expression:

$$\Delta E = hf = \frac{hc}{\lambda} \tag{11}$$

where c is the speed of light in a vacuum and h is the Planck constant.

QUESTIONS

The following data will be needed when you tackle Questions 11–13:

electron charge, $e = 1.60 \times 10^{-19}\,\text{C}$

$1\,\text{eV} = 1.60 \times 10^{-19}\,\text{J}$

Planck constant, $h = 6.63 \times 10^{-34}\,\text{J s}$

speed of light in vacuum, $c = 3.00 \times 10^{8}\,\text{m s}^{-1}$.

Q 11 The bandgaps for compounds of indium and antimony can be as small as 0.18 eV and those of gallium and nitrogen up to 3.4 eV. Calculate the wavelength of the radiations given off by the LEDS with each of those bandgaps.

Q 12 What bandgap energies would be needed to produce visible LEDs that emit light over the range 400 nm to 600 nm? Express your answer in joules and in eV.

Q 13 (a) A standard red LED was found to be passing a current of 15 mA when there was a pd of 2.15 V across it. What power was it drawing from the supply?

(b) An infrared-emitting diode, operating at a voltage of 1.3 V, passed a current of 100 mA. What power was it drawing from the supply?

(c) If you need 10^6 LEDs to produce a screen display, estimate the power required.

3.2 Liquid-crystal displays

Liquid-crystal displays (LCDs) are used in mobile phones, DVD players, hand-held games, calculators, on cookers, video-recorders, televisions, computer monitors and so on, and in aircraft cockpits. The production of LCDs is a multi-billion-pound industry.

A **liquid crystal** is a liquid in which the molecules arrange themselves in some sort of ordered pattern. Although we tend to think of molecules in liquids being completely randomly distributed, many liquids are in fact liquid crystals – these include DNA, RNA, soaps and detergents, and even the brain is about 70 per cent liquid crystal.

Liquid-crystal display technology has its origins in the 1970s, when George Gray of Hull University made the first stable synthetic liquid crystals, known as nematic alkylcyanobiphenyls. These have the important characteristic that their optical properties can be controlled electrically.

A so-called nematic liquid crystal consists of long thin molecules (the Greek word *nematos* means 'thread-like') that tend to arrange themselves with their long axes parallel to each other, as shown in Figure 2.29. This comes about because the molecules are **polar** – that is, their charge is neutral overall but not evenly distributed. Thus, one end is slightly positive and the other slightly negative, forming an **electric dipole** (Figure 2.30); they line up because the positive end of one molecule attracts the negative end of another.

Figure 2.30 An electric dipole

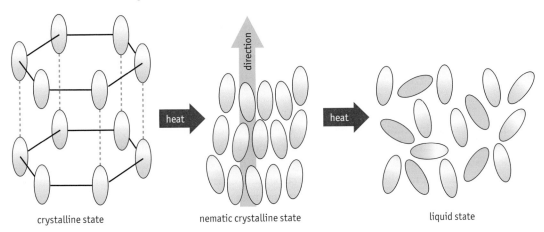

crystalline state   nematic crystalline state   liquid state

Figure 2.29 Molecules in a nematic liquid crystal

The LCDs in use today are of a type known as twisted nematic (TN) or super-twisted nematic (STN). A TN display consists of a thin layer of nematic liquid crystal between two glass plates whose surfaces have been coated with a very thin conducting layer and treated with a rubbed polymer layer; this has the effect of creating small parallel grooves. The molecules prefer to lie along, rather than across, the grooves. The glass plates are oriented with one set of grooves running at right angles to the other, which introduces a 90° twist in the molecular arrangement, as shown in Figure 2.31(a). (In an STN display, there is a twist of 270°.) This 'sandwich' is mounted between two pieces of polarising filter set in the 'crossed' position.

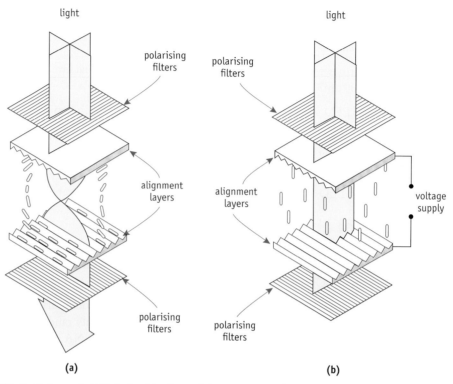

Figure 2.31 Twisted nematic LCD display: (a) voltage off, transparent, and (b) voltage on, opaque

STUDY NOTE

You met polarised light and polarising filters in the chapter *Good Enough to Eat*.

The light emerging through the first filter is polarised. As it travels through the liquid crystal, the twisted line of dipoles guides its plane of polarisation so that it rotates through 90° and can emerge through the second filter – the 'sandwich' is transparent. However, if a potential difference is applied across the liquid crystal, the molecules are forced to line up end-to-end between the plates, as shown in Figure 2.31(b). Then there is no longer any rotation of the plane of polarised light, so the light cannot emerge through the second filter – the 'sandwich' is opaque.

To make a display, the thin conducting coating is applied in a suitable pattern (for example, line segments to make up numbers), each connected to a separate small electrode that can be controlled by a digital on–off signal. Colours can be obtained by placing coloured filters in front of the display and shining white light through it. TV screens, computer monitors and various instrument displays can be made from arrays of many thousands of individual LCDs.

Electric field

To explain how the molecules are aligned by the applied potential difference, it is useful to introduce the idea of an **electric field**, which is defined as a region in which a charged object experiences a force. The **electric field strength**, E, is defined as the force experienced by a charge of 1 coulomb; in other words, the force per unit charge. If a charge q experiences a force of magnitude F, then the magnitude of field strength is:

$$E = \frac{F}{q}$$

(12)

The SI units of electric field strength are N C^{-1}. As force is a vector, electric field is also a vector; that is, it has direction as well as magnitude. The **direction of an electric field** is defined as that of the force experienced by a positive charge.

It is quite difficult to measure electric field strength by directly measuring the force on a charged object; this is mainly because in most practical cases the force is quite small. However, there is an alternative way of thinking about electric fields, which both provides an easier way to measure E and also gives some insight into the type of field that is present in an LCD.

Figure 2.32 shows an arrangement for producing an electric field between two parallel conducting plates separated by a distance Δx and with a potential difference ΔV applied between them.

Now think of placing a single small positive charge, q, on plate A in Figure 2.32. As A is connected to the positive terminal of the power supply, the charged particle will move straight across the gap to plate B – the field direction is straight across from A to B as shown in Figure 2.32. As the particle moves through the potential difference ΔV between the plates, it will gain energy ΔW, where:

$$\Delta W = q\Delta V \tag{13}$$

Another way to look at the same situation is to say that there is a force of magnitude F acting on the particle. As the particle crosses the gap, width Δx, this force does work ΔW, where:

$$\Delta W = F\Delta x \tag{14}$$

As Equations 13 and 14 are just two ways of describing the same thing, we can write:

$$F\Delta x = q\Delta V \tag{15}$$

As the electric field strength is defined as the force per unit charge, we can rearrange Equation 15 to get its magnitude:

$$E = \frac{F}{q} = \frac{\Delta V}{\Delta x} \tag{16}$$

In other words, the field strength is equal to the **potential gradient** – the rate at which the potential changes with distance across the gap. As field strength is a vector, we should really be careful with signs. The field is directed from A to B, which is the direction in which the potential is decreasing from positive to zero – that is, the potential gradient is negative. We can therefore write Equation 16 as:

$$E = -\frac{\Delta V}{\Delta x} \tag{16a}$$

For a uniform field between two parallel plates, the magnitude of the field is usually written as:

$$E = \frac{V}{d} \tag{16b}$$

where V is the potential difference between the plates and d their separation.

Equation 16 provides an alternative way to define and measure electric field strength. It also gives its alternative SI unit: V m^{-1}.

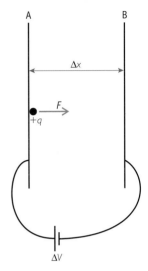

Figure 2.32 Electric field between two parallel plates

⚙ ACTIVITY 11 FORCE FIELD

Use a charged foil strip or a charged ping-pong ball to demonstrate the force on a charged object in an electric field between two parallel plates.

Electric field lines

An electric field can be represented diagrammatically by drawing **electric field lines**, which show the path and direction a free positive charge would take, as shown in Figure 2.33. The spacing between the electric field lines shows the strength of the field; for a strong electric field the electric field lines are close together. If voltage across the plates is increased, and the plates remain the same distance apart, the field strength will increase and the field lines will be drawn closer together, as in Figure 2.33(b). If the plates move closer together and the voltage across the plates remains the same, the field strength will increase and the field lines will be drawn closer together, as in Figure 2.33(c). The electric field between two parallel plates is called a **uniform electric field** because the electric field strength is the same at every point between the plates, and so all of the field lines are the same distance apart between the plates.

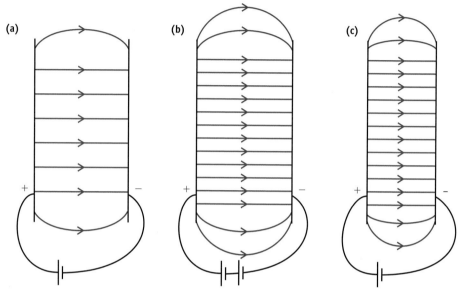

Figure 2.33 (a) Electric field lines for a uniform field between parallel plates; (b) when the pd between the plates is increased; (c) when the plates are moved closer together

STUDY NOTE

You met turning effects and moments in *Higher, Faster, Stronger.*

If an electric dipole (such as a polar molecule) is placed in such a field, then the positive and negative end will experience forces in opposite directions; in other words, there will be a turning effect due to the combined **moments** of the forces. The dipole will twist around until it is aligned with the field direction, as shown in Figure 2.34. In Activity 12, grass seeds become polarised in the electric field and then twist around until they align with the field.

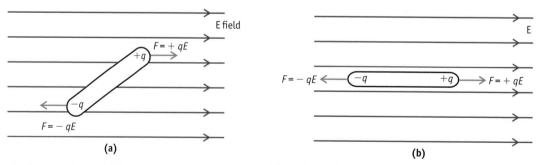

Figure 2.34 A dipole in an electric field experiences a turning effect and so (a) it turns until it is aligned with the field direction and (b) the resultant moment of the two forces becomes zero

The field around a point charge is a **radial electric field**. The field lines point towards or away from the centre of the charge because they show the direction a small positive charge would take. The field lines of a positive charge point outwards as they show the direction of the force on another small positive charge. The field lines of a negative charge (for example, that found on an electron) point inwards, as shown in Figure 2.35.

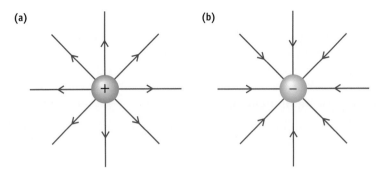

Figure 2.35 Electric field lines for (a) positive and (b) negative point charge

The field of a point charge is not uniform. It is strong near the point charge where the field lines are close together and weaker further away where there are bigger spaces between the field lines.

Figure 2.36 shows the apparatus that can be used to measure field strength: a double-flame probe. There are two tiny flames at the tips of two fine metal tubes, which are connected to a gas supply. A flame produces some ionisation in the air, and the ions move until there is no potential difference between the flame and its immediate surroundings. The two fine tubes are connected to the cap and case of an electroscope, where the deflection of the leaf depends on the potential difference between the cap and the case. The electric field strength close to the tips of the fine tubes can then be found from a measurement of the potential difference indicated by the electroscope and the physical separation of the tips.

Figure 2.36 Using a double-flame probe to explore an electric field

⚙ ACTIVITY 12 PROBING THE FIELD

Use grass seeds on oil to show the field lines in an electric field.

Use a double-flame probe to investigate the size and direction of the electric field between parallel plates and in some other arrangements.

PRACTICAL SKILLS REFERENCE

Carrying out practical work

See Practical skills notes 3.1–3.2

Lines of equipotential

Another way of representing an electric field is to draw lines to join up the points that are all at the same potential (relative to a chosen reference point defined as zero potential). These are called **lines of equipotential**. In three dimensions, there are equipotential surfaces – that is, all points that lie on an equipotential surface are at the same potential. Equipotential lines and surfaces are often referred to simply as **equipotentials**.

We draw equipotentials at uniform intervals, for example in Figure 2.37 the equipotentials are drawn every 100 V. For parallel plates, it is easy to work out that the equipotentials are parallel to the plates and equally spaced, which shows a uniform field.

Notice that *equipotentials are always at right angles to electric field lines.* Where the equipotentials are close together the field strength is high. Look at Equation 16a on page 81: if Δx becomes small for a given ΔV, then E is large.

In a radial electric field, the equipotential surfaces are concentric spheres that get further apart as the field gets weaker. A two-dimensional diagram of the field (Figure 2.38) shows equipotential lines that are concentric circles.

+300 V   0 V

+200 V   +100 V
equipotentials

Figure 2.37 Field lines and equipotentials in a uniform field

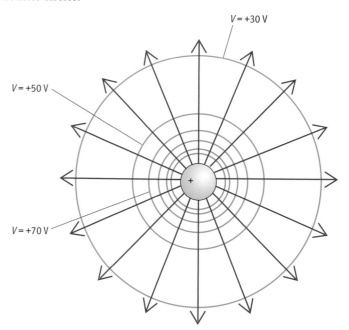
$V = +30\,\text{V}$
$V = +50\,\text{V}$
$V = +70\,\text{V}$

Figure 2.38 Field lines and equipotentials in a radial field

PRACTICAL SKILLS REFERENCE

Carrying out practical work

See Practical skills notes 3.1–3.2

ACTIVITY 13 PLOTTING LINES OF EQUIPOTENTIAL

Use conducting paper to plot lines of equipotential in various two-dimensional fields.

Alternatively, use a single-flame probe to find the equipotential surfaces between parallel plates and other arrangements. (You can use the equipment from Activity 12 but just light one of the probes and connect the earth of the electroscope to one of the plates.)

QUESTIONS

Q 14 Show that the SI units N C^{-1} and V m^{-1} are exactly equivalent.

Q 15 (a) If the voltage between two parallel plates is 1 kV and they are separated by 0.1 m, what is the magnitude of the electric field strength between them? Give your answer in both possible SI units.

(b) How large a force would this electric field exert on a charge of 3 μC?

Q 16 An LCD operating at a voltage of 2.5 V passed a current 140 pA ($1\,\text{pA} = 1 \times 10^{-12}\,\text{A}$). What power was it drawing from the supply?

Q 17 Figure 2.39 shows the electric field lines between two parallel plates. The potential difference, V, between the plates is 200 V and the distance, x, between the plates is 4.0 cm.

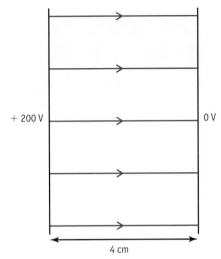

+ 200 V   0 V

4 cm

Figure 2.39 Diagram for Question 17

(a) What is the strength of the electric field between the plates?

(b) For each of the following, calculate the field strength and draw a diagram showing the electric field lines.
   (i) $V = 400$ V and $x = 4.0$ cm
   (ii) $V = 200$ V and $x = 8.0$ cm
   (iii) $V = 100$ V and $x = 2.0$ cm

(c) On each of your diagrams for part (b), add lines of equipotential at intervals of 50 V.

Q 18 Figure 2.40 shows some lines of electrical equipotential.

(a) Use the given distances between the equipotentials to calculate the field strength:
   (i)   between points X and Y, distance = 3.0 cm
   (ii) at point Z, distance between 50 V and 350 V = 6.0 cm.

(b) Copy the diagram and add electric field lines. Remember to put arrows on the field lines.

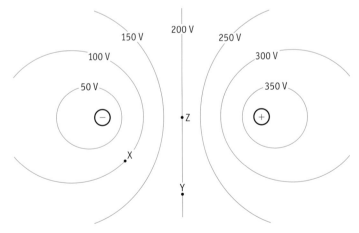

Figure 2.40 Diagram for Question 18

Q 19 The forward voltage required to operate a green LED is 2.3 V. The depletion region in the LED is 0.4 μm wide.

(a) Calculate the field strength in the depletion region of the LED.

(b) Calculate the wavelength of the light produced by the LED.

PRACTICAL SKILLS REFERENCE

Planning and experimental design

See Practical skills notes 2.1–2.4

Carrying out practical work

See Practical skills notes 3.1–3.2

Analysis and interpretation of data

See Practical skills notes 4.1–4.2

Conclusion and evaluation

See Practical skills notes 5.1–5.2

⚙ ACTIVITY 14 COMPARING LCD AND LED DISPLAYS

Compare the voltages and powers required to operate an LCD and an LED seven-segment display.

FURTHER INVESTIGATION

Measure and compare the response times of LEDs and LCDs. Explore the factors that affect response time.

Operating an LCD with an alternating supply

If an LCD is operated using a DC voltage for a long time, the display will be damaged. DC (direct current) means that the voltage and current are always in the same direction, though they may vary in size. A DC field will cause negative ions in the liquid between the electrodes to collect on the positive electrode and positive ions to collect on the negative electrode. This changes the electrodes and prevents the display from working correctly. To prevent electroplating, an LCD can be driven with an alternating (AC) voltage so that the electric field keeps changing direction and the electrodes are not damaged.

The **frequency** of the AC is important. Frequency of an alternating current or voltage is defined in the same way as the frequency of the waves you studied in *The Sound of Music*; it is the number of cycles per second, with SI units of Hz. Frequency is related to the **period**, which is the time for the current or voltage to go through one cycle (e.g. from maximum positive value to maximum negative, and back again to maximum positive).

The display must not be driven at very high frequency (usually no more than 100 Hz), otherwise the current charging and discharging the plates can damage the display. The greater the frequency, the more rapidly the display needs to be charged up and reach a given voltage, and therefore the larger the current needed. A large current can damage the display by heating it, and consumes unnecessary power. However, if the frequency of the signal driving the display is too low the image flickers, as we can see the segments in the display turn off and on.

The **amplitude** of an alternating current or voltage is its maximum positive (or negative) value. These values are also known as the **peak current** and **peak voltage**. The greater the amplitude of the applied voltage, the greater the current and the greater the heating effect. When operating an LCD, the current and voltage should usually be kept as low as possible in order to avoid damaging the device.

PRACTICAL SKILLS REFERENCE

Planning and experimental design

See Practical skills notes 2.1–2.4

Scientific background and information research

See Practical skills notes 1.2–1.3

Carrying out practical work

See Practical skills notes 3.1–3.2

⚙ ACTIVITY 15 OPERATING AN LCD WITH AN ALTERNATING SUPPLY

Use a signal generator and oscilloscope to find the minimum frequency and voltage amplitude required to operate an LCD without flicker.

QUESTION

Q 20 Figure 2.41 shows three different signals used to drive LCD displays. Find the amplitude, period and frequency of each signal.

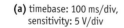

(a) timebase: 100 ms/div, sensitivity: 5 V/div

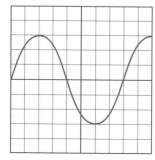

(b) timebase: 20 ms/div, sensitivity: 0.5 V/div

(c) timebase: 200 ns/div, sensitivity: 2 V/div

Figure 2.41 Diagrams for Question 20

3.3 Shine a light

Since 2012, the UK has had a total ban on making or importing incandescent light bulbs. Manufacturers and importers also have to comply with EU regulations concerning the efficiency of any bulbs they sell. The alternatives to incandescent bulbs (that is, those in which there is a heated wire filament) are either LED lighting or compact fluorescent light bulbs (CFL) (see Figure 2.42). Generally a CFL lasts 8–12 times longer than an incandescent bulb. It also uses less energy; for instance, a 15 W compact fluorescent could replace a 60 W incandescent bulb because it has greater efficiency.

(a)

(b)

Figure 2.42 Energy-efficient light bulbs: (a) LED (b) CFL

AC and DC

What does 60 W mean? Power, P, in a device such as a light bulb is calculated using:

$$P = IV = I^2R = \frac{V^2}{R} \tag{17}$$

where I is the current in a device, V the pd across it and R its resistance. Household lighting runs from an alternating supply where the voltage and current are constantly changing. The mean value of an alternating V, or I, is zero (it is negative for half the time). Yet clearly the mean power is not zero. It turns out that thinking about the mean power gives us a useful way to define the average value of an alternating current or pd.

STUDY NOTE

You used Equation 17 in the chapter *Technology in Space*.

From Equation 17:

$$\text{mean } P = \text{mean } I^2R = \text{mean } \frac{V^2}{R}$$

$$\langle P \rangle = \langle I^2 \rangle R = \frac{V^2}{R} \tag{17a}$$

STUDY NOTE

We use angle brackets $\langle$ and $\rangle$ to denote a mean value. Some texts use a bar over the averaged quantity e.g. $\overline{P}$.

The voltage and current of mains electricity vary as a sine wave, as shown in Figure 2.43(a). The mean value of the sine function is zero. However, the power varies the *square* of current or voltage, so it is always positive. As you can see from Figure 2.43(b), the mean value of the *square* of the sine function is 1/2. So, if the peak voltage and peak current are V_0 and I_0, we can write:

$$\langle I^2 \rangle = \frac{I_0}{2} \tag{18}$$

and:

$$\langle V^2 \rangle = \frac{V_0}{2} \tag{18a}$$

The quantities $\langle I^2 \rangle$ and $\langle V^2 \rangle$ are called the **mean-square current** and **mean-square voltage**.

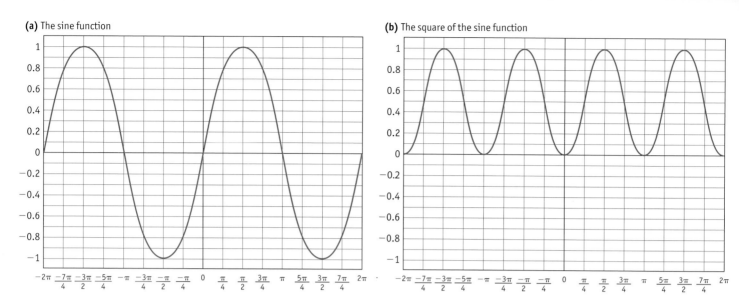

Figure 2.43 (a) The sine function (b) the square of the sine function

A useful way to define the average voltage or current is to take the square root of $\langle I^2 \rangle$ or $\langle V^2 \rangle$. This value is called the **root-mean-square** (rms) value, and given the subscript rms:

$$I_{rms} = \frac{I_0}{\sqrt{2}} \tag{19}$$

$$V_{rms} = \frac{V_0}{\sqrt{2}} \tag{19a}$$

The rms value is much more useful than the mean value when dealing with alternating currents and voltages. It is always positive, and there is a straightforward relationship between the mean power and the rms current and voltage:

$$\langle P \rangle = V_{rms} I_{rms} = I_{rms}^2 R = \frac{V_{rms}^2}{R} \tag{20}$$

It is always the rms voltage and not the peak voltage that is written on light bulbs and other consumer equipment. The rms voltage is the DC voltage that would produce the same electrical power as the AC supply. The peak voltage of the mains electricity supply in the UK is about 330 V, but the voltage written on most equipment is 230 V, which is the rms voltage. The 330 V peak AC mains supply produces the same brightness in an incandescent bulb as you would get if the bulb were attached to a steady 230 V DC supply.

PRACTICAL SKILLS REFERENCE

Carrying out practical work

See Practical skills notes 3.1–3.2

⚙ ACTIVITY 16 AC AND DC

Compare the peak AC voltage with the DC voltage required to produce the same brightness in a light bulb.

QUESTIONS

Q 21 The mains electricity supply in the USA is 120 V rms at 60 Hz. Calculate the peak mains voltage in the USA.

Q 22 An electric heater has a resistance of 0.5 Ω. Calculate the peak current in the heater when it is connected to the 230 V rms mains supply.

Q 23 A portable radio has a speaker with a resistance of 64 Ω. Calculate the power dissipated by the speaker when a sine wave of peak voltage 3.0 V is played.

Q 24 Figure 2.44 shows how the supply voltage to a bulb of resistance $10\,\Omega$ varies with time.

(a) Draw a graph to show how the current through the bulb varies with time.

(b) Calculate V_{rms} and I_{rms} for the bulb.

(c) Calculate the power dissipated by the bulb.

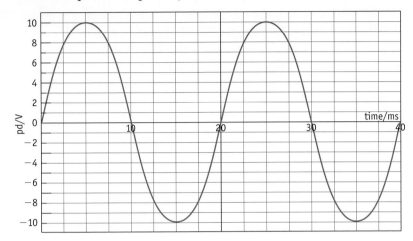

Figure 2.44 Diagram for Question 24

CFL

A compact fluorescent lamp (CFL) contains mercury vapour. At room temperature a few of the atoms are ionised, so as well as neutral atoms there are some free electrons and positive ions. When you apply a potential difference across the tube that contains the vapour, the resulting field accelerates the electrons and ions. Provided they can acquire enough kinetic energy between collisions, the free electrons cause further ionisation when they collide with the neutral atoms. The gas very rapidly becomes ionised (an ionised gas is called a **plasma**) and it is then a good electrical conductor.

Collisions between the electrons and the mercury atoms and ions excite the bound electrons to higher energy levels. The excited electrons lose energy by emitting ultraviolet photons as they fall to lower energy levels. The inside of the CFL is coated with phosphor, which absorbs the ultraviolet and emits visible radiation.

Although CFLs are more energy efficient, there is a downside: concerns have been raised about disposing of the CFLs due to the mercury vapour contained within them.

QUESTION

Q 25 The excited mercury vapour in a CFL emits ultraviolet radiation where the dominant photon energy is 4.9 eV. In order to excite the mercury ions so they give out this ultraviolet radiation, the free electrons in the plasma must acquire energy between collisions that is at least 4.9 eV.

If the average distance between collisions is 8.0×10^{-3} m, what must be the magnitude of the uniform electric field within the lamp?

 ACTIVITY 17 WHY SHOULD WE REPLACE TRADITIONAL LIGHT BULBS?

Prepare a report on the three different types of light bulbs – incandescent, compact fluorescent light bulbs and LEDs – and give your recommendations for replacement and use.

You should consider:

- how they work
- cost of purchase and use
- efficiency
- problems with pollution
- effectiveness.

Was the UK government correct to ban incandescent bulbs? What would you recommend?

3.4 Cathode-ray tube

The oldest type of electronic display is the cathode-ray tube (CRT), which was the most common type of display in use until LCD sales overtook CRT sales in 2004. CRTs have been very widely used in TVs, oscilloscopes and computer monitors, and still form the basis for many of the displays used in aircraft.

Figure 2.45 shows one type of cathode-ray tube, known as a Perrin tube after its designer Jean Baptiste Perrin (1870–1942). Figure 2.46 shows a CRT from a television. Inside each tube there is an **electron gun** in which electrons are released from a negative electrode (a **cathode**) and accelerated in an electric field between the cathode and **anode** (positive electrode). The resulting stream of high-energy electrons hits a phosphor screen, causing it to glow. (Originally the electron beam was called a 'cathode ray', before people knew what it consisted of.) There must be a vacuum inside the tube, otherwise air molecules would scatter the electron beam. Around the outside of the tube are deflection coils; current in these coils produces a magnetic field that deflects the electron beam.

Figure 2.45 Perrin tube

Figure 2.46 CRT used in a television

> ⚙ **ACTIVITY 18 USING A PERRIN TUBE**
>
> Use a Perrin tube to explore the operation of a cathode-ray tube. Adjust the tube so as to produce a 20 mm deflection of the spot on the screen.

We will now look in more detail at the operation of each part of a CRT and see how it contributes to the formation of an image.

Electron gun

Figure 2.47 shows details of an electron gun. An electric current heats the filament, F. During a process known as **thermionic emission**, electrons that have enough energy escape from the filament's surface, in a similar way to water evaporating from a puddle of rain. These electrons and the radiation from the filament heat up a nickel cathode (C) that is coated with a mixture of barium and strontium oxides, resulting in the thermionic emission of a large number of electrons from the cathode. A potential difference (often called the gun voltage) between the cathode (C) and anodes (A_1 and A_2) produces an electric field that accelerates the electrons. A_1 is either a hollow cylinder or a disc with a hole in it, which focuses the electrons into a narrow beam. A_2 is usually a cylinder with a disc with a hole at the screen end, through which the accelerated electrons pass on their way to the screen. Most guns also incorporate a grid (G), which is negative with respect to the cathode and so controls the overall rate at which electrons pass through the whole system.

The electrons hit the phosphor screen and are brought to rest. Their kinetic energy is transferred to the atoms and molecules in the phosphor screen coating, which become excited and then lose energy by emitting photons.

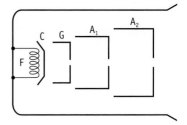

Figure 2.47 Electron gun

QUESTIONS

Q 26 **(a)** If a gun with voltage 5000 V accelerates an electron, how much energy does the electron acquire?

(b) What is the electron's:
  (i)   kinetic energy?
  (ii) speed when it emerges from the gun?

(c) How would the presence of air in the CRT affect your answer to **(b)**?

(Electron charge $e = 1.60 \times 10^{-19}$ C. Electron mass $m_e = 9.11 \times 10^{-31}$ kg.)

Q 27 Explain how each of the following would affect the brightness of the spot on the screen:

(a) increasing the potential of anode A_2

(b) increasing the voltage across the filament

(c) making the grid more negative with respect to the cathode.

Figure 2.48 Line scanning to build up an image

Deflecting the beam

To produce an image, the beam is swept rapidly across the screen in a scanning pattern (Figure 2.48). As you saw in Activity 18, current-carrying coils produce the deflection.

You may have wondered why an electron beam is deflected in a magnetic field. In the chapter *Transport on Track* you saw that a current-carrying wire in a magnetic field experiences a force at right angles to the wire and to the field direction. The magnitude of this force, F, is given by:

$$F = BIl \sin\theta \tag{21}$$

where B is the magnetic flux density, I the current, l the length of the wire and θ the angle between the wire and the field direction (see Figure 2.49). The direction of the force is given by Fleming's left-hand rule (Figure 2.50), by which the first finger points along the direction of the magnetic field, the second finger points along the direction of the (conventional) current, and the thumb points in the direction of the resulting motion – that is, the direction of the force.

This force arises because of the motion of the charged particles in the magnetic field, so electrons that are not confined to a wire will still experience a force when they move in a magnetic field. The size of the force can be related directly to the electrons' speed v.

Figure 2.49 Current-carrying wire in a magnetic field

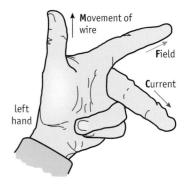

Figure 2.50 Fleming's left-hand rule

So what about the force on an electron, or indeed any other charge carrier? To deal with this we need to peer inside our current-carrying conductor (as in Figure 2.51) and consider the individual charge carriers.

Figure 2.51 Charged particles in a conductor

There are n charge carriers per unit volume. The volume of a piece of wire with length l and area A is Al, so the number of particles, N, in this length of wire is:

$$N = nAl \qquad (22)$$

In the chapter *Technology in Space* you used an expression that related the current, I, to the charge carriers' motion with average speed v:

$$I = nAqv \qquad (23)$$

Substituting this into Equation 21 gives:

$$F = BnAqvl \sin\theta \qquad (24)$$

Dividing this force by the number of particles N (Equation 22) gives an expression for the force on a *single* mobile charge carrier:

$$F = Bqv \sin\theta \qquad (25)$$

The direction of the force is still given by Fleming's left-hand rule. But beware: the direction of conventional current is that of the apparent movement of *positive* charge, i.e. in the opposite direction to the movement of electrons.

STUDY NOTE

The direction of conventional current is that in which positive charge appears to move, i.e. from the positive to the negative terminal of a power supply. Electrons, having negative charge, move in the opposite direction.

QUESTIONS

Q 28 What is the magnitude of the force that acts on an electron moving at speed $8.0 \times 10^7\,\mathrm{m\,s^{-1}}$ at right angles to a field whose flux density is 0.10 T? (Electron charge $e = 1.60 \times 10^{-19}$ C.)

Q 29 In Activity 18 you saw that a pair of coils, known as Helmholtz coils, placed either side of a CRT deflected the beam vertically. What can you deduce from this about the direction of the magnetic fields produced by the coils?

Helmholtz coils are named after their inventor, the Geman scientist Hermann von Helmholtz (1821–1894). They are rather special, being designed to produce a fairly uniform field in the region between them. This is achieved by having their centres separated by a distance equal to their radius (Figure 2.52). The magnitude, B, of the magnetic flux density so produced can be calculated using the following expression:

$$B = \frac{8\mu_0 nI}{5r\sqrt{5}} \qquad (26)$$

where I is the coil current, r the radius and separation of the coils, n the number of turns on each coil, and μ_0 is the permeability of a vacuum. $\mu_0 = 4\pi \times 10^{-7}\,\mathrm{N\,A^{-2}}$.

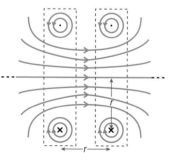

Figure 2.52 Magnetic fields produced by Helmholtz coils (**×** indicates current flowing into page; • indicates current flowing out of page)

In a CRT used to produce a display, such as that found on a TV screen (see Figure 2.53), the coils are mounted close to the electron gun, though they are not in fact Helmholtz coils. As you can see in Figure 2.46, two sets of coils are used; this is in order to scan the beam in two dimensions and build up an image. The coils and the current they carry must be designed to produce a deflection that matches the screen size. In Activity 18, you found the field that would deflect a particular electron beam through 20 mm, as would be required for a screen 40 mm from top to bottom. To produce a larger image, for example on a wide-screen TV, a larger field would be needed.

ACTIVITY 19 SCANNING THE SCREEN

Explain how the magnetic field must change in both direction and magnitude in order to produce the line scanning shown in Figure 2.48.

Figure 2.53 TV screen

The image on the screen

To produce a monochrome (single-coloured) image, all that is required is that signals fed to the gun control the brightness of the spot as it sweeps across the screen. A coloured image can be produced using three different phosphors, which glow red, green or blue when excited by an electron beam. Each pixel on the screen is a group of three dots, one of each phosphor. One way to excite the coloured phosphors is to use three electron guns, each targeting just one colour.

Equal exciting of each phosphor would cause white light to be emitted from the screen. Exciting just the blue and green phosphors equally would give cyan, the red and green equally would give yellow, and the red and blue equally would give magenta. Any colour and its intensity can be controlled by the speed and number of electrons hitting the combination of phosphors. These, and the position on the screen, are controlled by the signals fed to the electron gun and the deflecting coils.

Cathode-ray tubes produce good colour, brightness and definition. Their problem is their bulk and their power ratings, which are in the range 500 W and above.

3.5 Plasma screens

For many purposes, LCDs, LEDs or plasma screens have replaced CRT displays. Plasma displays were invented in 1964 at the University of Illinois in the USA. A plasma screen has about a million small **cells** that can emit spots of coloured light. Each cell is about 1 mm across and filled with gas (usually a mixture of neon and xenon). Each cell has electrodes on the top and bottom (Figure 2.54) and acts as a minute fluorescent light, creating a very bright display.

When a voltage is applied to the electrodes across the cell, this produces a strong electric field in the cell that causes ionisation as in a CFL (Section 3.3). The cell then contains a plasma (ionised gas), which gives the display its name.

glass front

front electrode

phosphor

rear electrode

Figure 2.54 Cell for a plasma display

When a free electron combines with a positive ion of neon or xenon, it gives off a photon of ultraviolet radiation. The photon may be absorbed by other atoms, causing them to ionise, which quickly leads to a very large number of ions in the cell. Many of the photons will hit the back of the cell, which has a fluorescent phosphor coating that gives off visible light when it absorbs UV radiation. The light comes out of the front of the cell, which is made of glass and is seen by the person viewing the screen. The screens are produced so that in each pixel there is a red cell, a green cell and a blue cell using different fluorescent phosphors, so that coloured images can be produced. Plasma screens are used for large displays to be viewed from some distance away, as each cell is relatively large.

Some displays use cells with two electrodes at the front (Figure 2.55). The field in the cell is not uniform; the electric field strength is not the same at all points in the cell. The gas in the cell is ionised where the electric field is strong. The electrodes can be close together if they are placed at the front of the cell, which means that a lower voltage can be used to get the same electric field strength.

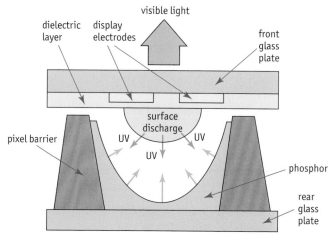

Figure 2.55 Cell with front electrodes

QUESTIONS

Q 30 The cells in a plasma screen are 150 μm high and have 200 V applied to them when they are turned on. What is the electric field strength in a plasma cell when it is turned on?

Q 31 When an electron combines with a xenon ion it produces an ultraviolet photon of wavelength of 147 nm. What is the energy of this photon?

Q 32 Sketch a copy of Figure 2.55 and on it draw a diagram of the electric field lines between the electrodes. Write a few sentences to explain where in the cell the gas is ionised.

3.6 Summing up Part 3

In Part 3 you have looked at instrument displays and light sources – cathode-ray tubes, light-emitting diodes, liquid-crystal displays, compact flourescent light bulbs, plasma displays – and the physics behind them. Magnetic and electric fields played their part and you saw how to calculate and measure these. You also extended your knowledge of electic circuit theory to include alternating current and voltage.

Use Activities 20 and 21 to check your progress and understanding.

⚙ ACTIVITY 20 SUMMING UP PART 3

Look back through your work and check that you know the meaning of all the terms printed in **bold**.

⚙ ACTIVITY 21 DISPLAY POWER

Small LCD displays require much less electrical power to operate than CRTs, but many consumers are buying large displays that have high power consumption. Compare the power consumption for televisions of similar display size that use different technologies to find the most efficient displays.

Suppose a member of your family wants to buy a new TV set, but has heard that some TVs have high power demands. Your relative is concerned about the environmental effects of electricity usage. What advice would you give him or her regarding this?

PRACTICAL SKILLS REFERENCE

Scientific questions and information research

See Practical skills notes 1.2, 1.3

QUESTION

Q 33 An LCD in a pocket calculator is operated by a 1.5 V power supply.

(a) If the conducting layers in the display are separated by 10 μm, what is the strength of the electric field within the display?

(b) What would be the magnitude of the force experienced by a free electron within this field? (Electron charge $e = 1.60 \times 10^{-19}$ C.)

LINK THE LEARNING

4 Message received

4.1 Delivering the message

In this chapter you have covered many aspects of modern communications technology. A substantial part of this chapter has been about displays.

In the 20th century, cathode-ray tubes (CRTs) were used for most domestic televisions. They produced good colour, brightness and definition, and had a quick response time. However, their power demands were large and they needed a lot of space for the three electron guns. As customers began to want larger televisions, the alternatives became more popular. LCD or plasma televisions are much thinner and allow for both wall mounting and greater screen size. The cost of these was initially high and did not meet the quality of CRTs. Furthermore, LCD television suffered from a slow refresh rate – the time it takes for the pixels to switch on and off to cope with a moving image. However, with improvements in technology and quality, CRTs are now rarely used as computer monitors, and LCDs are frequently used instead.

Plasma televisions have vastly increased in popularity. The size of a screen (i.e. its diagonal size) can be over a metre, and this large size means that plasma screens can contain 1920 × 1080 pixels, compared with the few hundred pixels along each dimension of a CRT screen. HDTV (high-definition TV) broadcast signals make use of this increase in pixel number, producing images that are sharper and more detailed than was possible with CRTs.

Another recent development is the organic light-emitting diode (OLED). This offers all of the benefits of a traditional LCD display, but each pixel radiates its own light. The main advantage is that you no longer need a backlight as in traditional LCD TVs, massively reducing power consumption and making the screen considerably thinner, allowing screens to be made even larger and promising ultra-high definition TV with screens containing 3840 × 2160 pixels.

In Activity 22, you will apply what you have learned about display technology and find out about some of the most recent developments.

PRACTICAL SKILLS REFERENCE

Scientific questions and information research

See Practical skills notes 1.2, 1.3

⚙ ACTIVITY 22 WHAT ELSE IS GOING ON?

Displays are very important to the electronics industry and many different types of display are being developed, including:

- OLED (organic light-emitting diode display)
- SED (surface-conduction electron-emitter display)
- FED (field emission display)
- ELV (electro-wetting light valve)
- EL (electro-luminescent).

Prepare a brief report on one type of display technology now available or currently being researched. Your report should outline how the display works, the advantages and disadvantages of the display technology and the main uses of the type of display.

Try typing 'display technology' into a search engine. Possible websites, accessible via **www.pearsonhotlinks.co.uk**, include Cambridge Display Technology.

4.2 Questions on the whole chapter

QUESTIONS

Q 34 **(a)** Plasmas have been studied using the fast discharge of banks of capacitors to deliver a burst of energy. If a bank of capacitors totalling 100 µF was to be charged to 50 kV and then discharged into the plasma in 1 ms, what mean power would be delivered?

(b) A 100 µF capacitor is charged up to just 10 V, and the energy stored is then discharged into a thermally insulated coil of wire of very low resistance. The temperature rise of the coil is recorded by a very sensitive thermometer as 0.005 °C. What rise in temperature would you expect if this same capacitor was charged to 20 V and then discharged into the same coil?

Q 35 Figure 2.56 shows electric field lines for two positive charges. Copy the diagram and add lines of equipotential.

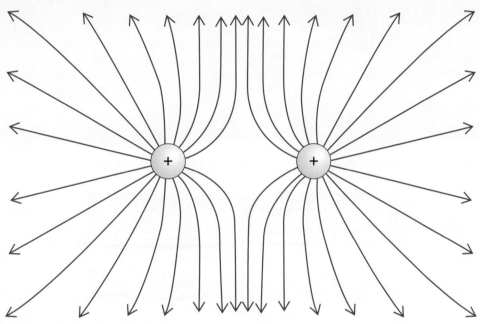

Figure 2.56 Diagram for Question 35

Q 36 In 1982 Sinclair Research Ltd launched the Sinclair Microvision – the first pocket-sized TV that used a flattened cathode-ray tube measuring only 10 cm × 5 cm × 2 cm. Compared with a normal size tube of that time, was much brighter and drew far less power from the supply.

(a)

(b)

Figure 2.57 Sinclair Microvision cathode-ray tube: (a) front view and (b) view from above

The tube consisted of an electron gun, collimator (to produce a narrow electron beam), vertical and horizontal deflection plates, a phosphor screen mounted on the rear electrode, a transparent front electrode, and a flat lens (a so-called Fresnel lens similar to those used on overhead projectors). The electron beam was fired in from the side of the screen.

The electric field between the deflection plates and another between the front tin oxide electrode and the rear electrode caused the electrons to be deflected towards the phosphor screen. The viewer looked through the flat lens at the screen at the back of the tube.

(a) (i) What must have been the direction of the electric field between the front tin oxide and rear electrodes?

(ii) The potential difference between these electrodes was 800 V and the separation between them 14 mm. What was the electric field strength between these electrodes?

(iii) Calculate the force on an electron in this electric field.
(Electron charge $e = 1.60 \times 10^{-19}$ C)

(b) The electrons were produced in a conventional electron gun by the process of thermionic emission.

  (i) If fewer electrons were required to be given off each second, what would have to be done to the filament?

  (ii) What effect would this have on the screen image?

  (iii) If the accelerating voltage of the electron gun were increased, what effect would this now have on the image of the screen?

(c) The phosphor screen was mounted on a metal plate. What was the advantage of this over the more usual glass?

(d) This television operated from a 6 V lithium battery with a current of approximately 0.08 A.

  (i) What was the power drawn from the supply?

  (ii) The battery was rated at approximately 1.2 Ah (ampere-hours). For how long could you have watched this television before the battery was discharged?

(e) Why do Smartphones have LCD or LED displays rather than a CRT?

Q 37 In 1897 Sir Joseph John Thomson (1856–1940), who believed that cathode rays consisted of a stream of charged particles, devised an experiment with which to measure their specific charge, i.e. the charge per unit mass. Thomson's result confirmed that the 'rays' were indeed streams of the particles we now call electrons; their specific charge is usually written as:

$$\frac{e}{m}$$

Thomson's apparatus is shown in Figure 2.58. Electrons were accelerated into the electric field between the parallel metal plates and deflected downwards from the initial straight-through path. A magnetic field was then applied across the tube, deflecting the electron beam back upwards and so nullifying the original deflection.

Figure 2.58 J.J. Thomson's apparatus

(a) (i) If the separation between the plates were 2.0 cm and the potential difference across the plates 2000 V, calculate the electric field, E, produced.

  (ii) If the charge on the electron is given the symbol e, write an expression for the magnitude of the force, F_{el}, on an electron in this electric field.

(b) Applying a magnetic field of flux density 3.125×10^{-3} T brings the electron beam back up to the straight-through line. If v is the speed of these electrons, write an expression for the magnitude of the force, F_{mag}, on these electrons due to this magnetic field.

(c) (i) From your answers to (a) and (b), obtain an expression for an electron's speed, v, in terms of E and B.

   (ii) Use your previous answers to obtain a numerical value for v.

(d) (i) If V is the accelerating potential difference, write an expression for the energy transferred to each electron as it is accelerated by the electron gun.

   (ii) Assuming that all the energy transferred by the gun is used to increase the electron's kinetic energy, write an expression relating an electron's kinetic energy to the accelerating voltage.

(e) (i) Use your previous answers to obtain an expression for $\dfrac{e}{m}$ in terms of E, B and V.

   (ii) The electrons in this apparatus had been accelerated through a potential difference of 3000 V. Calculate a value for $\dfrac{e}{m}$.

Q 38 'Which gives more dangerous electric shocks: AC or DC?' This is a popular discussion topic on some Internet forums. Here are some typical postings, which are all incorrect in some way.

(a) 'AC passes through zero, so if you touch a live wire from the mains you get about half a second to relax your muscles and let go.'

(b) 'AC is more dangerous because it delivers more amps.'

(c) 'DC is more powerful than AC because 230 V DC is 230 V all the time but 230 V AC mains varies from zero to 230 V.'

For each statement, explain why it is incorrect.

(Note: both AC and DC can kill. A current of only a few mA AC or DC can stop your heart, and this effect can be produced in the body by a few tens of volts.)

4.3 Achievements

Now you have studied this chapter you should be able to achieve the outcomes listed in Table 2.1.

TABLE 2.1 ACHIEVEMENTS FOR THE CHAPTER *THE MEDIUM IS THE MESSAGE*

	Statement from Examination Specification	Section(s) in this chapter
1	know and understand the distinction between base and derived quantities and their SI units	3.2
2	demonstrate their knowledge of practical skills and techniques for both familiar and unfamiliar experiments	1.1, 1.3, 3.1, 3.2
3	be able to estimate values for physical quantities and use their estimate to solve problems	3.1
4	understand the limitations of physical measurement and apply these limitations to practical situations	3.3
5	be able to communicate information and ideas in appropriate ways using appropriate terminology	1.4, 3.3, 3.4
6	understand applications and implications of science and evaluate their associated benefits and risks	3.3
8	understand the ways in which society uses science to inform decision making	3.3
108	understand that an electric field (force field) is defined as a region where a charged particle experiences a force	3.2
109	understand that electric field strength is defined as $E = \dfrac{F}{Q}$ and be able to use this equation	3.2
112	know and understand the relation between electric field and electric potential	3.2
113	be able to use the equation $E = \dfrac{V}{d}$ for an electric field between parallel plates	3.2
115	be able to draw and interpret diagrams using field lines and equipotentials to describe radial and uniform electric fields	3.2
116	understand that capacitance is defined as $C = \dfrac{Q}{V}$ and be able to use this equation	2.1
117	be able to use the equation $W = \dfrac{1}{2}QV$ for the energy stored by a capacitor, be able to derive the equation from the area under a graph of potential difference against charge stored and be able to derive and use the equations $W = \dfrac{1}{2}CV^2$ and $W = \dfrac{1}{2}\dfrac{Q^2}{C}$	2.2
122	be able to use the equation $F = Bqv\sin\theta$ and apply Fleming's left-hand rule to charged particles moving in a magnetic field	3.4
128	understand what is meant by the terms *frequency*, *period*, *peak value* and *root-mean-square value* when applied to alternating currents and potential differences	3.2, 3.3
129	be able to use the equations $V_{rms} = \dfrac{V_0}{\sqrt{2}}$ and $I_{rms} = \dfrac{I_0}{\sqrt{2}}$	3.3
132	understand that electrons are released in the process of thermionic emission and how they can be accelerated by electric and magnetic fields	3.4

Answers

Q 1 Think of the expression $V = IR$. If the cables have any resistance (R), then any current (I) must also be associated with a potential difference (V) between the ends.

Q 2 (a) 0 V   (b) 6 V   (c) 3 V

Q 3 Over a path of 100 m, a pulse will be spread by 0.1 ns. Each pulse has a length of 1 ns, so a spread of 0.1 ns would only make a slight difference to the pulse length and therefore dispersion will not be a major problem.

Q 4 (a) $I_x = \dfrac{I_0}{e}$

$$x = \frac{1}{\mu} = \frac{1}{2.0 \times 10^{-6}\,\text{m}^{-1}}$$

$$= 5.0 \times 10^5\,\text{m} = 500\,\text{km}$$

(b) From Equation 2:

$$\frac{I}{I_0} = e^{-\mu x}$$

When $x = 1.0 \times 10^5\,\text{m}$,
$\mu x = 2.0 \times 10^{-6}\,\text{m}^{-1} \times 1.0 \times 10^5\,\text{m} = 0.20$

$$\frac{I}{I_0} = e^{-0.2} = 0.82$$

i.e. the intensity has fallen to 0.82 of its initial value.

(c) From Equation 4, the maximum length, x, is given by:

$$\mu x = \ln(I_0) - \ln(I)$$

$$= \ln\left(\frac{I_0}{I}\right)$$

$$= \ln\left(\frac{200}{2}\right) = \ln(100) = 4.61$$

$$x = \frac{4.61}{\mu} = \frac{4.61}{2.0 \times 10^{-6}\,\text{m}^{-1}}$$

$$= 2.3 \times 10^6\,\text{m}$$

Q 5 From Equation 2 or Equation 4, we have:

$$\mu x = -\ln\left(\frac{I}{I_0}\right) = \ln\left(\frac{I_0}{I}\right)$$

Putting $I = \left(\dfrac{I_0}{2}\right)$,

$$\mu x = \ln 2$$

$$\mu = \frac{\ln 2}{x} = \frac{0.693}{x} = 1.386 \times 10^{-5}\,\text{m}^{-1}$$

With $x = 50\,\text{km} = 5.0 \times 10^4\,\text{m}$

$$\mu = \frac{0.693}{5.0 \times 10^4\,\text{m}} = 1.386 \times 10^{-5}\,\text{m}^{-1}$$

Q 6 Around 1.7 μm would be best, but other possibilities are near 1.35 μm and 1.22 μm – the other troughs on the graph.

Q 7 (a) Number of electrons $= \dfrac{Q}{e}$

$$= \frac{0.16 \times 10^{-12}\,\text{C}}{1.6 \times 10^{-19}\,\text{C}}$$

$$= 1 \times 10^6\,\text{electrons}$$

(b) From Equation 5:

$$V = \frac{Q}{C} = \frac{0.16 \times 10^{-12}\,\text{C}}{1 \times 10^{-12}\,\text{F}} = 0.16\,\text{V}$$

Q 8 From Equation 5, $Q = CV = 1\,\text{F} \times 0.45\,\text{V} = 0.45\,\text{C}$

Q 9 From Question 8, $Q = 0.45\,\text{C}$

Current $I = 50\,\text{mA} = 50 \times 10^{-3}\,\text{C s}^{-1}$

Time to charge $= \dfrac{0.45\,\text{C}}{50 \times 10^{-3}\,\text{C s}^{-1}} = 9.0\,\text{s}$

Q 10 The most convenient expression is Equation 8:

$$W = \frac{1}{2}CV^2$$

Substituting values:

(a) $W = \dfrac{1}{2} \times 10^4 \times 10^{-6}\,\text{F} \times (3\text{V})^2 = 4.5 \times 10^{-2}\,\text{J}$

For parts (b) and (c) you can multiply the answer to (a) by 4 and then by 9, since W is proportional to V^2. This gives:

(b) $W = 4 \times 4.5 \times 10^{-2}\,\text{J} = 0.18\,\text{J}$

(c) $W = 9 \times 4.5 \times 10^{-2}\,\text{J} = 0.405\,\text{J}\ (= 0.41\,\text{J}\ 2\,\text{s.f.})$

Q 11 From Equation 11: $\lambda = \dfrac{hc}{\Delta E}$

Indium/antimony:

$$\lambda = \frac{6.63 \times 10^{-34}\,\text{J s} \times 3.00 \times 10^8\,\text{m s}^{-1}}{0.18 \times 1.60 \times 10^{-19}\,\text{J}}$$

$$= 6.9 \times 10^{-6}\,\text{m}$$

Gallium/nitrogen:

$$\lambda = \frac{6.63 \times 10^{-34}\,\text{J s} \times 3.00 \times 10^8\,\text{m s}^{-1}}{3.4 \times 1.60 \times 10^{-19}\,\text{J}}$$

$$= 3.7 \times 10^{-7}\,\text{m}$$

Q 12 Using Equation 11: $\Delta E = \dfrac{hc}{\lambda}$

$\lambda = 400\,\text{nm}$

$\Delta E = \dfrac{6.63 \times 10^{-34}\,\text{J s} \times 3.00 \times 10^{8}\,\text{m s}^{-1}}{400 \times 10^{-9}\,\text{m}}$

$\quad = 5.0 \times 10^{-19}\,\text{J}$

$\quad = \dfrac{5.0 \times 10^{-19}\,\text{J}}{1.60 \times 10^{-19}\,\text{J eV}^{-1}}$

$\quad = 2.1\,\text{eV}$

Similarly for $\lambda = 600\,\text{nm}$:

$\Delta E = 3.3 \times 10^{-19}\,\text{J}$

$\quad = 3.1\,\text{eV}$

Q 13 (a) $P = VI = 2.15\,\text{V} \times 15 \times 10^{-3}\,\text{A} = 3.23 \times 10^{-2}\,\text{W}$

(b) $P = VI = 1.3\,\text{V} \times 100 \times 10^{-3}\,\text{A} = 0.13\,\text{W}$

(c) Taking the answer to (a) to be typical of visible LEDs, 10^{6} of them would have a power rating of:

$P = 3 \times 10^{-2} \times 10^{6}\,\text{W} = 3 \times 10^{4}\,\text{W} = 30\,\text{kW}$

Q 14 $1\,\text{V} = 1\,\text{J C}^{-1}$,
and $1\,\text{J} = 1\,\text{N m}$

so $1\,\text{V m}^{-1} = 1\,\text{J C}^{-1}\,\text{m}^{-1}$

$\quad = 1\,\text{N m C}^{-1}\,\text{m}^{-1}$

$\quad = 1\,\text{N C}^{-1}$

> **MATHS REFERENCE**
>
> Derived units
>
> See Maths note 2.3

Q 15 (a) From Equation 16b:

$E = \dfrac{V}{d} = \dfrac{1 \times 10^{3}\,\text{V}}{0.1\,\text{m}}$

$\quad = 1 \times 10^{4}\,\text{V m}^{-1}$

$\quad = 1 \times 10^{4}\,\text{N C}^{-1}$

(b) From Equation 12:

$F = qE = 3 \times 10^{-6}\,\text{C} \times 1 \times 10^{4}\,\text{N C}^{-1}$

$\quad = 3 \times 10^{-2}\,\text{N}$

Q 16 $P = IV = 140 \times 10^{-12}\,\text{A} \times 2.5\,\text{V} = 3.5 \times 10^{-10}\,\text{W}\ (= 0.35\,\text{nW})$

Q 17 (a) From Equation 16b:

$E = \dfrac{200\,\text{V}}{4.0 \times 10^{-2}\,\text{m}} = 5.0 \times 10^{3}\,\text{V m}^{-1}$

(b) (i) $E = \dfrac{400\,\text{V}}{4.0 \times 10^{-2}\,\text{m}} = 1.0 \times 10^{4}\,\text{V m}^{-1}$

(ii) $E = \dfrac{200\,\text{V}}{8.0 \times 10^{-2}\,\text{m}} = 2.5 \times 10^{3}\,\text{V m}^{-1}$

(iii) $E = \dfrac{100\,\text{V}}{2.0 \times 10^{-2}\,\text{m}} = 5.0 \times 10^{3}\,\text{V m}^{-1}$

The field lines should be drawn as in Figure 2.59 so that their spacing indicates the field strength.

(c) See Figure 2.59. Notice that the field strength is indicated by the spacing of the field lines and by the spacing of the equipotentials.

(i)

(ii)

(iii)

Figure 2.59 Answers to Question 17(b) and (c)

Q 18 **(a) (i)** Between X and Y, $\Delta V = 100\,\text{V}$, $\Delta x = 3\,\text{cm}$ (measured from diagram)

$$E = \frac{100\,\text{V}}{3\,\text{cm}} = 33\,\text{V cm} = 3.3 \times 10^3\,\text{V m}^{-1}$$

(ii) Measuring between the 50 V and 350 V equipotentials along a line joining the two charges and passing through Z: $\Delta V = 300\,\text{V}$, $\Delta x = 6\,\text{cm}$

$$E = \frac{300\,\text{V}}{6\,\text{cm}} = 50\,\text{V cm}^{-1} = 5 \times 10^3\,\text{V m}^{-1}$$

(b) See Figure 2.60.

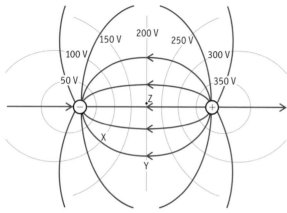

Figure 2.60 The answer to Question 18(b)

Q 19 **(a)** Using Equation 16b:

$$E = \frac{2.3\,\text{V}}{0.4 \times 10^{-6}\,\text{m}}$$

$$= 5.8 \times 10^6\,\text{V m}^{-1}$$

(b) From Equations 10 and 11:

$$\frac{hc}{\lambda} = q\Delta V$$

$$\lambda = \frac{hc}{q\Delta V}$$

$$= \frac{6.63 \times 10^{-34}\,\text{J s} \times 3.00 \times 10^8\,\text{m s}^{-1}}{1.60 \times 10^{-19}\,\text{C} \times 2.3\,\text{V}}$$

$$= 5.40 \times 10^{-7}\,\text{m} = 540\,\text{nm}$$

Q 20 **(a)** Amplitude 15 V, period 800 ms, frequency 1.25 Hz

(b) Amplitude 2 V, period 120 ms, frequency 8.3 Hz

(c) Amplitude 4 V, period 400 ns, frequency 2.5×10^6 Hz

Q 21 From Equation 19a:

$$V_0 = V_{\text{rms}} \times \sqrt{2} = 120\,\text{V} \times \sqrt{2} = 170\,\text{V}$$

Q 22 $I_{\text{rms}} = \dfrac{V_{\text{rms}}}{R}$ and $I_0 = I_{\text{rms}} \times \sqrt{2}$

so $I_0 = \dfrac{V_{\text{rms}} \times \sqrt{2}}{R} = \dfrac{230\,\text{V} \times \sqrt{2}}{0.5\,\Omega}$

$$= 650\,\text{A}$$

Q 23 Using any version of Equation 20:

$$P = V_{\text{rms}}\,I_{\text{rms}} = I_{\text{rms}}^2 R = \frac{V_{\text{rms}}^2}{R}$$

e.g. $V_{\text{rms}} = \dfrac{V_0}{\sqrt{2}}$, $P = \dfrac{V_0^2}{2R} = \dfrac{(3.0\,\text{V})^2}{2 \times 64\,\Omega} = 0.070\,\text{W}$

Q 24 **(a)** $I_0 = \dfrac{V_0}{R} = \dfrac{10.0\,\text{V}}{10.0\,\Omega} = 1.00\,\text{A}$

See Figure 2.61. Note that the voltage is in phase with the current.

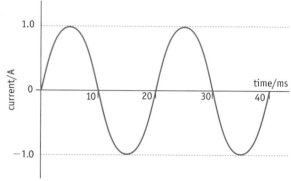

Figure 2.61 The answer to Question 24(a)

(b) From Equation 19:

$$V_{\text{rms}} = \frac{V_0}{\sqrt{2}} = 7.07\,\text{V}$$

$$I_{\text{rms}} = \frac{I_0}{\sqrt{2}} = 0.707\,\text{A}$$

(c) $P = V_{\text{rms}}\,I_{\text{rms}} = \dfrac{V_0 I_0}{2} = 5.00\,\text{W}$

Q 25 $E = \dfrac{V}{d} = \dfrac{4.9\,\text{V}}{8.0 \times 10^{-3}\,\text{m}} = 610\,\text{V m}^{-1}$

Q 26 **(a)** Using Equation 13, $\Delta W = q\Delta V$, so energy transferred to the electron is

$$\Delta W = 1.60 \times 10^{-19}\,\text{C} \times 5000\,\text{V} = 8.0 \times 10^{-16}\,\text{J}$$

(b) (i) Assuming that the kinetic energy gained by the electron is equal to the energy transferred by the electric field in the gun, $E_k = 8.0 \times 10^{-16}\,\text{J}$.

(ii) $E_k = \dfrac{1}{2}mv^2$, and so:

$$v = \sqrt{\frac{2E_k}{m}}$$

$$= \sqrt{\frac{2 \times 8.0 \times 10^{-16}\,\text{J}}{9.11 \times 10^{-31}\,\text{kg}}}$$

$$= 4.19 \times 10^7\,\text{m s}^{-1}$$

(c) Electrons would collide with air molecules, transferring energy to them, so the electrons' kinetic energy would be greatly reduced. (They would also be scattered and the beam would not remain focused.)

Q 27 **(a)** The spot would be brighter as the kinetic energy of each electron would be increased, so each electron would transfer more energy to the phosphor screen.

(b) The spot would be brighter as more electrons would be produced by thermionic emission, more would be accelerated, and so there would be a greater overall rate of energy transfer to the phosphor.

(c) The spot would be less bright as the electrons would be decelerated to some extent by the grid.

Q 28 From Equation 25 with $\theta = 90°$, $\sin\theta = 1$:

$$F = 0.10\,\text{T} \times 1.60 \times 10^{-19}\,\text{C} \times 8.0 \times 10^7\,\text{m s}^{-1}$$

$$= 1.3 \times 10^{-12}\,\text{N}$$

Q 29 From Fleming's left-hand rule, the field must be horizontal and directed across the tube.

Q 30 From Equation 16b:

$$E = \frac{200\,\text{V}}{150 \times 10^{-6}\,\text{m}} = 1.3 \times 10^6\,\text{V m}^{-1}$$

Q 31 From Equation 11:

$$E = \frac{hc}{\lambda} = \frac{6.63 \times 10^{-34}\,\text{J s} \times 3.00 \times 10^8\,\text{m s}^{-1}}{147 \times 10^{-9}\,\text{m}}$$

$$= 1.35 \times 10^{-18}\,\text{J}$$

Q 32 See Figure 2.62. The gas will be ionised near the top of the cell where the field strength is high, shown by the field lines being close together.

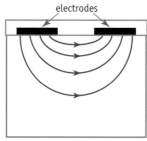

electrodes

Figure 2.62 Answer to Question 31

Q 33 **(a)** From Equation 16b:

$$E = \frac{1.5\,\text{V}}{10 \times 10^{-6}\,\text{m}}$$

$$= 1.5 \times 10^5\,\text{V m}^{-1} = 1.5 \times 10^5\,\text{N C}^{-1}$$

(b) From Equation 12, $F = eE$

$$F = 1.60 \times 10^{-19}\,\text{C} \times 1.5 \times 10^5\,\text{N C}^{-1}$$

$$= 2.4 \times 10^{-14}\,\text{N}$$

PROBING THE HEART OF MATTER

Why a chapter called *Probing the Heart of Matter?*

Science has followed a long quest to probe the structure of matter in an attempt to discover what we are all made of. In recent years this story has taken a new twist. Two different strands of physics have come together to provide a picture of how the fundamental structure of matter and the Universe itself are linked.

In the 1970s, theoreticians working on the physics of fundamental particles developed theories that were far ahead of the experimental physicists' means to test them out, so they started searching for new ways of seeing if their theories made sense. At the same time, physicists working on the relatively new subject of cosmology (the study of how the Universe came into being) discovered that they needed to understand particle physics in order to chart the very early history of the Universe.

The particle physicists realised that one way to test their theories was to see if they made sensible predictions about the way in which the early Universe evolved. This helped the cosmologists to solve some of their most difficult problems regarding the development of the Universe after the Big Bang. As a consequence, physicists think that they have a fairly complete theory telling them how the Universe evolved from 10^{-35} seconds after the Big Bang right up to the present day. After 10^{-12} s, theory is supported by experimental evidence.

You may have read about the latest developments in this field in newspapers and science magazines, or you may have seen television documentaries on the subject. In 2012, scientists working at CERN's Large Hadron Collider (LHC) announced the discovery of the Higgs boson, the particle responsible for giving other particles their mass. The Higgs boson is named after Peter Higgs (Figure 3.1a) who, as a young researcher in the 1960s, was one of a handful of people who predicted its existence. Higgs and the Belgian physicist François Englert shared the 2013 Nobel Prize in Physics for their work. After a two-year closure for maintenance and upgrading, the LHC reopened in April 2015, ready to do more ground-breaking work on the particles that make up our world (Figure 3.1b).

Figure 3.1 (a) Peter Higgs at the Large Hadron Collider (LHC)

Figure 3.1 (b) The CERN control room in April 2015

Particle physics is a subject that is alive with new possibilities, and many of the most able young physicists starting their research careers choose to work in this exciting field. In this chapter you will study some of the ideas that underpin this theory. You will also see where some of the problems lie, and the work that is being done by experimenters and theoreticians to try to produce solutions. In doing so, you will be introduced to some aspects of modern particle physics and will also study the physics of circular motion and particle acceleration that underlies many of the experiments.

OVERVIEW OF PHYSICS PRINCIPLES AND TECHNIQUES

You will begin by looking at the Big Bang theory of the origin of the Universe and how it manages to take us from a short time after the creation event right up to the present day. Along the way, we will introduce some of the physics of fundamental particles, together with some key ideas about conservation laws and forces. Next, you will look at the use of beams of particles to probe matter, which will involve learning about electrostatic and magnetic forces and about circular motion. In the final part of the chapter, you will see some of the most important results that these devices have produced. In the course of the chapter, you will also carry out several activities that involve communication and IT skills: researching information, and discussing and communicating what you have found.

In this chapter you will extend your knowledge of:

- momentum from *Higher, Faster, Stronger* and *Transport on Track*
- kinetic energy and work from *Higher, Faster, Stronger* and *Transport on Track*
- wave–particle duality from *The Sound of Music, Technology in Space* and *Digging Up the Past*
- electric and magnetic fields from *Transport on Track* and *The Medium is the Message*
- vectors from *Higher, Faster, Stronger* and *Transport on Track*
- using graphs from *Higher, Faster, Stronger, Digging Up the Past* and *The Medium is the Message*.

Elsewhere, you will do more work on:

- forces and motion in *Build or Bust?*
- motion in a circle in *Reach for the Stars*
- kinetic energy and work in *Reach for the Stars*
- inverse-square law fields in *Reach for the Stars*
- nuclear physics in *Reach for the Stars*.

1 In the beginning

1.1 Very large and very small

You will have read in the introduction about the **Big Bang theory** of the origin of the Universe. The idea is that at some time in the past, about 13 800 million years ago, the Universe came into being. All of the matter and energy in the observable Universe, which is now spread out to the furthest distance we can see (about 10^{20} km), was compressed into a very small volume. From that tiny, super-dense state at the instant of creation, the Universe grew. As it expanded, it cooled, and matter clumped together to form galaxies, stars and planets. The evidence for this model of the history of the Universe comes from observations of the light from distant galaxies.

> **ACTIVITY 1 THE EXPANDING UNIVERSE**
>
> What ideas do you already have, from GCSE science and from general reading, about the history of the Universe? Share your ideas with a partner, and then with the rest of your class. Here are some phrases that may help you to recall what you have studied or read about previously:
>
> - galaxies moving apart
> - red shifts
> - distant galaxies – looking into the past
> - the age of the Universe.

You know that we think of matter as being made up of atoms – the building blocks of the chemical elements (see Figure 3.2). But atoms are not the smallest, most basic particles of matter, and not all matter is made up of atoms. For example, there is a beam of electrons inside old-style television sets (when they are producing a picture); electrons are particles of matter, much smaller than atoms.

Figure 3.2 Model of a DNA molecule showing the atoms from which it is made

STUDY NOTE

For a reminder about atoms, nuclei, protons, neutrons and electrons, see PRO Additional Sheet 1.

> **ACTIVITY 2 THE PARTICLES OF MATTER**
>
> In the next part of this chapter, you will learn about the current picture that physicists have developed of the sub-atomic particles of matter. What ideas do you already have about particles? Share your ideas with a partner, and then with the rest of the class. Here are some phrases that may help you to recall what you have studied or read about previously:
>
> - atoms, nuclei, protons, neutrons and electrons
> - quarks
> - particle accelerators – 'atom smashers'
> - radioactivity, fission and fusion.

2 Theories of everything

Scientists like to develop theories that account for as many different observations as possible. For example, Newton developed a theory of gravity that explained how things fall on the Earth and how the Moon orbits the Earth. Until then, people felt that there must be one set of laws for what happens on Earth, and another for what happens in the heavens. So Newton's theory was a dramatic advance, unifying two previously separate realms of ideas.

Today we know much more about the laws of physics, especially about the particles from which matter is made and the forces that act between them. Physicists would like to come up with a single theory to explain everything we know about particles and forces: it would be a 'theory of everything'. It is perhaps a rather arrogant idea. Such a theory would still leave a lot of important aspects of life yet to explain: evolution, human consciousness, poetry and love, for example. However, such a theory would still be a remarkable achievement. In this part of the chapter you will learn something about how physicists have made progress towards a 'theory of everything' during the last few decades.

2.1 Ideas in cosmology

Late on 6 December 1979, Alan Guth (Figure 3.3), then a researcher at SLAC (the Stanford Linear Accelerator Center in California), sat down to start some work. He was accustomed to working late at night, once his young son had gone to bed, as he could concentrate better. He had been working on a paper about the early Universe with some colleagues and had decided to check to see if their ideas had any influence on the way the Universe expanded at about 10^{-35} seconds after the Big Bang. Guth had been doing similar calculations for some time and the work that night was quite routine. The results he discovered were not routine, however. They were to change thinking about cosmology completely.

Figure 3.3 Alan Guth, American cosmologist

Not yet realising the full implications of what he had done, Guth cycled into the lab the next morning. In his book *The Inflationary Universe*, Guth records that he set a personal best time of 9 minutes, 32 seconds on that ride. Arriving at this office he opened his notebook, and at the top of one of the pages wrote:

> *Spectacular realization*

He then set out to refine the rough work that he had done the night before. Guth records that he cannot find any other instances in his notes where he has used a double box to emphasise a point.

At the time Guth was coming to the end of his period of employment at SLAC and was looking for another job. He decided to take the risk of not immediately setting down his new ideas in a published paper. Instead he travelled round several research centres in America, presenting his ideas in seminars and seeing if he could impress people enough to offer him a job. The risk paid off. At one conference he was given 10 minutes to explain his theory to the attending physicists. Part way through Guth's presentation, Murray Gell-Mann, a Nobel Prize winner and important character in this chapter, jumped up and shouted, 'You've solved the most important problem in cosmology!' On the Monday after, Guth took the plunge and rang MIT (the Massachusetts Institute of Technology) to ask whether they were interested in offering him a job. Twenty-four hours later he was offered an assistant professorship.

What Guth had done was to apply some of the very latest ideas in particle physics to the Universe shortly after the Big Bang. Particle physicists had started working on **Grand Unified Theories** (GUTs), which brought the fundamental forces that act between fundamental particles together into one theory. Unfortunately, these theories could not be directly checked by experiments. The effects that they predict can only be seen if the particles involved have very high energies – much higher than could be produced in experiments. However, particles with high enough energies would have existed early in the history of the Universe when all matter and energy were compressed into a tiny volume, so GUTs should have something to say about what was going on at the time. What Guth's calculations told him was that, instead of expanding at a relatively gradual rate, as everyone had assumed, the Universe must have exploded in size at a gigantic rate for a short period of time: in a time interval of 10^{-35} seconds, the Universe would become 10^{50} times bigger. This is like magnifying a proton to the current size of the visible Universe in less time than it takes for a light ray to travel from one side of a proton to another. Guth named this extraordinary expansion **inflation**, and it has become a central part of our thinking about the early history of the Universe.

2.2 Building blocks

You probably picture an atom as being like a miniature Solar System, with a nucleus made up of protons and neutrons, surrounded by orbiting electrons. This is the picture that scientists developed in the early decades of the twentieth century, and for many purposes it is still a very useful model. Electrons, protons and neutrons were at first thought to be **fundamental particles**, i.e. particles that cannot be divided into smaller constituents. However, it was inevitable that physicists would question whether, just as the atom was found to be divisible, these particles too could be subdivided.

Later in this chapter you will learn something of how this question was investigated. In the meantime, we will look at some of the results of these investigations.

The **standard model of particle physics** (Figure 3.4), which is currently our best theory of how the Universe works, identifies 12 fundamental particles from which all matter is made and four fundamental forces that govern reactions between the particles. The 12 fundamental particles divide into two distinct groups according to their properties. There are:

- six types of **quark**, the particles of which protons and neutrons (and some other particles) are made; the name *quark* was coined by Murray Gell-Mann

- six types of **lepton**, including the electron; the name *lepton* comes from Greek and means 'light'; not all leptons are 'light', but the name has stuck.

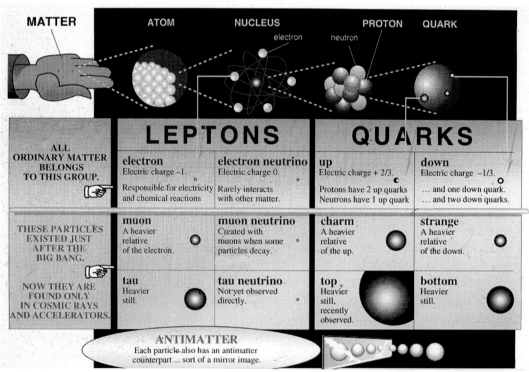

Figure 3.4 Constituents of matter

Although electrons are thought to be truly fundamental, physicists have shown that protons and neutrons are not – each is made up of three quarks.

The particles are generally grouped into three **generations** – so-called because there is some 'family' resemblance between particles of one generation and another (not because one group precedes another). Table 3.1 lists the names (which are often slightly weird), symbols and some properties of the quarks. All three quarks in a horizontal row have the same electrical charge, shown in units of the proton charge e. There are also three generations of lepton (Table 3.2). In each generation, one of the leptons is a neutrino.

Many of the names in Tables 3.1 and 3.2 will be new to you. Probably the only familiar one is the electron. As this chapter progresses you will start to see how the others fit into our description of the Universe. Do not get worried if you feel that you are not entirely sure what these particles are – this was the problem that physicists faced as they started to discover them.

1st generation	2nd generation	3rd generation	Charge
up, u	charm, c	top, t	$+2e/3$
down, d	strange, s	bottom, b	$-1e/3$

Table 3.1 Quarks

1st generation	2nd generation	3rd generation	Charge
electron, e	muon, μ	tau, τ	$-1e$
electron-neutrino, v_e	muon-neutrino, v_μ	tau-neutrino, v_τ	0

Table 3.2 Leptons

Table 3.3 shows where leptons are found in nature. We cannot provide a similar table for quarks, because it is believed that quarks always occur bound together in twos or threes, never separately.

Lepton	Where it is found
Electron	found in atoms important in electrical currents produced in beta radioactivity
Muon	produced in large numbers in the upper atmosphere by 'cosmic rays'
Tau	so far only seen in lab experiments
Electron-neutrino	produced in beta radioactivity produced in large numbers by atomic reactors produced in huge numbers by the nuclear reactions in the Sun
Muon-neutrino	produced by atomic reactors produced in upper atmosphere by cosmic rays produced in the Sun by nuclear reactions
Tau-neutrino	so far only seen in lab experiments

Table 3.3 Leptons in nature

It is worth mentioning something about the masses of the fundamental particles.

● The top quark is the heaviest (most massive) of all fundamental particles. Its mass is almost 200 times that of a proton. That makes it much heavier than many molecules; for example, it is heavier than a $C_{14}H_3O$ hydrocarbon.

● An electron has a mass roughly $\dfrac{1}{2000}$ times that of a proton.

● For many years it was suggested that electron-neutrinos (which appear during beta decay of a radioactive material) have absolutely no mass at all. However, sensitive experiments have shown that this is not the case and that the neutrinos have a very tiny mass, less than 10^{-9} of the mass of a proton.

⚙ ACTIVITY 3 CHARGE AND NO CHARGE

A proton is made up of three quarks: two up quarks and a down (written uud). A neutron is also made of three quarks, a different combination of up and down quarks.

● Show that the charge of a proton is correctly given by adding the charges of two up quarks and one down quark.

● What combination of up and down quarks make up a neutron?

A proton is made of two up quarks and a down (uud); a neutron is an up and two downs (udd). All other possible collections of three quarks also correspond to particles, although all of them apart from the proton (which has the lowest mass) are unstable. This means that they will decay into other (less massive) particles. That is why they are unfamiliar to you. Particles such as the Δ^{++} (pronounced 'delta plus plus' and made from three up quarks, uuu) and the Δ^{-} ('delta minus', ddd) typically last about 10^{-25} seconds before they decay, so they are not seen in the everyday world. We only understand their properties to the extent that we do because we have been able to make and study them in experiments. However, in the early moments after the Big Bang such particles were much more common than they are now, and so our attempts to discover their properties also help us to understand what was going on in the early Universe.

Symmetry and prediction

You have probably noticed similarities between the patterns of the lepton and quark families. Each has six members in three generations. The particles' masses increase as you move up the generations. The model is said to have **symmetry**.

When the American physicist Murray Gell-Mann (Figure 3.5) first proposed the quark model in 1964, he only included three quarks (up, down and strange) as they could account for all the properties of the known particles. At the time, only the first two generations of leptons were known. During the next ten years, physicists gradually became convinced on theoretical grounds that a fourth quark (the charm quark) should also exist. Partly this was due to an appreciation of the symmetry that would be established between four leptons and four quarks. More convincing grounds for looking for the charm quark came from patterns in the manner in which certain particles decayed. Particles containing the charm quark were discovered in 1974. Evidence for the tau lepton accumulated between 1974 and 1977. Straightaway physicists suspected that the symmetry between charged leptons and their neutrinos implied the tau neutrino's existence. However, it was not confirmed until 2000.

Figure 3.5 Murray Gell-Mann

1973 saw the first suggestion that the number of quarks should be extended to include the top and bottom quarks. Their suspected hidden roles in particle reactions was the primary reason, although as the tau lepton and the charm quark became established, symmetry between the quarks and leptons certainly led physicists to suspect that a fifth and sixth quark were waiting to be found. The bottom quark was discovered in 1977. The huge mass of the top quark required several advances in experimental technology before enough energy could be put into a reaction in order to create this quark. It finally turned up in 1995.

The elegant symmetry between the number of quarks and leptons has been a subtle motivator in the path to discovering the fundamental particles, but it has always required the support of other theoretical evidence, and finally the crucial confirmation of experimental results.

⚙ ACTIVITY 4 TOP QUARK

Research the story of how the bottom and top quarks were predicted and discovered. Identify the key events in the story and produce an annotated timeline showing what happened.

PRACTICAL SKILLS REFERENCE

Scientific questions and information research

See Practical skills notes 1.2, 1.3

2.3 Fundamental forces and interactions

Physicists have identified four fundamental forces that operate in nature. Two of these, gravity and the electromagnetic force, will be familiar to you from your studies in physics so far. By a fundamental force we mean a force that cannot be explained as the action of some other force. For example, the force of friction is not a fundamental force as the electrostatic forces between atoms cause it when one object rubs against another. The four fundamental forces are shown in Table 3.4.

Force	Range	Relative strength	Acts between …
Gravity	no limit	10^{-34}	all objects
Electromagnetic	no limit	10	charged objects
Strong force	10^{-15} m	10^3	quarks
Weak force	10^{-18} m	10^{-10}	fundamental particles

Table 3.4 The four fundamental forces of nature

Range and strength

The **range** of a force is the maximum distance by which two objects can be separated and still feel the force acting. The **relative strength** of the force is a way of comparing how big an influence each force would have on the same pair of objects. The relative strengths quoted correspond roughly to the force in newtons between two protons separated by 10^{-15} m.

The so-called strong and weak forces have probably not come into your study of physics before, as they only act over very short distances – smaller than the diameter of a nucleus. For this reason physicists did not discover them until they started to experiment with nuclei.

As this chapter progresses you will start to get a feel for the properties of the strong and weak forces. For the moment, just think of them as new forces that we do not see in our everyday life, as their size is so small at the sort of distances we deal with (though as you can see from Table 3.4, over small enough distances even the weak force is not as weak as gravity). You would not worry about the force of gravity pulling you towards the Moon as you cross the road because it is so small; likewise, when we hit a tennis ball the strong and weak forces do not influence its motion.

STUDY NOTE

The electromagnetic force combines both magnetic and electrostatic effects; in the chapters *Transport on Track* and *The Medium is the Message*, you will have seen that these effects are closely interconnected.

⚙ ACTIVITY 5 FORCES – FUNDAMENTAL OR NOT?

Make a list of the forces you have come across in physics so far – Figure 3.6 illustrates several examples. Divide them into two categories – contact forces, which require two objects to be touching each other for the force to exist (such as friction), and force fields, which do not require contact (such as gravity). Try to explain each of the forces you have listed in terms of a fundamental force from Table 3.4. Which category of force do you think is the more fundamental – force fields or contact forces?

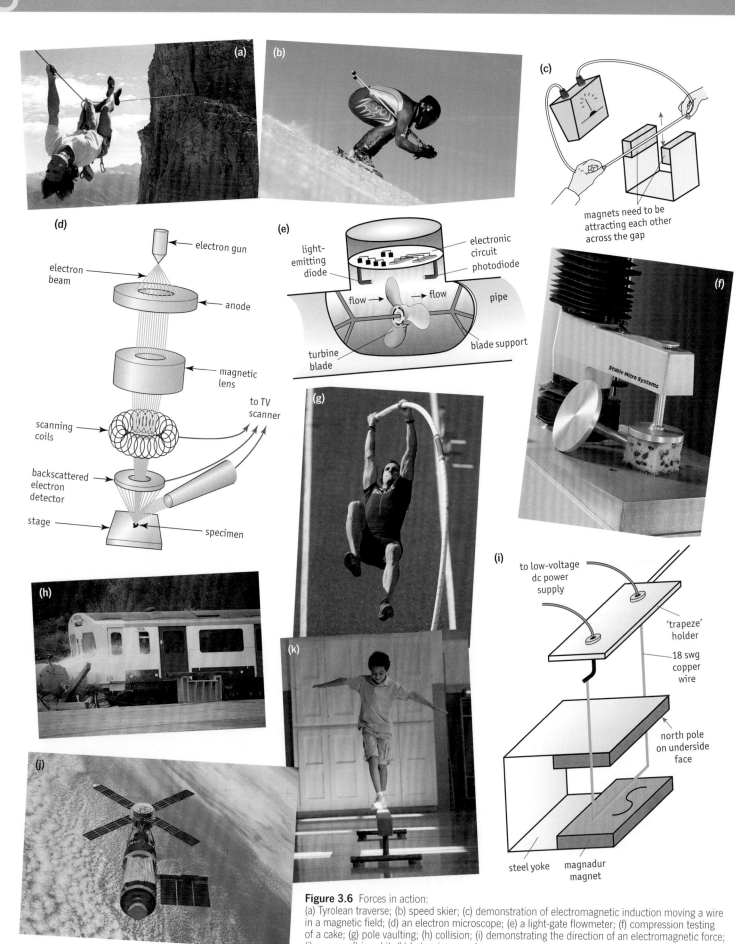

Figure 3.6 Forces in action:
(a) Tyrolean traverse; (b) speed skier; (c) demonstration of electromagnetic induction moving a wire in a magnetic field; (d) an electron microscope; (e) a light-gate flowmeter; (f) compression testing of a cake; (g) pole vaulting; (h) collision; (i) demonstrating the direction of an electromagnetic force; (j) spacecraft in orbit; (k) balancing on a beam

Binding particles together

Not all of the fundamental forces act on each of the fundamental particles. For example, quarks experience the strong force but leptons do not. As you can see from Table 3.4, between two quarks the strong force would completely dominate over the weak force. In fact the strong force is so strong that it pulls quarks together into groups from which they can never be separated. These groups are essentially particles in their own right, called hadrons (see Figure 3.7). As far as we can tell, quarks are never found on their own – they are always bound together into hadrons.

Figure 3.7 Two types of hadron

The strong force allows only two types of quark combination to exist. One type, the **baryon**, is always composed of three quarks. The most familiar examples of such a combination are the particles found in the nucleus – the proton and the neutron. The second type of hadron is the **meson**. Mesons are made of a quark and an anti-quark, which are bound together (see Section 2.4).

QUESTION

 In each of **(a)** to **(e)** below, state which fundamental forces are involved:

(a) a quark and an anti-quark are held together to make a meson

(b) the Earth is held in its orbit around the Sun

(c) a compass needle points north

(d) protons and neutrons are close together in an atomic nucleus

(e) an electron-neutrino is absorbed by an atomic nucleus.

2.4 Anti-matter

Physicists have known about the existence of anti-matter since the positron (the anti-matter version of the electron – see below) was discovered in 1932. Since then, anti-matter versions of all the fundamental particles have been discovered. Anti-matter has become standard stuff used in experiments.

The first thing to understand about anti-matter is that it is not very special stuff. Despite what you might have read in science fiction, it does not have anti-gravity properties or anything weird like that. For every type of particle of matter that exists, there is a corresponding particle of anti-matter with the same mass but with the opposite electrical charge (if it is electrically charged). So, as well as there being electrons in the Universe, there are also objects called anti-electrons, which are also known as positrons. These have the same mass as electrons, but they have a positive charge identical in size to that of a proton. For every quark that we mentioned earlier, there is an anti-quark; for every lepton, there is an anti-lepton. These are shown in Tables 3.5 and 3.6. Each anti-particle is represented by the same symbol as its corresponding particle, with a bar above it – except for the positron, which has its own symbol, e^+.

At first glance, there is something of a mystery surrounding the neutrinos and their anti-particles. After all, if the neutrinos have the same mass as their equivalent anti-neutrinos, and both are neutral, then there does not seem to be any physical property that distinguishes them. However, they certainly react differently when they pass through matter.

STUDY NOTE

The distinction between leptons and anti-leptons is discussed further in Section 2.5.

STUDY NOTE

The $\bar{\mu}$ and $\bar{\tau}$ are also written μ^+ and τ^+, respectively.

1st generation	2nd generation	3rd generation	Charge
$\bar{u}$	$\bar{c}$	$\bar{t}$	$-\dfrac{2}{3}e$
$\bar{d}$	$\bar{s}$	$\bar{b}$	$+\dfrac{1}{3}e$

Table 3.5 Anti-quarks

1st generation	2nd generation	3rd generation	Charge
positron $\bar{e}$ or e^+	$\bar{\mu}$	$\bar{\tau}$	$+1e$
electron-anti-neutrino $\bar{\nu}_e$	muon-anti-neutrino $\bar{\nu}_\mu$	tau-anti-neutrino $\bar{\nu}_\tau$	0

Table 3.6 Anti-leptons

Figure 3.8 CERN experiment that produced the first atoms of anti-hydrogen

There is nothing in the laws of physics to suggest that a position in orbit around an anti-proton will not be as stable as a familiar hydrogen atom (an electron in orbit around a proton). The problem is what happens when this hydrogen anti-atom meets a hydrogen atom: the two annihilate one another, leaving only a flash of gamma rays. Physicists believe that matter and anti-matter behave in exactly the same manner in most circumstances. Yet, as always, physicists like to check such things, just in case they happen to be wrong. A few atoms of anti-hydrogen were first produced in an experiment called PS210 at CERN in 1995 (Figure 3.8). More recent experiments have produced tens of thousands of anti-hydrogen atoms. PS210 was designed to produce and detect such anti-atoms as a first step to confirming their properties.

PRACTICAL SKILLS REFERENCE

Scientific questions and information research

See Practical skills notes 1.2, 1.3

> ⚙ **ACTIVITY 6 MAKING ANTI-ATOMS**
>
> Read about the PS210 experiment and write a summary of between 80 and 100 words, explaining what was done and why it was significant.

Mesons

In Section 2.3, we discussed the action of the strong force on quarks and said that it is so strong that it binds them together into particles. One such grouping combines three quarks together (such as in the proton and neutron) to form a baryon. We also mentioned another possible combination; a quark can bind to an anti-quark to make a meson. An example of such a quark–anti-quark pairing is a u quark with a $\bar{d}$ anti-quark, $\bar{u}$. Such an object would have a charge equal to the sum of the two quark charges:

$$u \text{ charge} = +\frac{2}{3}e$$

$$\bar{d} \text{ charge} = +\frac{1}{3}e$$

$$\text{total charge} = +1e$$

Any pairing is possible. Table 3.7 shows the possibilities if we just take the lightest quarks, the up and the down.

STUDY NOTE

The π^0 is alternately composed of $u\bar{u}$ or $d\bar{d}$. As a π^0 flies through space it continually oscillates from one combination to the other.

Quarks	Total charge	Name of bound particle	Symbol
$u\bar{u}$	0	pi zero	π^0
$d\bar{d}$	0	pi zero	π^0
$u\bar{d}$	$+1e$	pi plus	π^+
$\bar{u}d$	$-1e$	pi minus	π^-

Table 3.7 Combining up and down quarks to make pions

All these combinations correspond to one of a family of three particles called pi-mesons or **pions** (notice that there are two ways to make a pi-zero meson). Pions are the lightest of the meson family of particles and are very commonly produced in particle physics experiments. They are also produced high in the Earth's atmosphere when cosmic rays (high-energy protons) from the Sun and beyond the Solar System hit the nuclei of atoms in the atmosphere.

Figure 3.9 shows a schematic diagram of a typical shower of particles produced by a single cosmic ray particle, together with some of the detectors and observations used to study the effects of cosmic rays. N is used to represent a light nucleus. Notice that the shower contains many muons, neutrinos and positrons, as well as pions.

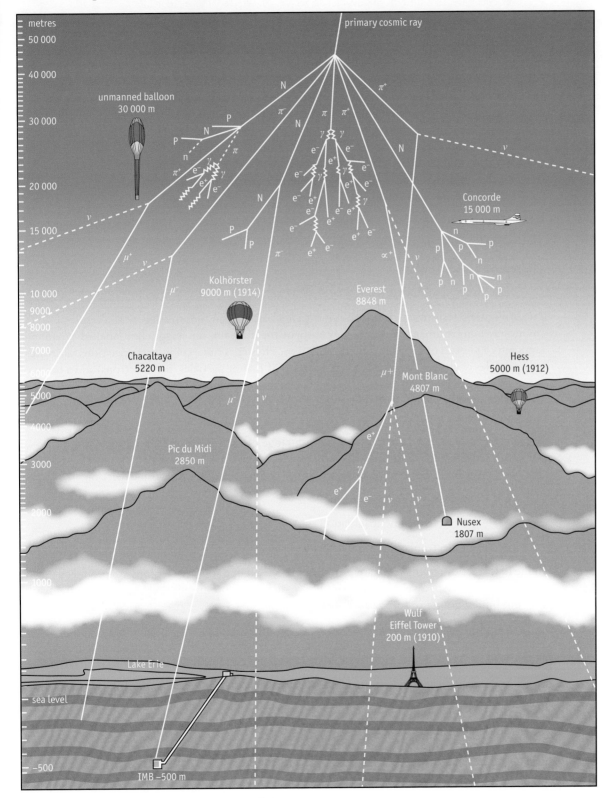

Figure 3.9 Shower of particles produced by cosmic radiation

STUDY NOTE

The K^0 and the $\overline{K^0}$ are different particles, but the π^0 and the $\overline{\pi^0}$ are the same. Can you see why?

ACTIVITY 7 MAKING MESONS

All mesons are made up of a quark and an anti-quark.

- A phi meson (ϕ) is made of a strange quark and a strange anti-quark. What will be its charge?
- Four types of K meson are possible: $u\overline{s}$, $\overline{u}s$, $d\overline{s}$, $\overline{d}s$. Which of these are neutral?

Playing with fire

It is very difficult to experiment with anti-matter. If an anti-matter particle comes into contact with a matter particle of the same type (for example, an electron and a positron or a proton and an anti-proton), then they will react with each other and turn into electromagnetic radiation (usually gamma-ray photons). This is called an **annihilation reaction**. For instance, an up quark and its anti-particle may come together to form a meson, but this is very unstable and decays very rapidly; a π^0 meson decays in about 10^{-17} s. The amount of energy released in a single reaction of this sort is tiny and poses no safety risk – the problem is keeping the anti-matter particles away from the matter particles for long enough to experiment on them. (Remember, any equipment used will be made of matter particles.)

Here is the equation that represents an electron and an anti-electron (positron) annihilating:

$$e^- + e^+ \rightarrow \text{electromagnetic radiation}$$

The reverse can also occur:

$$\text{electromagnetic radiation} \rightarrow e^- + e^+$$

Part of Figure 3.9 shows such reactions occurring naturally in the atmosphere as a result of cosmic-ray bombardment. Towards the centre of the diagram, the wiggly line labelled γ (gamma) represents a photon of electromagnetic radiation. Physicists can exploit these reactions to create new particles to study. The idea is to take a beam of matter particles (say, electrons) and a beam of anti-matter particles (positrons in this case) and accelerate them to very high energy. The beams are then allowed to collide with one another. This process sounds simple, but steering the beams is actually a highly complex affair.

When the matter and anti-matter come within range of each other, they annihilate, producing a burst of electromagnetic radiation that will rapidly materialise into new particles, perhaps ones that have never been seen before. Here are three equations that show possible reactions:

$$e^- + e^+ \rightarrow \text{electromagnetic radiation} \rightarrow \mu^- + \mu^+$$

$$e^- + e^+ \rightarrow \text{electromagnetic radiation} \rightarrow \tau^- + \tau^+$$

$$e^- + e^+ \rightarrow \text{electromagnetic radiation} \rightarrow u + \overline{u}$$

In the first reaction, two muons are produced. In the second, two tau leptons are produced – this is how the tau was discovered. In the third, two quarks are produced; they cannot exist freely like this, but bind together to form a pion (see Table 3.7).

QUESTIONS

Q 2 Write an equation to show what happens when a muon annihilates with its anti-particle.

Q 3 Some mesons are made from a quark bound to its own anti-quark. Explain why such a meson has no electric charge.

Mass from energy

It is surprising to find that particles of matter can disappear, leaving only electromagnetic radiation, which has energy but no mass. It is also surprising to find that particles can appear where before there was only radiation. This sounds as if the laws of conservation of energy and mass are both being broken, and in a sense that's true. However, mass m and energy E are deeply interrelated, as described by Einstein's equation:

$$E = mc^2 \tag{1}$$

where c is the speed of light in a vacuum (also called 'free space'). When a particle–anti-particle pair is created, we can use the expression to deduce the possible mass that might be created. To produce particles 'out of nothing', the photons of electromagnetic radiation must have a great deal of energy; the more massive the particles produced, the greater the energy of the photons.

Equation 1 has a more general application. If ever the mass of particles changes by an amount, Δm, there is a corresponding change in the amount of energy, ΔE, and these two are related by:

$$\Delta E = c^2 \Delta m \tag{1a}$$

QUESTIONS

Use the following data to answer Questions 4–6:

mass of positron = mass of electron, $m_e = 9.11 \times 10^{-31}\,\text{kg}$

mass of proton, $m_p = 1.67 \times 10^{-27}\,\text{kg}$

speed of light in vacuum $c = 3.00 \times 10^8\,\text{m s}^{-1}$

Q 4   Explain why electron and positron beams must be accelerated to high energies in order to produce particles such as tau leptons, which are more massive than electrons.

Q 5   Using Equation 1, calculate the energy equivalent to the mass of a proton.

Q 6   An electron–positron pair can be created from a single photon, if it is energetic enough. Any excess energy is shared equally between the two particles as kinetic energy.

    (a) What energy of photon is required to create an electron–positron pair?

    (b) A photon of energy $2 \times 10^{-13}\,\text{J}$ disappears, and an electron–positron pair is created. Calculate the kinetic energies of the electron and the positron.

Anti-matter space fuel?

It has often been suggested that anti-matter might be the answer to providing fuel for future space missions. The idea is that an engine in which matter and anti-matter annihilated each other would accelerate a spacecraft. This would power it so that it could flash across the galaxy, just like the Starship *Enterprise*, at speeds approaching the speed of light (Figure 3.10). Scientists estimate that just 20 kg of anti-matter would do the job. The problem is making the stuff, and containing it. With present technology, physicists at CERN can now produce about 10 ng of anti-matter particles a year, so it would take 100 million years to make 1 g. So anti-matter drives are still a long way off.

MATHS REFERENCE

SI Prefixes

See Maths note 2.4

Figure 3.10 The Starship *Enterprise*

⚙ ACTIVITY 8 PHYSICS IN FICTION: ANGELS AND DEMONS

Angels and Demons by Dan Brown is a detective story about a secret society that wants to destroy the Vatican using an anti-matter bomb. In the book, the anti-matter has been stolen from CERN.

Chapter 21 of the book contains lots of physics. Read the chapter, identify the physics and write an explanation for a reader who has studied physics to AS level. (You might need to refer to work that you did for earlier chapters, including *The Medium is the Message*.) Comment on whether the situation described in the book is plausible.

2.5 Particle reactions

'Everything not forbidden is compulsory.' While the exact attribution of this expression is not clear, it is suspected that Murray Gell-Mann borrowed it from T.H. White's version of the legend of King Arthur, *The Once and Future King* (1958), where it is used to describe a system of government.

Applied to particle physics, the quote draws attention to our ability to learn as much from reactions that do *not* happen as we can from those that do. Nature will explore every pathway open to it, so failure to observe a reaction that appears possible points to some physical law, previously unknown, which would be violated if the reaction were to take place.

Conservation laws

Consider a reaction such as this:

$$p + p \rightarrow p + n \qquad\qquad (R1)$$

When reading an equation of this sort, we are following a convention:

reacts to produce

stuff going in

stuff coming out

Physicists have discovered **conservation laws** that dictate what can happen in these reactions. As you saw in Section 2.4, Einstein's equation $E = mc^2$ gives a certain amount of freedom, in that the particles coming out do not have to be the same mass, or number, as those going in. The energy in the reaction can be deployed to create new particles. However, it is not a totally free ride.

Reaction R1 does not happen because it violates the law of **conservation of charge.**

On the left-hand side, the total charge entering the reaction = $+2e$ ($+e$ for each proton)

On the right-hand side, the charge is only $+e$, as the neutron is neutral.

The charge going in is not the same as that emerging – *the charge is not conserved*, and the reaction is never observed in nature.

However, the following reactions all conserve charge and are commonly observed in the right sort of experiments:

$$p + p \rightarrow p + p \qquad\qquad (R2)$$
$$p + n \rightarrow p + n \qquad\qquad (R3)$$
$$p + p \rightarrow p + n + \pi^+ \qquad\qquad (R4)$$

STUDY NOTE

Section 3.4 discusses how particle physicists use energy and momentum conservation to analyse reactions.

Alongside conservation of charge sit the very important laws of **conservation of energy** and **conservation of momentum**. These are always obeyed in particle reactions, but this can't be read directly from simple reaction equations such as the ones above. Further information would have to be provided.

Baryon number

Consider the following reactions

$$p + p \rightarrow p + \pi^+ \qquad\qquad (R5)$$
$$p + p \rightarrow p + \pi^0 \qquad\qquad (R6)$$
$$p + p \rightarrow p + p + n \qquad\qquad (R7)$$

Each of them satisfies conservation of charge and energy, but they have never been observed in nature. This is a classic example of how we can learn something from the absence of a reaction. Physicists established **baryon number conservation** by considering apparently possible reactions such as these alongside others that evidently worked perfectly well, such as:

$$p + p \rightarrow p + p + \pi^0 \tag{R8}$$

$$p + p \rightarrow p + n + \pi^+ + \pi^- \tag{R9}$$

Baryons – hadrons composed of three quarks – have baryon number $B = 1$.

Mesons – hadrons composed of a quark anti-quark pair – have baryon number $B = 0$.

Now, the total number of baryons emerging from a reaction has to be the same as the total number entering (that is the conservation law). Reactions R5 and R6 have two baryons entering ($B = 2$ initially) and only one baryon emerging ($B = 1$ for the proton + 0 for the pion = 1 in total). Reaction R7 has $B = 3$ on the right-hand side, which is clearly not the same as the $B = 2$ on the left.

Reactions R8 and R9 obey baryon number conservation. As pions, like all mesons, have $B = 0$ they can be created in any number during the reaction, subject to conservation of charge, energy and momentum.

Baryon number of quarks

Conservation of baryon number was established before the notion of quarks had been conceived. Nevertheless, by establishing the following rules it is easily extended:

All quarks have $B = \dfrac{1}{3}$

All anti-quarks have $B = -\dfrac{1}{3}$

From this it follows:

Baryon   contains 3 quarks so: $B = 3 \times \dfrac{1}{3} = 1$

Anti-baryon   contains 3 anti-quarks so: $B = 3 \times \left(-\dfrac{1}{3}\right) = -1$

Meson   contains a quark and an anti-quark: $B = \dfrac{1}{3} + \left(-\dfrac{1}{3}\right) = 0$

> **STUDY NOTE**
>
> As you saw in Section 2.4, pions are mesons.

Particle decays

The various conservation laws apply when particles react with each other, and also when a given particle decays. In Section 2.2 we came across members of a short-lived family of particles called the deltas. The Δ^{++} (uuu combination) decays as follows:

$$\Delta^{++} \rightarrow p + \pi^+ \tag{R10}$$

This conserves charge and conserves baryon number.

ACTIVITY 9 DELTA DECAY

Given that all the deltas (Δ) decay into a nucleon and a pion, work out decays for the Δ^+, Δ^0 and Δ^-.

Lepton number

Given what we have said previously about the apparent symmetry between the quarks and the leptons, it's not surprising that there is an equivalent conservation of lepton number.

A classic example of conservation of lepton number at work is the beta decay of a neutron:

$$n \rightarrow p + e^- + \bar{v}_e \qquad \text{(R11)}$$

Lepton number going into the reaction is 0, on the reasonable grounds that there are no leptons. However, there are two leptons emerging. This happens to be fine, as the electron has lepton number $L = 1$, but the anti-electron neutrino has $L = -1$ (compare with an anti-quark having $B = -\frac{1}{3}$), so the total is zero.

Lepton number works just like baryon number, with one minor complication: there are three separate lepton numbers, accounting for each generation of lepton separately, as shown in Table 3.8.

	First generation	Second generation	Third generation
	Electron	Muon	Tau
L_e	1	0	0
L_μ	0	1	0
L_τ	0	0	1
	electron neutrino	muon neutrino	tau neutrino
l_e	1	0	0
L_μ	0	1	0
L_τ	0	0	1

Table 3.8 Lepton numbers

This looks somewhat complicated at first, but simply reflects the idea that the number of particles in each generation is separately conserved. This is seen in the decay of the muon:

$$\mu^- \rightarrow e^- + \bar{v}_e + v_\mu \qquad \text{(R12)}$$

The electron and its anti-neutrino ensure that $L_e = 0$ on the right-hand side, as it is on the left. The muon neutrino is there to balance the muon decaying on the left, ensuring that $L_\mu = 1$ on both sides.

Neutral anti-matter

As we know, each matter particle has an anti-matter equivalent with the same mass. If the matter particle has charge, then the anti-matter version will also be charged, but the charge will be the opposite sign (electron and positron, for example). However, if the matter particle has no charge (such as the neutrinos) the matter and anti-matter version would appear to be identical, were it not for the reversal of lepton number, which still serves to distinguish the two particles in their reactions.

> ### ⚙ ACTIVITY 10 REACTION GAME
> Use a set of reaction cards to challenge your classmates to a duel. Each group presents the other with a set of reaction equations, including some that are forbidden by conservation laws. The group that can correctly identify the most allowed and forbidden reactions (and state which conservation law is violated) wins.

2.6 History of the Universe

We started this part of the chapter by discussing the Big Bang and the idea of an expanding Universe. You may have felt that we have taken a bit of a diversion by considering the current view of the fundamental particles and forces that govern the Universe. Now we need to bring these two topics together.

As the Universe expands, it cools downs; energy and matter are spread more thinly, and temperature is related to the concentration of energy. Out in space, it's cold. The COBE satellite (Cosmic Background Explorer, Figure 3.11) measured the temperature out there to be about 2.7 K. (Stars are, of course, much hotter; so is some of the gas between them. The 2.7 K refers to the temperature of 'empty space', where there is essentially no matter, just radiation.) Matter is very thinly spread – on average, less than one proton per cubic metre.

When the Universe was much younger than it is today, the energy was concentrated in a smaller volume, and the temperature was much higher. At high temperatures, the particles of matter have much higher energies, and they rush around, colliding and interacting. (These are the conditions that physicists try to reproduce in their giant accelerators.) These temperatures are too high for atoms and molecules to exist – they smash one another into their constituent parts.

So the history of the Universe can be summarised as follows:

- in its early stages: small, hot, dense; matter in the form of quarks, hadrons, leptons
- today: vast, cold, matter spread thinly; matter in the form of atoms and molecules.

Using a log–log graph

Figure 3.12 shows in more detail how the temperature of the Universe has changed with time. Here you can see the different phases that the Universe has gone through since the end of the inflationary phase. Notice that *both* scales are logarithmic. This is necessary in order to show the great ranges of these quantities.

Figure 3.11 The COBE (Cosmic Background Explorer) satellite

> **STUDY NOTE**
> 2.7 K = −270.3 °C
> The Kelvin temperature scale is discussed in the chapter *Reach for the Stars*.

> **STUDY NOTE**
> You met logarithmic scales in the chapter *Digging Up the Past*.

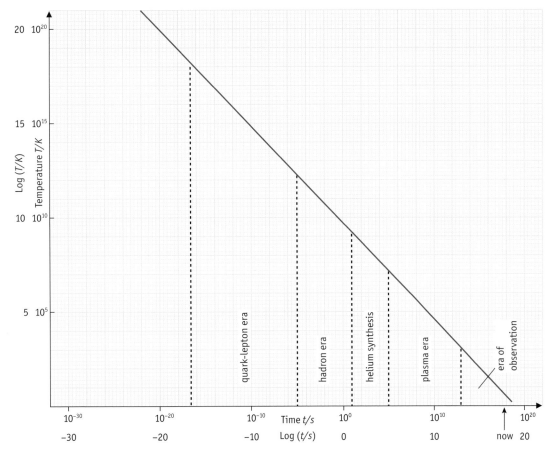

Figure 3.12 Fall in temperature of the Universe with time

> **STUDY NOTE**
> Figure 3.12 is an example of a log–log graph. Both scales are logarithmic. For more on log–log graphs, see PRO Additional Sheet 2.

The shape of the graph has been worked out from theoretical calculations that predict a relationship between temperature and time:

$$T = \frac{K}{\sqrt{t}} = kt^{-\frac{1}{2}}$$

(2)

where T is the absolute temperature (in kelvin), t is the time since the Big Bang, and k is a constant.

Equation 2 is an example of a **power-law relationship** – where one variable (T in this case) is directly proportional to another variable (t) raised to a power or **exponent** (here, the exponent is $-\frac{1}{2}$).

When numbers related by a power law are plotted in a **log–log graph** (that is, one where both axes are logarithmic), then the resulting graph is a straight line. We can illustrate this by taking logs of Equation 2:

$$\log(T) = \log(k) + \log(t^{-\frac{1}{2}})$$

$$\log(T) = \log(k) - \frac{1}{2}\log(t)$$

(3)

MATHS REFERENCE

Index notation and powers of 10

See Maths note 1.1

Powers that are not whole numbers

See Maths note 1.5

Gradient of a linear graph

See Maths note 5.3

Logs and powers

See Maths note 8.4

Log scales

See Maths note 8.6

Using log graphs

See Maths note 8.7

Plotting log (T) on the y-axis against log (t) on the x-axis gives a graph of the type $y = mx + c$, where the gradient, m, is equal to the exponent.

ACTIVITY 11 THE UNIVERSAL STORY

Use Figure 3.12 to answer the following questions.

● How many seconds is it since the origin of the Universe?

● At what time did the 'hadron era' start? What was the temperature at that time?

● According to the graph, what is the temperature (of 'space') now?

Find the gradient of the graph in Figure 3.12 and check that the graph is described by Equations 2 and 3.

Use Figure 3.12 to find the value of the constant, k, in Equations 2 and 3, and use Equation 2 to deduce its SI units.

Step-by-step history

Now we can look in more detail at the different periods in the history of the Universe, as shown in Figure 3.12. Physicists like to divide up the history of the Universe into distinct periods, depending on what was happening at the time.

After inflation

We will start just after the end of the period of inflation, 10^{-35} s after the origin of the Universe. At this time, the particles of matter and anti-matter that were in the Universe had been scattered over vast distances. If nothing had happened to fill the Universe again then matter would now be very scarce indeed. Certainly life could never have evolved, as it is difficult to see how stars could ever form in such a 'thin' Universe. However, at the end of the inflation period the Universe was topped up with matter and anti-matter by the decay of particles called 'Higgs particles'. These are very important in the theory of the Universe and particle physics in general. (In 2012 physicists working at CERN announced very good evidence for the existence of the long-sought Higgs boson. One of the tasks of the LHC accelerator had been to search for Higgs. As with all scientific discoveries, further confirmatory evidence is being assembled.)

After inflation had stopped, the Universe was filled with matter, anti-matter and radiation again. You might think that nothing important has therefore happened, but that is not true. The amount of new matter and anti-matter that appeared seems to be just enough to slow the Universe down to a gentle expansion – its gravity stops it from expanding so fast that stars and galaxies would never get the chance to form. But with much more matter, the Universe would have collapsed under its own gravity back to a 'big crunch' long ago. Inflation seems to have been just what was needed to get the right amount of matter into the Universe.

Soon after inflation ended, most of the anti-matter in the Universe disappeared. This is a very interesting puzzle that has had physicists thinking for decades. Where did the anti-matter go? You may ask how we know that it vanished at all. The answer is simple – we are here (see Figure 3.13). Everything that we touch is made of matter (even the Moon – we know because people have walked on it). In reactions such as those you saw in Section 2.4, matter and anti-matter are created and destroyed in equal amounts. But to account for the predominance of matter, anti-matter must in fact decay more rapidly than matter, so physicists have to think how their ideas might need to be modified or even overturned completely.

Quark-gluon plasma

Between 10^{-18} and 10^{-12} seconds after the Big Bang, no hadrons existed. The Universe was composed of a soup of quarks, leptons and various particles involved with the four fundamental forces (see Figure 3.4). As the Universe expanded, so the average distance between the quarks got bigger until at the end of this period in history they started to group into the hadrons that we find now.

The hadron era

Unsurprisingly, the next period, 10^{-12} seconds to 3 minutes after the Big Bang, is called the hadron era. As well as leptons, the Universe was now full of hadrons – mostly protons and neutrons. Various reactions were going on and the numbers of protons and neutrons were being kept in balance. However, the protons and neutrons were not able to join together to make nuclei, as the electromagnetic radiation in the Universe was hot enough (had enough energy) to break them apart again straightaway. As the end of this era drew near, the radiation had cooled down enough to let hadrons start to combine.

Nucleosynthesis

Three to 13 minutes after the Big Bang, nuclei started to form. Protons and neutrons came together to make deuterium, an isotope of hydrogen. Deuterium nuclei came together to make helium nuclei. As this process took place, all the free neutrons were effectively swept up into nuclei. A detailed calculation shows that at the end of nucleosynthesis, the Universe had about 78% hydrogen and 22% helium (measured by mass), which is almost exactly what we see now.

Plasma era

For the next 3000 years, the Universe was basically made up of leptons (especially electrons), electromagnetic radiation, helium nuclei and protons. The electrons could not stick to the nuclei to make atoms, as the energy in the radiation was great enough to blast them away again. The Universe was effectively an ionised gas – a plasma.

After about 3000 years, the temperature had dropped (the radiation was less energetic) to such an extent that electrons that did stick to nuclei could stay there. This is a very important step in the history of the Universe. Until then, the electromagnetic force had been very important as all the objects in the Universe were charged. After the loose charges had been stuck together and the Universe had become neutral, gravity became the most important of the fundamental forces.

Figure 3.13 Earth and the living creatures that inhabit it are made from matter

STUDY NOTE

A very similar nuclear process goes on in the Sun now – you will learn more about this in the chapter *Reach for the Stars*.

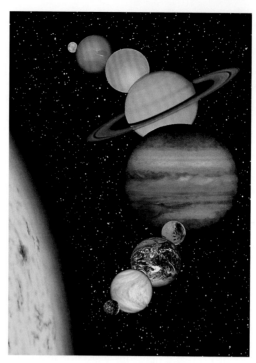

Figure 3.14 Planets of the Solar System. (Pluto is classified as a dwarf planet.)

Towards the present

After the plasma era, stars and galaxies formed, planets revolved in orbit about some stars (Figure 3.14). One planet (that we know of) took the radical step of evolving life. After some time, one complex form of life started to wonder about the Universe it lives in. And that's where you come in . . .

2.7 Summing up Part 2

In this part of the chapter you have learned about two important areas of physics:

● cosmology – the way the Universe has evolved since the Big Bang

● particle physics – the standard model of the fundamental particles and the forces that act between them.

You have also seen how these two areas, which might at first seem unrelated, are in fact intimately connected.

> ### ⚙ ACTIVITY 12 SUMMING UP PART 2
> Look back at your notes on Activities 1 and 2 in Part 1 of this chapter. How might you respond to these activities, now that you know a lot more about cosmology and particle physics?

QUESTIONS

Q 7 (a) What two particles are represented by e^+ and e^-?

(b) What difference is there between them?

(c) What happens when these two particles meet?

Q 8 For each of the particles in the list below, say whether it is a hadron, meson, baryon, quark or lepton (it may fall into more than one category).

proton	pion	electron
charm	neutron	anti-electron
up	electron-neutrino	

3 Towards the standard model

The standard model of particle physics, with its 12 fundamental particles and four fundamental forces, has been very successful. As we have seen in Part 2, it has allowed physicists to link their picture of the tiniest particles to their picture of the history of the whole Universe – particle physics has been united with cosmology. This has been one of the great achievements of physics in the second half of the 20th century. However, the standard model hasn't always been with us. In this part of the chapter, we will look at what went before.

3.1 The discovery of the atomic nucleus

In the 19th century, most scientists came to accept that matter is made of atoms, although they never expected to see any. The idea of atoms allowed scientists to explain all sorts of things – the molecular kinetic theory, for example, in which the behaviour of gases was explained in terms of the movement of the particles that make up a gas. Much of chemistry depends on our understanding of the ways in which atoms arrange and rearrange themselves. It was thought that atoms were indivisible, the fundamental building blocks of matter. Then came two discoveries that changed all that.

- In 1896, Henri Becquerel discovered radioactivity. Atoms of some elements, including uranium and radium, were capable of giving out smaller particles. Atoms were not indivisible.

- In 1897, Joseph (J.J.) Thomson discovered the electron. These were tiny particles, much lighter than atoms, and many different metals could emit them; this suggested that they must be more fundamental than atoms.

Thomson devised a model of the atom based on his discovery of the negatively-charged electron. He suggested that each atom was made up of thousands of electrons, spinning in such a way that they stuck together to form a neutral atom. This was not a very successful model. You have probably come across two other models of the atom: the plum-pudding model and the **nuclear model**, like a mini solar system. These are shown in Figure 3.15.

STUDY NOTE

You will study the molecular kinetic theory in the chapter *Reach for the Stars*

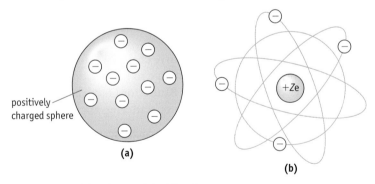

positively charged sphere

(a)

(b)

Figure 3.15 (a) Plum pudding and (b) nuclear models of the atom

STUDY NOTE

Z (the atomic number) represents the number of protons in the nucleus. You will learn more about the constituents of the nucleus in *Reach for the Stars*.

⚙ ACTIVITY 13 MODEL ATOMS

Discuss the two models shown in Figure 3.15, and J.J. Thomson's all-electron model. Here are some points to consider:

- How are the positive and negative electric charges arranged in each model?
- What can you say about the attraction and repulsion between electric charges in each model?
- What do we mean by the term 'model' here?

The discovery of radioactivity was important not just because it showed that atoms could fall apart; it also became a useful research tool. The radiation produced by radioactive substances could be used as a probe to find out more about matter on the scale of individual atoms. Someone who made profound use of this tool was Ernest Rutherford (1871–1937).

Figure 3.16 Geiger and Marsden's apparatus (based on a drawing in the paper that reported their results)

> **STUDY NOTE**
>
> An alpha particle is made up of 2 protons and 2 neutrons.

Figure 3.16 shows the apparatus used by Rutherford's co-workers German physicist Hans Geiger (1882–1945) and graduate student Ernest Marsden (1889–1970) to investigate the scattering of alpha radiation by a thin foil of gold. It was Rutherford who suggested the experiment and who interpreted the results. The source of radiation, R, is a thin-walled glass tube containing purified radium. The gold foil, F, is less than a thousandth of a millimetre thick. The tube, T, is used to pump the air out of the chamber. Radiation passing through the foil hits the scintillator, S, and produces a flash of light. The experimenter counts the flashes while looking through the microscope, M.

When the microscope was aimed directly at the source of radiation, Geiger and Marsden saw frequent flashes of light, showing that most of the alpha particles passed straight through the gold foil. When they moved the microscope round through a small angle, they found that the number of flashes became much less; relatively few alpha particles were deflected by the foil. This was what they had anticipated. However, when they moved the microscope round through large angles, they were surprised to find that they could still see some flashes, even when the microscope was on the same side of the foil as the source. A very few alpha particles were being reflected back from the gold foil towards their original source. It was this 'back-scattering' that Rutherford had to explain.

Rutherford explains back-scattering

To understand how alpha particles are scattered by gold foil, we need to know something about electrostatic forces. Alpha particles are positively charged, and so they are repelled by other positive charges and attracted by negative charges. Rutherford pictured an alpha particle as a tiny, positively charged bullet moving at high speed. If it encountered a gold 'plum-pudding atom', it would be deflected slightly as the positive and negative charges of the atom pushed and pulled the alpha particle. Yet because the positive charge of the pudding is spread out, and the embedded electrons have very small mass, it would not be deflected very much.

To explain the back-scattering of alpha particles, Rutherford came up with the nuclear model of the atom. He suggested that nearly all the atom's mass was concentrated at the centre of the atom in a tiny nucleus. To use an anology: a fast-moving bullet will not bounce back off a low-mass object in its path; it will simply knock it out of the way. Rutherford realised that the nucleus of the atom must be very tiny compared to the whole atom – most alpha particles whiz straight through without ever coming close to a nucleus.

Rutherford also proposed that all of the atom's positive charge was in the nucleus. The electrons orbited around this nucleus, providing a negative charge to balance the positive charge of the nucleus. Assuming an inverse-square electric field, Rutherford modelled the paths of charged alphas to predict the fraction of alphas as a function of scattering angle.

Geiger and Marsden measured the numbers of alpha particles scattered through different angles, and Rutherford's model was able to explain the pattern of their results. His model also predicted the results that would be obtained if the alpha particles had different energy, and if foils of different thickness, and made from different metals, were used. Geiger and Marsden carried out a series of experiments to test Rutherford's model – and every time the results were as predicted. This was enough to convince most physicists that the nuclear model of the atom was correct. Table 3.9 shows some results obtained using gold foils of different thickness.

Thickness, t, of gold foil/10^{-6} m	Number, N, of scintillations per minute
0.23	21.9
0.46	38.4
1.07	84.3
1.70	121.5
1.89	145.0

Table 3.9 Some of Geiger and Marsden's results

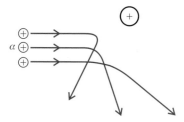

Figure 3.17 Alpha particles (α) are deflected as they pass close to a nucleus

Table 3.10 summarises the experimental results and Rutherford's conclusions and Figure 3.17 shows how Rutherford's model explained Geiger and Marsden's results.

Observation	Conclusion
Most of the alpha particles passed straight through the gold foil without any deflection.	Most of the atom is empty space. Chance encounters with electrons are not enough to deflect the alpha particles, as the electrons are much lighter than the alpha particles ($m_{alpha} \sim 8000\,m_e$).
The number of deflected particles is small.	The volume occupied by the central massive region (nucleus) is very small.
A very few of the alpha particles (less than 1 in 1000) were scattered through angles greater than 90°.	The atom must include a small dense region containing most of the atom's mass to cause such large changes in direction for even a few alpha particles, as they have such high momenta when striking the foil.
The fraction of particles scattered at different angles agreed with mathematical models that assumed an inverse-square field.	The small highly dense region within the atom must also carry a large electrical charge. (The positive sign of the charge was inferred from the existence of negatively charged electrons, not from this experiment.)

Table 3.10 Summary of Geiger and Marsden's results and Rutherford's conclusions

⚙ ACTIVITY 14 A MODEL FOR RUTHERFORD SCATTERING

Use a model such as the one shown in Figure 3.18 to investigate Rutherford scattering. A marble runs down a ramp, and then past a plastic or metal 'hill'. The closer it gets to the hill, the more it is deflected. Replace the hill with a metal or paper cylinder to show that only a large mass can produce large-angle deflections and that only an inverse-square force field will produce the scattering pattern observed.

Figure 3.18 Model to represent the scattering of an alpha particle by a nucleus

⚙ ACTIVITY 15 ANALYSING THE RESULTS

By drawing suitable graphs using the data from Table 3.9, find a mathematical relationship between N, the number of scintillations per minute, and t, the thickness of the foil.

MATHS REFERENCE

Testing mathematical relationships
See Maths note 5.5
Using log graphs
See Maths note 8.7

QUESTIONS

Q 9 Look at Figure 3.17. How does the degree of deflection change as the alpha particle passes closer to the nucleus? At what point on the path is the alpha particle changing direction most rapidly? What do these observations tell you about the nature of electrostatic force?

Q 10 Alpha radiation is easily absorbed. An alpha particle has a range of about 5 cm in air, because it collides with innumerable molecules of the air and loses its energy. Alpha particles travel even shorter distances through solid materials, because they are much denser than air. Use these ideas to explain some of the features of the design of Geiger and Marsden's apparatus.

Q 11 Geiger and Marsden tested Rutherford's model using thicker gold foils. Would this make it more or less likely that an alpha particle would score a direct hit on a gold nucleus? How would you expect their results to change?

Q 12 Geiger and Marsden used thin sheets of mica to slow down the alpha particles so that they would not pass so rapidly through the gold foil. How would you expect their results to change?

STUDY NOTE

Mica is a silicate mineral that naturally forms very thin sheets.

Q 13 Silver atoms are the same size as gold atoms, but their nuclei are smaller and have lower charge. Geiger and Marsden tried silver foil instead of gold. What do you think they observed?

Figure 3.19 A fast-moving alpha particle slows down as it approaches a gold nucleus, until it goes into reverse

Charge repelling charge

Rutherford used his model to say something about the size of the nucleus of the gold atom. To see how he did this, think about an alpha particle that approaches the nucleus of a gold atom head-on, as shown in Figure 3.19. As the alpha particle approaches, it begins to feel the repulsion of the gold nucleus. (The two positive charges repel one another.) The repulsive force slows it down. The closer the alpha particle gets, the greater the force, and so the greater its deceleration. Eventually it comes instantaneously to a halt, and then starts to move back towards where it came from.

You can compare this motion with that of a ball thrown up in the air. The ball moves fast as it leaves your hand; it gradually slows as it rises in the air; it stops instantaneously, and then it accelerates downwards. In this case, the force is caused by the gravitational attraction between the ball and the Earth; for the alpha particle, the force is the electrostatic repulsion.

QUESTION

Q 14   Use the information in the above paragraph and the theory already studied in *Higher, Faster, Stronger* to sketch the shape of a velocity–time graph for a ball thrown in the air and for an alpha particle as it approaches a nucleus head-on. State and explain two similarities and two differences between the graphs.

If we can find out how close the alpha particle gets to the gold of the nucleus, this could give us an idea of how big the nucleus is. To do this, Rutherford needed to know about the law that describes the electrostatic force between two charged particles. He was able to show that the alpha particles reached within about 10^{-13} m of the centre of the atom. (This is the distance shown as d in Figure 3.19.) So, although he could not say how big the nucleus was, he could deduce that its diameter must be no greater than about 10^{-13} m; that's no more than a thousandth of the diameter of an atom. (Today we know that the diameter of a nucleus is even smaller, about 10^{-15} m across.) In the next section we will look at the law that allowed Rutherford to make his estimate.

3.2 Electrical forces

No doubt you have carried out some simple experiments on static electricity. If you rub a balloon on your jumper, it becomes charged and you can do interesting things with it – stick it to the wall, for example. Here you are making use of the attractive force between opposite charges. Suppose the balloon gains a positive charge when you rub it; it will then attract negative charges in the wall, and the two stick together.

If you rub two balloons on your jumper so that they both become charged, you can observe the repulsive force between like charges (charges with the same sign). Hang the balloons side by side and they repel one another, as shown in Figure 3.20(a).

(a)   (b)

Figure 3.20   (a) Like charges repel one another; (b) unlike charges attract one another

The fact that there are two types of electric charge, and the rules of attraction and repulsion, were known in the 18th century. Today we understand that a balloon becomes charged when you rub it because electrons are transferred from one object to another. If electrons are transferred from the balloon to your jumper, the balloon will become positively charged because it has lost electrons, which are negatively charged.

The French physicist Charles Coulomb (1736–1806) went beyond these simple rules of attraction and repulsion. He wanted to know more about the nature of the force, and in particular how the force between two charged objects depended on their separation. He used balls of pith, because

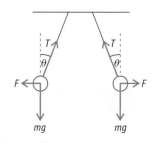

Figure 3.21 Two identical charged balls suspended by insulating threads

pith is a good insulating material that retains an electric charge long enough for measurements to be made. (Today we might use expanded polystyrene.) Figure 3.21 shows two charged balls suspended so they repel one another. This is a clever arrangement because, provided we know the weight of each ball, we can determine the force between them.

Note that Figure 3.22 is symmetrical. Each ball feels a horizontal electrostatic force of magnitude F regardless of the magnitude of the charge it carries because the force is proportional to the product of the charges on each ball. The pair of electrostatic forces is an example of Newton's third law of motion:

- they are equal in magnitude and opposite in direction
- they act on different bodies (the two balls)
- they are the same type of force (electrostatic).

Now look at the forces acting on just one of the balls, as shown in the vector diagram in Figure 3.22. There are three forces:

- the ball's weight mg
- the electrostatic force F
- the tension of the string T.

STUDY NOTE

You met Newton's third law and force vector diagrams in the chapter *Higher, Faster, Stronger*.

The easiest of these to measure is the weight mg. We can use this, along with the angle θ, to find the electrostatic force F. The ball is stationary; that is, it is in equilibrium, so we know that the forces are balanced. The weight mg downwards is balanced by the vertical component of the tension T acting upwards:

$$mg = T \cos\theta \tag{4}$$

The electrostatic force F to the right is balanced by the horizontal component of the tension T acting to the left:

$$F = T \sin\theta \tag{5}$$

We can eliminate T from these equations by dividing Equation 5 by Equation 4:

$$\frac{F}{mg} = \frac{T \sin\theta}{T \cos\theta} = \tan\theta$$

Rearranging gives:

$$F = mg \tan\theta \tag{6}$$

So, if we can measure mg and the angle θ, we can calculate F. Alternatively, you could find F from a scale drawing of Figure 3.22(b).

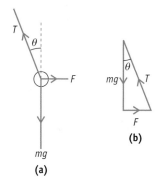

Figure 3.22 The three forces acting on one of the balls shown in Figure 3.21

QUESTIONS

Q 15 Figure 3.21 shows two charged balls repelling one another. Sketch diagrams to show how the situation would change **(a)** if the balls had more charge, and **(b)** if the balls had opposite charge.

Q 16 Sketch two charged balls hanging close to one another, one having twice as much positive charge as the other. (Hint: remember Newton's third law!)

Q 17 Calculate the electrostatic force on each of the balls shown in Figure 3.23, using data from the diagram.

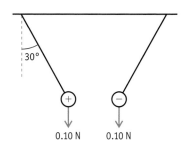

Figure 3.23 Diagram for Question 17

Figure 3.24 Defining quantities: the force between two charged spheres

Coulomb's law

Coulomb investigated the force between two charged spheres. Two spheres with charges Q_1 and Q_2 are separated by a distance r; note that r is measured from the centre of one sphere to the centre of the other (Figure 3.24). It is not surprising to find that the force depends on the size of each of the two charges, and that the force decreases as the spheres are moved apart. Coulomb found that the force is proportional to each of the charges:

$$F \propto Q_1 \text{ and } F \propto Q_2 \tag{7}$$

and that the force is inversely proportional to the square of the distance r between the two charges – the force obeys an inverse square law:

$$F \propto \frac{1}{r^2} \tag{8}$$

These relationships can be combined:

$$F \propto \frac{Q_1 Q_2}{r^2}$$

$$F = \frac{k Q_1 Q_2}{r^2} \tag{9}$$

where k is a constant of proportionality. Equation 9 represents Coulomb's law of electrostatic force. In SI units, with F in newtons, r in metres and Q in coulombs:

$$k = 9.0 \times 10^9 \, \text{N m}^2 \, \text{C}^{-2}$$

The constant k is usually written in a different form as:

$$k = \frac{1}{4 \pi \varepsilon_0}$$

where the quantity ε_0 is known as the **permittivity of free space** and has the value $8.854 \times 10^{-12} \, \text{F m}^{-1}$ (farads per metre). Coulomb's law is usually written as:

$$F = \left(\frac{1}{4 \pi \varepsilon_0} \right) \frac{Q_1 Q_2}{r^2} \tag{9a}$$

Coulomb's law tells us about one of the four fundamental forces of nature, part of the standard model. One remarkable feature of this law is that, while Coulomb found out about it by making measurements on charged pith balls in his laboratory, it applies equally to the charged particles of the sub-atomic world – protons, electrons and the other members of the sub-atomic menagerie.

QUESTIONS

Q 18 Show that the SI units of k and ε_0 are consistent with one another, i.e. that the units F m^{-1} and $(\text{N m}^2 \, \text{C}^{-2})^{-1}$ are equivalent.

Q 19 Calculate the force that acts between two charges, each of $10 \, \text{nC}$, separated by a distance of $5 \, \text{cm}$.

Q 20 The average separation of a proton and an electron in a hydrogen atom is about $0.037 \, \text{nm}$.

(a) Calculate the force that each exerts on the other.

(b) Draw a diagram to show the directions in which these forces act.
($e = 1.60 \times 10^{-19} \, \text{C}$, $k = 9.0 \times 10^9 \, \text{N m}^2 \, \text{C}^{-2}$)

> **STUDY NOTE**
>
> In the chapters *Transport on Track* and *The Medium is the Message*, you will have met the farad as a unit of capacitance. $1 \, \text{F} = 1 \, \text{C V}^{-1}$

> **STUDY NOTE**
>
> Although the symbol k is used in Equation 9 it is important to note that this is *not* the same as k_B, the Boltzmann constant $(1.38 \times 10^{-23} \, \text{J K}^{-1})$, which is used in thermal physics.

PRACTICAL SKILLS
REFERENCE

Planning and experimental
design

See Practical skills notes
2.1–2.4

Carrying out practical work

See Practical skills notes
3.1–3.2

Analysis and interpretation
of data

See Practical skills notes
4.1–4.2

Conclusion and evaluation

See Practical skills notes
5.1–5.2

⚙ ACTIVITY 16 EXPLORING COULOMB'S LAW

Explore the inverse-square relationship of the force between two charged objects (Figure 3.25). This requires care, as well as dry atmospheric conditions.

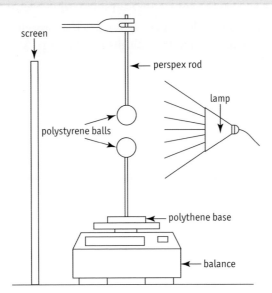

Figure 3.25 Apparatus for Activity 16

Force, work, energy and potential

Now that we know an expression for the electrostatic force between two charged particles, we can go back to Rutherford's problem. How close can an alpha particle get to the nucleus of a gold atom?

Figure 3.26 shows, in graphical form, how the force depends on the distance of the alpha particle from the gold nucleus. As an alpha particle (Q_1 = +2e) speeds directly towards a gold nucleus (Q_2 = +79e), it feels a force that is at first small but which increases more and more rapidly. When the values are plotted on a logarithmic graph, the gradient is −2, showing that there is an inverse square relationship between F and r. The force slows the alpha particle until it stops, goes into reverse and accelerates back towards where it came from.

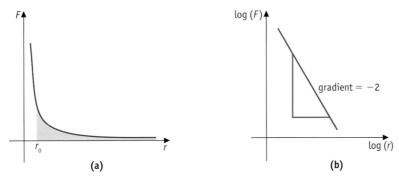

Figure 3.26 (a) The force F between an alpha particle and a gold nucleus changes with their separation r; (b) the same values plotted on a logarithmic graph

As the alpha particle approaches the nucleus, it loses kinetic energy and gains electrostatic potential energy; the repulsive force does work. At the point of closest approach, in a head-on encounter, the kinetic energy is zero; all of the initial kinetic energy has been transformed to potential energy. This is similar to the situations that you met in the earlier chapter *Higher, Faster, Stronger*, where a vertically launched projectile gains gravitational potential energy at the expense of its kinetic energy, and where a bungee jumper loses kinetic and gravitational potential energy as the cord gains elastic potential energy.

The work done on the alpha particle as it approaches to a distance r_0 is represented by the shaded area under the graph in Figure 3.26. The alpha particle stops and goes into reverse when the area is equal to its initial kinetic energy. Knowing the initial kinetic energy of the alpha particles, Rutherford was able to calculate how closely they must have approached the nucleus.

Using calculus, it is possible to show that, as an alpha particle approaches from a very large distance to a distance r, the increase in electrostatic potential energy ΔE, is given by:

$$\Delta E = \frac{Q_1 Q_2}{4\pi\varepsilon_0 r} \tag{10}$$

STUDY NOTE

To see how to derive the equation for potential energy in an inverse-square field, refer to PRO Additional Sheet 3.

As you saw in the chapters *Technology in Space* and *The Medium is the Message*, electrostatic potential energy is closely related to the idea of potential difference (voltage) and potential at a point relative to some chosen zero level. In a circuit, the zero of potential is usually chosen to be the negative terminal of the power supply (so that all other potentials are positive). When we are dealing with an isolated charge, the zero of potential is usually chosen to be at an infinite distance from the charge. Then if the charge is positive, the potential is positive everywhere around it, and if the charge is negative the potential is negative everywhere.

Remebering that potential is the potential energy per unit charge, we can divide Equation 10 by Q_2 (and drop the subscript from Q_1) to define the electrostatic potential, V, at a distance r from a charge Q:

STUDY NOTE

You will do more on potential in an inverse-square field in *Reach for the Stars*.

$$V = \frac{\Delta E}{Q_2}$$

$$= \frac{Q}{4\pi\varepsilon_0 r} \tag{11}$$

QUESTIONS

Q 21 (a) Sketch and explain how the graphs shown in Figure 3.26 would change if the nucleus used was that of a copper atom, containing 29 protons.

(b) Sketch graphs to show how the electroctatic potential, V, varies with distance, r, from a gold nucleus and a copper nucleus.

Q 22 A student carries out Activity 16 (Figure 3.25) to test Coulomb's Law on a good, dry day and obtains the results recorded in Table 3.11. (For details of the experiment, see Activity sheet 16.)

(a) Explain how the student calculated the column headed 'Force F/mN'.

(b) Plot a graph of $\log_{10}(r/m)$ against $\log_{10}(F/mN)$ and discuss whether there is enough evidence in these results to confirm that the force between charges obeys an inverse square law.

Separation r/m	Balance reading/g	Force F/mN
0.049	0.130	1.27
0.061	0.084	0.82
0.074	0.058	0.57
0.107	0.027	0.26
0.149	0.013	0.13

Table 3.11 Data for Question 22

Q 23 Alpha particles typically have a kinetic energy of 8.0×10^{-13} J. If such an alpha particle travelled directly towards a gold nucleus ($Q = +79e$), what would be its distance of closest approach? ($e = 1.60 \times 10^{-19}$ C, $k = 9.0 \times 10^9$ N m^2 C^{-2})

3.3 Force fields

Coulomb's law is an example of an inverse-square law, so called because the force decreases in inverse proportion to the square of the distance between the two charged spheres. At twice the distance, the force drops to one-quarter its previous value; at three times the distance the force drops to one-ninth its previous value – and so on.

Figure 3.27 shows the electric field lines of the field around a charged sphere. The lines get further apart as they spread out, showing that the force gets weaker with distance. At twice the distance the lines are spread out over four times the area, showing that the force is a quarter of the strength. This is an inverse-square relationship – doubling the distance gives a quarter of the force.

An electric field is defined as a region in which a charged object experiences a force, and the electric field strength, E, is defined as the force per unit charge. If a charge q experiences a force of magnitude F, then the magnitude of field strength is:

$$E = \frac{F}{q} \tag{12}$$

The direction of an electric field is defined as that of the force experienced by a positive charge. A diagrammatic representation of a field using electric field lines tells us two things about the electric field near a charged object:

● the strength of the field (closer together means stronger)

● the direction of the field (a small positive charge will feel a force in the direction of the arrow; note that the arrows go from positive to negative charge).

Figure 3.28 shows how field lines can be used to represent the electric field in a variety of situations. Notice that we can still say that a field exists around an isolated charged sphere (Figure 3.28), even if it is not exerting a force on another charged object. The field lines show us that there would be a force if we placed another charged object nearby.

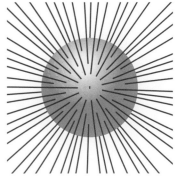

Figure 3.27 Electric field around a charged sphere

STUDY NOTE

Electric field strength and field lines were introduced in the chapter *The Medium is the Message*.

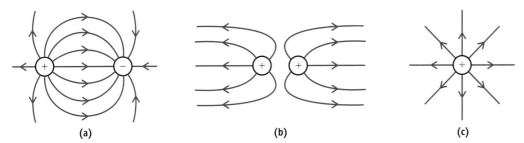

(a)   (b)   (c)

Figure 3.28 Electric fields around (a) two charged spheres with opposite charges, (b) two charged spheres with like charges, and (c) an isolated positively-charged sphere

To find out about the field strength at point X, a distance r from the charge Q, we place a small test charge q at X (see Figure 3.29). Outside a charged sphere, we have to measure r from the centre of the sphere. The force F on q is given by Coulomb's law, from Equations 9 and 11:

$$F = \frac{kQq}{r}$$

and electric field strength E is then:

$$E = \frac{kQ}{r} \tag{9b}$$

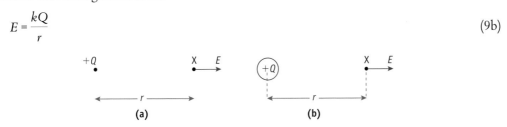

(a)   (b)

Figure 3.29 Calculating the electric field strength near (a) a point charge Q and (b) a charged sphere

⚙ **ACTIVITY 17
PROBING THE FIELD**

Investigate the size and direction of the electric field around a charged sphere.

PRACTICAL SKILLS REFERENCE

Planning and experimental design

See Practical skills notes 2.1–2.4

Carrying out practical work

See Practical skills notes 3.1–3.2

Analysis and interpretation of data

See Practical skills notes 4.1–4.2

Conclusion and evaluation

See Practical skills notes 5.1–5.2

QUESTIONS

Q 24 Draw a diagram to show the field lines around an isolated *negatively* charged sphere. Explain why your diagram is different from Figure 3.28(c).

Q 25 Copy the diagram of Figure 3.28(a). Mark and label a point where the electric field is strong, and another point where the field is weak. Draw arrows to show the electrostatic forces acting on the two spheres. Repeat for Figure 3.28(b).

Q 26 Calculate the electric field strength at a point where a charge of 20 mC feels a force of 10 N.

Q 27 An electron, charge $e = 1.60 \times 10^{-19}$ C, passes between two charged plates; the electric field strength between the plates is 2000 N C^{-1}.

What force acts on the electron and what is its acceleration?
(Electron mass $m_e = 9.11 \times 10^{-31}$ kg)

Q 28 The approximate distance of closest approach of an alpha particle and a gold nucleus (charge = +79e) is 1.0×10^{-13} m. What is the electric field strength at this distance from the nucleus?
($e = 1.60 \times 10^{-19}$ C, $k = 9.0 \times 10^9$ N m^2 C^{-2})

3.4 Collisions

Rutherford used collisions to probe the structure of atoms. Since then, particle physicists have used collisions as one of their main ways to deduce the nature of sub-atomic particles.

A gold nucleus is much more massive than an alpha particle – about 50 times as massive. So in Rutherford's experiments, a small particle was colliding with a much bigger one that remained more or less at rest. If there is less of a difference in the masses, the 'target' particle is set in motion; this is what happens in most particle-physics experiments. The diagram in Figure 3.30 shows what happens when an alpha particle collides with a much smaller nucleus – that of a helium atom.

Figure 3.30 was obtained using a cloud chamber, which is a device that can show up the tracks of charged sub-atomic particles. In this case, the cloud chamber was filled with helium gas, and a source of alpha radiation was then inserted. The tracks of alpha particles appear as straight lines extending out from the source. One alpha particle has hit a helium nucleus, and the two particles have gone off at right angles to each other.

Figure 3.30 Collision between an alpha particle and a helium nucleus

The nucleus of a helium atom consists of two protons and two neutrons, so it is identical to an alpha particle. You can investigate a collision between two identical particles using pucks containing dry ice, so that they glide on a layer of carbon dioxide gas, or two identical 'hover-footballs'. When a moving puck (or football) collides with an identical stationary one, they move off at 90° to one another, as shown in Figure 3.31, just like the alpha particle tracks in Figure 3.30.

Momentum and energy in collisions

By analysing the tracks left in particle detectors, particle physicists can make deductions about the mass and energy of sub-atomic particles. In order to do this, they use some fundamental physical laws – the conservation of momentum and of energy.

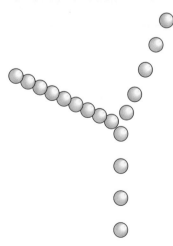

Figure 3.31 Diagram drawn from a multiflash photo of a collision between two identical pucks

In the chapters *Higher, Faster, Stronger* and *Transport on Track*, you explored one-dimensional collisions. You saw that, in a collision (or an explosion), momentum (mass × velocity) is always conserved; that is, the total momentum of the interacting objects is unchanged. We can now generalise this to all collisions. As momentum is a vector, we can draw a vector diagram to represent the momentum of the objects before and after they collide. Figure 3.32 shows such a diagram drawn from a multiflash photo, such as that shown in Figure 3.31; the speed was deduced from the spacing of the images in the multiflash photographs. You can see that the vector sum of momenta afterwards is equal to the momentum of the single moving puck before the collision. Momentum is conserved in *all* collisions.

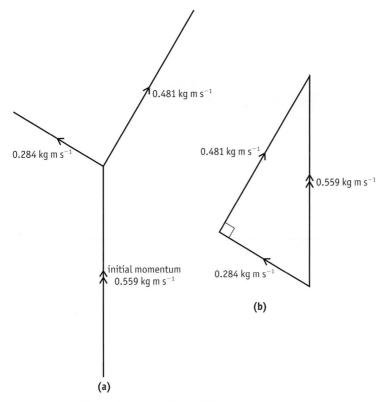

Figure 3.32 Momentum vectors of the pucks shown in Figure 3.31

As you saw in the chapter *Transport on Track*, an elastic collision is defined as one in which kinetic energy is conserved. By analysing the momentum and energy in a collision, particle physicists can make deductions about the nature of the particles involved and the interactions between them. You will explore some aspects of this in Activities 18 and 19. For these activities, it is useful to be able to relate kinetic energy directly to momentum without needing first to find the speeds of the moving objects.

An object of mass m moving at speed v has momentum of magnitude p given by:

$$p = mv \tag{13}$$

The same object's kinetic energy, E_k, is given by:

$$E_k = \frac{1}{2}mv^2 \tag{14}$$

To express E_k in terms of p, square Equation 13:

$$p^2 = m^2v^2$$

then divide both sides by $2m$:

$$\frac{p^2}{2m} = \frac{m^2v^2}{2m} = \frac{mv^2}{2}$$

and so we have:

$$E_k = \frac{p^2}{2m} \tag{15}$$

It turns out that an *elastic* collision between a moving and a stationary object of equal mass always results in the two moving at right angles to each other. Analysis of Figure 3.32 illustrates why this is the case. Since momentum is conserved, we can write the vector sum:

$$\mathbf{p} = \mathbf{p}_1 + \mathbf{p}_2 \tag{16}$$

PRACTICAL SKILLS
REFERENCE

Scientific questions and
information research

See Practical skills notes
1.1–1.4

Planning and experimental
design

See Practical skills notes
2.1–2.2

Carrying out practical work

See Practical skills notes
3.1–3.2

Analysis and interpretation
of data

See Practical skills notes
4.1–4.2

Conclusion and evaluation

See Practical skills notes
5.1–5.2

Since both particles have the same mass, m, conservation of kinetic energy gives:

$$\frac{p^2}{2m} = \frac{p_1^2}{2m} + \frac{p_2^2}{2m} \tag{17}$$

Multiplying Equation 17 by $2m$ leads to a relationship between the magnitudes of the vectors (in other words, between the lengths of the sides of the triangle):

$$p^2 = p_1^2 + p_2^2 \tag{18}$$

Equation 18 is Pythagoras's relationship between the sides of a *right-angled* triangle – so the angle θ must be 90°, i.e. in this particular case the particles must move at right angles to one another after they collide.

In Activities 18 and 19, you will explore collisions that give rise to different angles between the tracks.

ACTIVITY 18 ACCOUNTING FOR MOMENTUM AND ENERGY

Explore how to show that momentum is conserved in collisions by making measurements on multiflash photographs and also investigate how the kinetic energy changes in collisions.

ACTIVITY 19 NUCLEAR DODGEMS

Use marbles or ball bearings on a level glass surface to model collisions between sub-atomic particles. Explore what happens when a moving marble collides with a stationary marble whose mass is the same as, greater than, or less than its own.

Use ICT to analyse the collisions andto investigate the conservation of linear momentum and kinetic energy.

QUESTIONS

Q 29 Estimate your own mass, and the fastest speed at which you can run. Use Equations 13 and 14 to calculate your momentum and kinetic energy. Confirm that your answers are related by Equation 15.

Q 30 Calculate the kinetic energy of an alpha particle (mass 6.7×10^{-27} kg) whose momentum is 5.0×10^{-20} kg m s^{-1}.

Q 31 Many particle collisions result in the creation of new particles; for example when an electron and positron collide and interact to produce a muon/anti-muon pair. Muons have greater mass than electrons.

(a) Would momentum be conserved in such a collision?

(b) Would such a collision be elastic? If not, would the total kinetic energy increase or decrease?

Q 32 Particles with no electric charge (such as neutrons and neutrinos) leave no tracks in particle detectors. How might particle physicists deduce the presence of such particles from a record of tracks left by charged particles before and after a collision?

3.5 Particle diffraction

Rutherford's alpha-particle scattering experiment, carried out in 1908, showed how the underlying structure of matter could be investigated by firing a beam of charged particles at the matter and looking at the pattern that the particles formed after they had interacted with the matter. This is the basic approach used in much of particle physics; an approach that was developed and extended throughout the 20th century. In Section 3.4, we analysed the results of such experiments in terms of collisions. In this section we see how ideas about diffraction and wave–particle duality can be brought to bear on such experiments.

In the earlier chapter *Digging Up the Past*, you saw how diffraction of X-rays enables archaeologists to probe the structure of materials, and you saw that electron beams, too, behave like waves and can be diffracted. Electron diffraction is used to probe matter on an even smaller scale than X-rays – electron beams are used to measure the size of nuclei and to reveal the presence of quarks within protons. After a brief review of diffraction in Activity 20, this section looks in more detail at electron diffraction, and you will see how it comes about that electron beams can be used to probe matter on such a minute scale.

⚙ ACTIVITY 20 DIFFRACTION OF LIGHT

Observe what happens when light is diffracted by:

● silk, chiffon or a piece of net curtain

● a grating of parallel lines

● a regular grid of crossed lines

● a random array of fine dust particles.

Explore the effects of changing the separation of the lines and the wavelength of the light.

Figure 3.33 shows apparatus that you can use in the laboratory to demonstrate electron diffraction by graphite, and Figure 3.34 shows the pattern produced on the screen with such apparatus when polycrystalline graphite is placed in the electron beam. With a single crystal the pattern is an array of dots, like the pattern you get if you shine a laser beam through crossed diffraction gratings. Just as the grating diffracts laser light, so the beam of electrons is diffracted by the regular array of atoms that make up the graphite crystal. The photograph in Figure 3.34 shows what happens when a beam of electrons shines through a piece of polycrystalline graphite. (It's difficult to obtain a single crystal.) This material is made of many tiny crystals, all at different orientations. Because of this, the spots of the diffraction pattern are smeared out into rings.

> **STUDY NOTE**
>
> You also met electron beams ('cathode rays') in the chapter *The Medium is the Message*.

Figure 3.33 Using an electron diffraction tube to demonstrate electron diffraction by graphite

Figure 3.34 Diffraction pattern produced when an electron beam passes through polycrystalline graphite

STUDY NOTE

Wave–particle duality was introduced in *The Sound of Music*.

Wave–particle duality

Diffraction is something that happens to waves. When waves pass through a narrow gap, they spread out. The bright spots you see in the laser-light diffraction pattern are formed when many waves arrive at the same point in step with one another.

The fact that electrons can be diffracted tells us something striking about electrons: sometimes they seem to behave as particles, sometimes as waves. When they are moving freely as a beam, and when they strike the screen at the end of the tube, they behave as particles. When they pass between the atoms of a crystal of graphite, they behave like waves. This double behaviour is not like anything we experience on the macroscopic scale of things, but it is a phenomenon we have to come to terms with when considering the behaviour of matter and energy on the microscopic scale.

The rule is that sometimes we have to use wave ideas to explain our observations, and sometimes particle ideas. We can not say that under certain circumstances electrons are waves; they are neither waves nor particles, and all we can do is use our ideas of waves or particles to explain their behaviour as best we can. What we need is a way of translating back and forth between the two ways of describing electrons. In the earlier chapter *Digging Up the Past*, you met this 'translation' – it is the de Broglie equation:

$$\lambda = \frac{h}{p} \tag{19}$$

where p is the electron's momentum and λ is its wavelength (more correctly, its de Broglie wavelength, after Louis de Broglie (1892–1987) who found this relationship). The quantity h is the Planck constant:

$$h = 6.63 \times 10^{-34}\,\text{J s}$$

Equation 19 thus allows us to translate from a particle property (momentum) to a wave property (wavelength). Figure 3.35 shows a log–log graph relating λ to p. Notice that the gradient of the graph is negative (because the two quantities vary in inverse proportion). The gradient is −1 (because $\lambda \propto p^{-1}$).

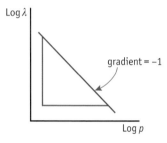

Figure 3.35 Logarithmic graph of de Broglie wavelength λ and momentum p for an electron

MATHS REFERENCE

Using log-graphs

See Maths note 8.7

In Activity 20 of this chapter and in the chapter *Digging Up the Past* you saw that, in order to analyse the structure of a material using a diffraction technique, the wavelength of the radiation should be comparable with the separation of the diffracting objects. So probing matter on a sub-atomic scale requires a shorter wavelength than that used to investigate structure on the scale of whole atoms or molecules. In Activity 21, you will explore how the de Broglie wavelength of electrons can be controlled. You will need to draw on ideas about charge and voltage that you met in earlier chapters (such for instance, the chapter *Technology in Space*).

The electron beam is produced in an electron gun (Figure 3.36). Electrons are first released from a heated cathode by thermionic emission, and then accelerated by a high voltage applied between cathode and anode. They pass through a hole in the anode and then travel at constant speed as they pass through the diffracting object (a graphite crystal, for example) and then strike the fluorescent screen.

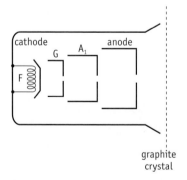

Figure 3.36 Electron gun

So long as there is no other energy transfer taking place, the kinetic energy, E_k, acquired by an electron as it is accelerated in the gun is given by:

$$E_k = eV \tag{20}$$

where e is the electron's charge and V the potential difference between cathode and anode. Provided the electron's speed does not approach the speed of light, we can use the non-relativistic Equations 15, 19 and 20 to derive an expression for the de Broglie wavelength. From Equations 15 and 20:

$$p = \sqrt{2mE_k} = \sqrt{2meV}$$

and so, using Equation 19:

$$\lambda = \frac{h}{p} = \frac{h}{\sqrt{2meV}} \tag{21}$$

STUDY NOTE

Electron guns were described in more detail in *The Medium is the Message*.

ACTIVITY 21 DIFFRACTION OF ELECTRONS

Use apparatus such as that shown in Figure 3.33 to explore electron diffraction and the relationship between the de Broglie wavelength and accelerating voltage.

PRACTICAL SKILLS REFERENCE

Carrying out practical work
See Practical skills notes 3.1–3.2

Analysis and interpretation of data
See Practical skills notes 4.1–4.2

Conclusion and evaluation
See Practical skills notes 5.1–5.2

QUESTIONS

Q 33 Calculate the kinetic energy, the speed and the momentum of an electron which is accelerated through a pd of 500 V.
(electron mass $m_e = 9.11 \times 10^{-31}$ kg; electron charge $e = 1.60 \times 10^{-19}$ C)

Q 34 What potential difference is required to accelerate an electron to a speed of 1.0×10^6 m s^{-1}?

Q 35 As electrons are accelerated from the cathode to the anode of an electron beam tube, does their de Broglie wavelength increase or decrease?

Q 36 If the cathode–anode voltage is increased in an electron beam tube, does the de Broglie wavelength of the electrons in the beam increase or decrease?

Q 37 Calculate the de Broglie wavelength of an electron moving at 1.0×10^6 m s^{-1}.
(electron mass $m_e = 9.11 \times 10^{-31}$ kg; Planck constant $h = 6.63 \times 10^{-34}$ Js)

Q 38 Calculate the momentum of an electron whose de Broglie wavelength is 3.0×10^{-10} m.

Q 39 Estimate your own mass and your speed when walking. Find the order of magnitude of your own de Broglie wavelength, and hence explain why you do not notice the wave-like aspect of your nature – for example, you are not noticeably diffracted when walking through a metre-wide doorway.

Faster electrons

When electrons are diffracted as they pass through a graphite crystal, they are interacting with the electrons of the carbon atoms that make up the graphite. Electrons can be used to look at the nuclei of atoms, too, but if they are to be diffracted by something as small as an atomic nucleus, they must have a very short wavelength, similar to the dimensions of the nucleus itself.

Since atomic nuclei are of the order of 10^{-15} m in diameter, we need electrons whose wavelength is of this order too. To achieve this, the electrons must be accelerated through many millions of volts.

At this point, we need to be aware of the limitations of some of the equations used earlier. In particular, consider what happens if we try to calculate the speed of an electron that has been accelerated through 100 MV (= 1.00×10^8 V):

$$E_k = eV = 1.60 \times 10^{-19} \text{ C} \times 1.00 \times 10^8 \text{ V}$$

$$= 1.60 \times 10^{-11} \text{ J}$$

(No problem here.)

Assuming $E_k = \frac{1}{2}mv^2$:

$$v = \sqrt{\frac{2E_k}{m}}$$

$$= \sqrt{\frac{2 \times 1.60 \times 10^{-11} \text{ J}}{9.11 \times 10^{-31} \text{ kg}}} = 5.9 \times 10^9 \text{ m s}^{-1}$$

This is greater than the speed of light ($c = 3.00 \times 10^8$ m s^{-1}), which is impossible (according to the theory of relativity). For speeds approaching the speed of light, we need to use a different expression for the particle's energy, derived from relativity. Particles that are moving at speeds that are greater than, say, 0.1c are described as relativistic, and some of the relationships between quantities such as velocity, kinetic energy and momentum have to be revised. In particular, we can no longer calculate a particle's kinetic energy using $E_k = \frac{1}{2}mv^2$.

Diffraction patterns

Figure 3.37(a) shows the diffraction pattern of electrons that have been diffracted by polycrystalline graphite. You saw a similar pattern in Activity 20 when light is diffracted by tiny dust particles; there is a bright central spot, surrounded by rings of decreasing brightness. The graph in Figure 3.37(b) shows how the brightness varies across the centre of this diffraction pattern. The central hump of the graph corresponds to the bright central spot; then there is a dip where the pattern is dark, then a lower peak for the first bright ring, and so on.

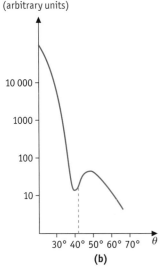

Figure 3.38 Electron diffraction patterns for (a) carbon nuclei and (b) oxygen nuclei

(a)

(b)

Figure 3.37 (a) Diffraction pattern produced when light is diffracted by a random array of fine particles; (b) graph showing how the brightness varies along the diameter of the pattern

The diffraction pattern can change in two ways: the electrons' de Broglie wavelength can be changed, as can the size of the particles that are causing the diffraction. We will concentrate on the size of the particles. With smaller and smaller particles, the diffraction rings get wider and wider. So, measuring the diameter of the rings can tell us about the size of the tiny particles that are doing the diffracting.

Figure 3.38 shows what happens when high-energy electrons are diffracted by samples of carbon and oxygen. The de Broglie wavelength of the electrons is similar to the diameter of the nuclei, so the electrons are strongly diffracted. In this case, the results don't appear as a set of rings on a screen. Instead, the detector is moved round to different positions to measure the intensity of the diffracted electrons. The result is a graph of intensity against angle. You can see the similarity in shape between these graphs and the graph for the diffraction of light (Figure 3.37(b)). In Figure 3.38(a) there is a central peak, then a minimum at an angle of about 50°, and then a smaller peak. From this, the diameter of the carbon nuclei can be determined.

PRACTICAL SKILLS REFERENCE

Analysis and interpretation of data

See Practical skills notes 4.1–4.2

Conclusion and evaluation

See Practical skills notes 5.1–5.2

⚙ ACTIVITY 22 INTERPRETING ELECTRON DIFFRACTION PATTERNS

Use the electron diffraction patterns shown in Figure 3.38 to determine the diameters of carbon and oxygen nuclei.

'Seeing' protons

The smallest and lightest of all atoms is the hydrogen atom. Its nucleus is a single proton. In the 1950s, experiments were done using electron diffraction with liquid hydrogen as the target. With electrons accelerated through nearly 1000 MV (10^9 V), Robert Hofstadter of Stanford University in California was able to estimate the diameter of the proton as 5.6×10^{-15} m. This result gained him the Nobel Prize for Physics in 1961.

As far as Hofstadter could tell, a proton was simply a single particle, not much different from a rather fuzzy billiard ball. The electrons he used interacted with the protons via the Coulomb force – their charges interacted. Hofstadter could find no evidence that the proton was anything other than a sphere of charge.

Today, as we have seen in Part 2 of this chapter, our picture of protons is rather different. High-energy electron accelerators have been used to probe right inside protons, and this has shown that the charge inside a proton is not uniformly distributed; each proton is made of three quarks tightly bound together. This three-fold structure was revealed by 'deep inelastic scattering' experiments in the late 1980s in which beams of high-energy electrons were fired at protons, rather as Rutherford had fired alpha particles at gold foil many years previously.

⚙ ACTIVITY 23 DEEP INELASTIC SCATTERING

Write a paragraph to explain why very high-energy electrons were needed to reveal the quarks that make up a proton. Include the following terms:

energy   wavelength   diffraction.

3.6 Summing up Part 3

In this part of the chapter, you have seen that the story of our understanding of the particles that make up matter developed rapidly during the 20th century. This has been achieved by probing matter with increasingly energetic particles – electrons, alpha particles and others. The electrons used in deep inelastic scattering experiments had energies equivalent to being accelerated through tens or hundreds of billions of volts. In Part 4, you will see how such high energies can be achieved.

You have looked at the development of our understanding of the structure of matter, starting from Rutherford's alpha particle scattering experiment; an understanding of electrostatic forces, developed in the 19th century, helped make this possible. You have seen how the conservation of momentum (a fundamental law of physics) is brought into the analysis of collisions, and how knowledge of the wave-like nature of electrons (and other particles) explains their usefulness as a tool for probing the heart of matter.

⚙ ACTIVITY 24 SUMMING UP PART 3

During the course of the 20th century, experiments using a variety of sub-atomic 'probes' were carried out to investigate the structure of atoms, and to find out about other aspects of the fundamental nature of matter. Draw a timeline to show this, giving dates and brief descriptions of the experiments, along with any other details you wish to include. Include as many as possible of the terms printed in bold – and make sure you know their meanings. Refer back to Part 2 for information about some other experiments that you might show on your timeline.

4 Particle beams and accelerators

In this part of the chapter, you will find out how charged particles are accelerated to high energies, so that they can be used in experiments to recreate the conditions in the early stages of the history of the Universe.

4.1 Big experiments

CERN is the European centre for particle physics research. It sits between the Alps and the Jura mountains, near Geneva in Switzerland. It's a great place to work if you're a particle physicist and you enjoy skiing. The photograph in Figure 3.39 shows an aerial view of CERN, with the positions of the underground tunnels marked. The main particle accelerator is circular in shape. It is so large – 8.6 km across – that it extends under the border into France.

Figure 3.39 The CERN lab near Geneva

One reason for building this vast particle-physics laboratory was political. During the Second World War, scientists from opposing sides had been fighting each other, competing to develop the nuclear bomb. After the war, it was hoped that a European centre could be built so that scientists from different countries could work together for non-military research. The Conseil Européen pour le Recherche Nucléaire (CERN's original name) was established in 1954. It was the first international organisation that Germany joined after the war. There are now (2015) 21 member states. CERN employs around 2400 people. The laboratory's scientific and technical staff design and build the particle accelerators, ensure their smooth operation, and also help prepare, run, analyse and interpret the data from complex scientific experiments. Some 10 000 visiting scientists – half of the world's particle physicists – come to CERN for their research. They represent over 600 institutions and 113 nationalities.

Particle accelerators such as those at CERN are so expensive that they can only work if many countries contribute. The member states' contribution to the CERN budget in 2014 was about £780 million, of which the UK contributed 13.88%. It sounds a lot, but works out at about £1 per person in Europe. Other non-member countries such as the USA, China and Japan also contribute in exchange for their scientists taking part in the experiments. The CERN council has to work hard to justify the money spent and tries to keep costs down. An example of this is the closing down of the Large Electron Position collider (LEP) in order use the tunnel for the Large Hadron Collider (LHC).

Discuss the following in small groups. You may find that you have opinions but need to look up information to back up your arguments. A visit to the *Physics and Ethics Education Project* website might help (see www.pearsonhotlinks.co.uk).

● Is it ever right for politics to interfere with scientific research?

● Are there any examples from the 20th or 21st centuries where politics has played a part in the development of physics research?

● Do the developments of physics ever affect political decisions? Try to use some contemporary examples.

People and ideas

Physicists like to develop theories that are simple. They are currently trying to come up with one theory that can explain the four fundamental forces (electromagnetic, weak, strong and gravitational) in terms of a single theory – rather as, in the 19th century, electrostatics and magnetism were unified through the theory of electromagnetism. In the 1960s, the electromagnetic and weak forces were unified in this way. This theory then predicted the existence of so-called W- and Z-boson particles. These particles have a lot of mass. This led teams of scientists to try to accelerate particles to high energies, collide them and try to create the Z and W particles. In 1983, Italian physicist Carlo Rubbia headed a team of 100 physicists at CERN that detected these particles. This helped to confirm the electroweak unified theory. Carlo Rubbia and his colleague Simon van de Meer were awarded the Nobel Prize for this work.

Scientists are now using the LHC to help them probe the mysterious dark matter and dark energy of the Universe, study the properties of the Higgs Boson and explore possible extra dimensions of space–time.

If you ever visit CERN, you will see people from many different countries. Because particle-physics research involves using expensive machines, single universities and research institutes rarely do research in isolation. Organisations collaborate not just within one country but internationally. If you look at scientific papers written today, they list all the scientists who have contributed to the project, and in particle physics this may mean scores of names (Figure 3.40).

Figure 3.40 The first eight pages of this article list the numerous authors who have contributed to this research. Particle-physics research involves collaboration between physicists of many different nationalities.

Figure 3.41 Discussions in the CERN cafeteria

Someone once joked that broken English, not English, is the international language at CERN. In the cafeteria (Figure 3.41) you can hear physicists arguing about the meaning of the results they have just analysed. Much of this work involves mathematical modelling, computer programming and careful, detailed discussion. All of the skills that they learn and practise are transferable to other disciplines. Many young physicists do not remain in physics after their PhD at CERN. Some of them go on to work in the City as merchant bankers, stockbrokers, managers, etc. In these jobs they are using many of the same skills that they needed in their scientific research.

Over the years, it became increasingly important to be able to send data from CERN back to the scientists' home universities very quickly to help speed up analysis. The World Wide Web was established by scientists at CERN to help them communicate across different continents as well as countries. With the LHC, the CERN experiments have petabytes (10^{15} bytes) of information to analyse, and as a result they have developed a so-called Grid, which allows access to the processing power of many computers distributed around the world. The Enabling Grids for E-sciencE (EGEE) infrastructure has been developed for Europe and there is another Grid system for the USA. As the Web was CERN's response to a new wave of scientific collaboration at the end of the 1980s, so the Grid is the answer to the need for increased data analysis required by the world particle-physics community.

Physicists tend to be just as passionate about their other interests – many are accomplished musicians, and concerts are frequently held at CERN. The mountains and lakes around CERN attract physicists who are keen skiers, mountaineers and sailors.

> ⚙ **ACTIVITY 26 PARTICLE PHYSICS TODAY**
>
> Use the Internet to find some information about current and planned particle-physics experiments.
>
> With a small group of students, share your opinions on the way current particle-physics research is organised and funded. Is this different from other areas of scientific research?

4.2 Achieving high energy

In the 1960s, as particle physics developed, there grew a demand for higher and higher energy accelerators. Partly this demand was fuelled by the desire of engineers and practical physicists to see what could be built and discovered at higher energies. At the same time, theoreticians wanted experiments that could test the predictions of the standard model of fundamental particles and forces. The theory predicted the existence of quarks; it also suggested that there were much more massive particles. To discover these particles required bigger accelerators capable of delivering more energy.

Particle physics is often known as high-energy physics (HEP). As you have seen in Parts 2 and 3, there are three main reasons why particle physicists use high energies.

- If you want a positively charged particle to get closer to a nucleus of an atom, a lot of energy must be supplied.

- The more energy that can be given to particles, the shorter their wavelength and the smaller the detail that can be investigated using them as a probe.

- By colliding particles together, the energy is re-distributed, producing new particles. The higher the collision energy, the larger the mass of particles that can be produced.

As theory predicted different particles, this drove the advancement of the technology – more energetic machines were needed. Sometimes unexpected results were found and this led to refinements in the theory, which sometimes needed even higher energies in order to be tested.

You have already encountered particle accelerators that operate on essentially the same principle as those used in particle-physics research. In the electron gun that forms part of any cathode-ray tube, electrons are accelerated in an electric field produced by applying a potential difference between the cathode and anode. In such a tube used in an old-style TV set, or in demonstrating electron beams in the laboratory, electrons are accelerated through a few thousand volts at most. In particle-physics experiments, much higher energies are required. This has several consequences.

One is that the required energies cannot be achieved all in one go; accelerators have to be designed that 'kick' the particles many times over, increasing their energy each time. Another consequence is that the equations of classical physics break down at speeds close to that of light, and to describe and predict the behaviour of such energetic particles, the equations of special relativity must be used. The second of these is related to some of the units that particle physicists use, and which you are likely to come across in your reading about particle physics.

Relativistic effects

In Section 3.5, you saw that, according to classical physics, an electron accelerated through 100 MV would travel faster than light. This is actually impossible, and the incorrect prediction arose because we used inappropriate equations. In particle-physics accelerators, electrons (and other charged particles, such as protons and positrons) are accelerated through potential differences of several gigavolts (1 GV = 10^9 V) or even teravolts (1 TV = 10^{12} V).

Accelerators give particles energy. When particles are accelerated they go faster, but they can never reach the speed of light. Einstein was able to show that any particle that has mass when at rest can never be accelerated to the speed of light. Even if we keep giving the particle energy when it is close to the speed of light, its speed will keep increasing but it will never reach that elusive 3×10^8 m s^{-1}. The reason for this is that the particle's mass keeps increasing the more energy it is given. (You might guess this from the famous equation $E = mc^2$.) And so the particle becomes more and more massive, and more and more difficult to accelerate. That's why, unless you can find a loophole in Einstein's theory of relativity, faster-than-light travel will remain in the realm of science fiction.

In 1964, American physicist William Bertozzi conducted an experiment to find the speeds of electrons accelerated by electric fields. Figure 3.42 shows Bertozzi's apparatus. The electrons were effectively accelerated through potential differences up to 15 MV. The 'time-of-flight' of the electrons between the two detectors was measured using an oscilloscope, enabling their speeds to be found. To check that their energies were as expected, the electrons were allowed to strike a target, which heated up. Bertozzi measured their speeds and found that the electrons did not exceed the speed of light and the relationship between their speed and energy was as Einstein predicted.

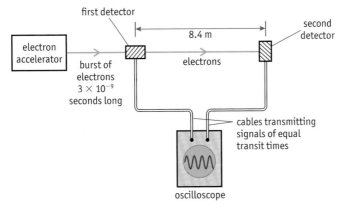

Figure 3.42 Bertozzi's apparatus

⚙ ACTIVITY 27 THE UNIVERSAL SPEED LIMIT

By analysing some of Bertozzi's results, demonstrate that electrons behave as Einstein predicted and do not obey the equations of classical physics when they reach high energies.

PRACTICAL SKILLS REFERENCE

Analysis and interpretation of data

See Practical skills notes 4.1–4.2

Conclusion and evaluation

See Practical skills notes 5.1–5.2

Time distortion

Fast-moving particles do not only put on mass compared to their stationary colleagues; they also appear to have their own distorted view of time.

According to Einstein's theory, if we were able to mount a clock on a fast-moving electron and (even more unlikely) were able to watch it ticking as it flew past, then compared to a stationary clock we happened to have to hand, the clock on the electron would appear to tick more slowly. Several seconds might pass on our stationary clock while we waited for the clock to tick over on the electron's clock.

This was such a bizarre claim of Einstein's theory that careful experiments were needed to confirm that the effect was taking place. One of the most direct of these was carried out in 1941 when Bruno Rossi and D. B. Hall compared the number of cosmic ray muons per second passing Echo Lake (3240 m above sea level) with those at Denver in Colorado (1616 m). Muons typically decay with a half-life of 2.2 μs. Consequently, although they were travelling in excess of 99% of the speed of light, very few of them should have covered more than 600–700 m in that time. In fact Rossi and Hall found that most of the muons were making it down to Denver. This can be explained if their internal clocks tick more slowly when viewed from Earth, so that 2.2 μs for the muon is actually more like 10 μs for us on the ground.

When Einstein refined his theory in 1915 to take gravity into account, he predicted that the strength of a gravitational field also had an effect on the passage of time. A clock at the top of Mount Everest ticks over more rapidly compared to one at sea level in the stronger gravity field there. This effect has been confirmed by flying an atomic clock at high altitude. It also affects satellites in orbit in the weaker gravity field at that height. However, this is more complex to work out, as satellites orbiting the Earth are moving at high speed, which tends to slow the clocks down. This is a bigger effect than the gravitational 'speeding up', so overall the clocks run slightly slow. The very accurate clocks running in the orbiting Global Positioning Satellites (GPS) have to take the effect of gravity and speed into account in their software.

In the 2014 Science Fiction film *Interstellar* (Figure 3.43), an extreme version of gravitational time distortion was portrayed. In the film, astronauts were required to land on a planet orbiting a black hole. Close to the black hole's immense gravity, time on the surface of the planet was greatly distorted compared to that experienced by their colleague in orbit. Although they spent a time on the surface measured in hours, 21 years had passed for the scientist waiting for them on their mother ship.

STUDY NOTE

Half-life and decay are treated in more detail in *Reach for the Stars*.

Figure 3.43 The science fiction film *Interstellar* involves time distortion

Units

When a particle has been accelerated to a very high energy, it is more useful to talk about its energy than its speed – its speed is, in any case, very nearly the speed of light, regardless of any additional energy it acquires, and if you want to know what a particle can do (for example, whether it can create new particles in a collision) then its energy is the most important thing to know. For this reason, particle physicists nearly always use a system of units that relates directly to particle energies.

When talking about the energy of an electron (or other sub-atomic particles), it's rather awkward to work in joules, with all those negative powers of ten. Instead, we often state its energy in units of electronvolts (eV).

The kinetic energy of an electron accelerated through 1 V is 1 electronvolt (1 eV). The magnitude of electron charge, $e = 1.60 \times 10^{-19}$ C, so:

$$1\,eV = 1.60 \times 10^{-19}\,J$$

The most energetic accelerators can produce particles with energies in the giga-electronvolt (GeV) or even tera-electronvolt (TeV) range. Remember that the electronvolt is a unit of energy, not of voltage. It can be used for any particle, not just electrons – and they do not have to be electrically accelerated.

Particle physicists commonly express mass in units that, at first, look a little odd, but do in fact make their life simpler. In Part 2 you met Einstein's equation:

$$E = mc^2 \qquad\qquad \text{(Equation 1)}$$

We can use this to find the energy associated with a particle when it is at rest – that is, the energy that would be liberated if its entire rest mass were to dematerialise or, equivalently, the energy needed to create the particle 'from nothing'. This energy, E_0, is called the particle's **rest-mass energy** (or, sometimes, just its rest energy). An example shows how this leads to new (non-SI) units for mass.

An electron at rest has mass $m_0 = 9.11 \times 10^{-31}$ kg

Its energy equivalent is:

$$E_0 = 9.11 \times 10^{-31}\,kg \times (3.00 \times 10^8\,m\,s^{-1})^2$$
$$= 8.199 \times 10^{-14}\,J$$
$$= 5.12 \times 10^5\,eV$$
$$= 0.512\,MeV$$

We can express the mass in terms of its energy equivalent, again using Equation 1:

$$m_0 = \frac{E_0}{c^2} = 0.512\,MeV/c^2$$

Rather than putting in numbers and doing any calculations, particle physicists leave this expression just as it is. The mass of the electron is now expressed in the rather odd-looking units of MeV/c^2.

Now imagine the electron has been accelerated so that its kinetic energy is 1 GeV. Its mass is still given by Equation 1, where E is its total energy (rest energy and kinetic energy combined) – though now its kinetic energy vastly exceeds its rest energy:

$$m = \frac{E}{c^2} = 1\,GeV/c^2$$

Without needing to convert back to kg, you can see that the mass equivalent to the electron's total energy is now about two thousand times its rest mass (as $1\,GeV = 10^3\,MeV$).

Now suppose this 1 GeV electron were to collide head-on with a positron that had also been accelerated to 1 GeV. Given that a muon has rest mass $0.106\,GeV/c^2$, would the collision be able to produce a muon-anti-muon pair? We can see straightaway that it would be possible to produce *many* such pairs, as the energy of the electron and positron greatly exceeds the energy associated with a muon and an anti-muon at rest.

STUDY NOTE

You met the eV in the chapter *Technology in Space*.

STUDY NOTE

Particle physicists sometimes omit the c^2 from the mass unit, so you might, for example, see the electron rest mass quoted as 0.512 MeV.

So by expressing particle masses in units of MeV/c^2 or GeV/c^2, particle physicists can tell whether there is enough energy in a given collision to produce a given particle, without having to do any arithmetic at all.

Another non-SI unit that particle physicists find convenient to use is the **atomic mass unit** (u). This is defined as one-twelfth of the mass of a carbon-12 atom. The hydrogen atom was the first choice for the standard of atomic mass, but as there are several isotopes of hydrogen that are difficult to separate out, carbon-12 was chosen. The atomic mass unit is useful because experimentally it is much easier, and more precise, to compare the masses of atoms and molecules (relative masses) than to measure their absolute masses.

The mass of a ^{12}C atom is 1.99×10^{-26} kg, so:

$$1\,u = 1.99 \times \frac{10^{-26}}{12} = 1.66 \times 10^{-27}\,kg$$

QUESTIONS

Q 40 A proton has rest mass $m = 1.67 \times 10^{-27}$ kg.

  (a) What is its rest energy **(i)** in J **(ii)** in GeV?

  (b) What is its rest mass expressed in units of GeV/c^2?

Q 41 A pi-zero meson has a rest mass of 0.14 GeV/c^2. How many such particles could, in principle, be produced in a collision between an electron and a positron, each with energy 1 GeV?

Q 42 Suppose an electron is accelerated to an energy of 20 GeV. Express its mass in units of GeV/c^2.

Q 43 Show that $1\,u$ is equivalent to 931 MeV.

Q 44 Express the rest mass of an electron in atomic mass units.

Designing accelerators

In 1928, a Norwegian, Rolf Wideroe, invented the linear accelerator (LINAC). A LINAC accelerates charged particles to very high energies without the need for high voltages. Instead of being accelerated once, charged particles travel along a series of tubes separated by gaps, and are given voltage kicks at each gap.

In a LINAC there is a series of tubular electrodes (see Figure 3.44). These are connected to an alternating voltage supply, so that the voltage of each electrode switches back and forth between positive and negative. There is actually no electric field *inside* the tubes, so the particles experience no force there. The aim is that the particle should emerge from the end of one tube and find that the next tube attracts it.

Figure 3.44 Construction of a linear accelerator

A proton leaves the ion source when electrode A is negative. It accelerates towards A, and then travels at a steady speed inside the tube. While it is inside the electrode, the voltage switches so that A is positive and B is negative. Now, when the proton enters the gap between A and B, it accelerates towards B. When it is inside B, again the voltage is switched on so that, on exiting, B is positive and C is negative.

Each time, the same magnitude of voltage V is used and so the energy of the particle is built up in steps. After n electrode gaps, the proton's kinetic energy is equal to $n \times e \times V$.

As an alternating voltage must be used, it can easily be stepped up using transformers. The frequency of the oscillating voltage is 'radio frequency' (i.e. a few MHz) and is kept constant. What does this mean for the design of the electrode drift tubes? They need to get longer as the proton gains speed so that the time to move between gaps is constant.

However, to achieve very high energies, very long accelerators are needed. The longest linear accelerator is at Stanford, California (Figure 3.45). The Stanford Linear Accelerator (SLAC) is 3 km long and can produce 20 GeV–50 GeV electrons. This has been used to smash electrons into protons and neutrons to help probe their structure. It was using this accelerator that physicists managed to show that protons and neutrons have a substructure of three quarks (see Section 3.5).

High-energy physicists are now collaborating on a new project: the International Linear Collider (ILC). The site has yet to be decided (2015) with Japan considered to be the most likely candidate due to the government being willing to contribute substantially to the costs. It may go active in 2026. It will consist of two linear accelerators facing each other and producing collisions between beams of electrons and positrons at energies around 500 GeV, initially.

The main limitation of a LINAC is that each accelerating section is used only once, and so to achieve higher energies the machines must be made longer and so are more expensive.

In a circular accelerator, charged particles cycle round and round the same accelerating sections again and again, giving the particles more energy each time they come around. In this way, not only can the energy of a particle be increased but two particles (oppositely charged) can be accelerated in opposite directions around the circle and then made to collide – increasing the total energy of interactions further.

The first circular accelerator was the cyclotron, invented by the American Ernest Lawrence (1901–1958) in 1929. He used the idea (developed by Wideroe for the linear accelerators) of accelerating charged particles in a gap between electrodes then allowing them to drift to the next gap. However, rather than let the particles drift to the next gap along a straight line, requiring more electrodes, he used a magnetic field to bring the particles round in a circle to be accelerated through the same gap every half-turn. Figure 3.46 shows a simplified illustration of the cyclotron.

Figure 3.45 It is claimed the Stanford Linear Accelerator is the world's straightest object

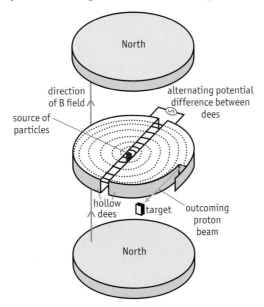

Figure 3.46 Principle of the first cyclotron

The electrodes were two hollow semi-circular pieces of metal – called 'dees' because of their shape – inside which the particles orbited. In 1931 Lawrence achieved a proton energy of 80 keV with a cyclotron 11 cm in diameter. The following year he achieved 1 MeV with a 26 cm diameter instrument. To reach energies much beyond 25 MeV, a new type of cyclotron had to be developed. As they reach higher kinetic energies, the particles' mass increases, which means that they take longer to complete each orbit. In a synchro-cyclotron (synchrotron), the frequency of the accelerating voltage is reduced as the kinetic energy of the particle increases – the accelerating 'kicks' are synchronised with the particles' orbital motion. To reach still higher energies, the design has to be modified again, and in modern large accelerators such as those at CERN, the particles are steered around in a path of constant radius as their energy is increased.

4.3 Circular motion

To see how particles can be steered in a circular path, we need to consider circular motion in general and also to see how charged particles behave in magnetic fields. These two aspects of particle accelerators are the subjects of this section.

Finding the force

How can an object be made to travel in a circular path at constant speed? In the absence of any resultant force, a moving object continues along a straight line at constant velocity, as described by Newton's first law of motion. To produce circular motion, then, a resultant force must be acting. An object in circular motion is constantly changing its direction – in other words, it is accelerating. Its velocity is constantly changing, but the change is one of direction not of speed.

STUDY NOTE

You met Newton's first law of motion in the chapter *Higher, Faster, Stronger*.

A few examples will help to identify the forces that produce motion in a circle. Imagine that you are the hammer thrower shown in Figure 3.47. The hammer is massive. You have to whirl it around so that it builds up speed before you release it. If your grip isn't strong enough, the hammer will fly off before you mean to let go. To keep the hammer moving round, you need to keep pulling on it. You can feel the hammer pulling on you outwards as you pull it inwards. So an inwards force is needed to keep the hammer moving along its circular path.

As a second example, consider the motion of the Moon or an artificial satellite around the Earth (Figure 3.48). (The motion is not quite circular, but it is very nearly so.) What forces are acting on a satellite as it orbits? There is no air resistance slowing the satellite down, as space is essentially a vacuum. The only force is the gravitational attraction between the Earth and the satellite. If this is the only force acting on the satellite, then it must be the force causing the satellite to accelerate as it orbits. The force is directed towards the Earth – towards the centre of the circle. In fact, for any object to move in a circle, a force towards the centre of the circle is required.

Figure 3.47 Throwing the hammer

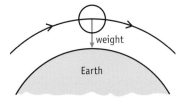

Figure 3.48 Satellite orbiting Earth

Centripetal, not centrifugal

Any force that acts towards the centre of a circle is described as a centripetal force. The word 'centripetal' means 'directed towards the centre' or 'centre seeking'.

Because of our experience of circular motion, we tend to believe that such motion involves a force acting radially outwards from the centre of a circle – a centrifugal ('centre-fleeing') force. When you are in a car that is turning a bend quickly (Figure 3.49), you feel thrown outwards, away from the centre of the circle. But there is no force actually pushing you outwards. The illusion arises because you are experiencing your motion relative to the car. Seen relative to a car cornering, you may seem to be moving outwards, but relative to the ground, you are actually trying to continue in a straight line – just as Newton's first law states. You will continue to move in a straight line until a resultant force acts on you that provides enough centripetal force to move you in the same circle as the car. This is normally provided by a seat-belt and friction from the seat, or by the side of the car if you're not belted up.

Figure 3.49 Car rounding a bend on a banked track

Table 3.12 lists some further examples of circular motion, with the name of the force (or forces) that act as the centripetal force in each case. In Activities 28 and 29, you explore two more examples of circular motion.

In the chapter *The Medium is the Message*, you saw that a charged particle moving in a magnetic field experiences a force at right angles to its direction of motion. This is just the sort of force that is required to produce motion in a circle. In Activity 29, you can see how a magnetic field can be used to steer an electron beam into a circular path as required in a particle accelerator.

Example of circular motion	Inward force
Earth orbiting the Sun	Gravitational attraction between the Sun and Earth
Car rounding a bend on a flat road	Frictional force of the road on the tyres
Plane turning in a horizontal circle	Lift force on the plane's tilted wings
Car rounding a bend on a banked track (Figure 3.49)	Normal reaction of the road (and friction)

Table 3.12 Some examples of circular motion

⚙ ACTIVITY 28 WHIRLING A BUNG

Take a rubber bung and tie it to one end of a piece of string about 1.5 m long. Find a suitable safe space and whirl the bung at a controlled speed in a horizontal circle around your head, initially with a radius of about 1 m. You can feel a force on your hand through the string as the bung whirls around. The string is pulling outwards on you because you are pulling inwards on the string.

List the forces that are acting on the bung as it moves and describe the direction it acts on the bung. Which of these forces is affecting the horizontal motion of the bung?

Investigate qualitatively how different factors affect the size of the force on your hand from the string. You could start by changing the radius of the circle, or by using bungs of different mass. How do they affect the size of the force? What else affects the size of the force?

⚙ ACTIVITY 29 ELECTRONS IN ORBIT

Use a fine-beam tube (Figure 3.50) to observe electrons moving in circular paths under the influence of a magnetic field.

We will return to magnetic fields and particle accelerators later. In order to understand the situation properly, we must first take a more precise look at circular motion.

Describing circular motion

If we want to analyse motion in a circle, it is useful to be able to describe the position of a body around a circle and how fast it is moving. Figure 3.51 shows how we do this.

The important quantities are as follows:

- The radius of the circle, r, which has SI units of metres, m.

- The object's **angular displacement**, $\Delta\theta$. This is the angle through which the object has moved, relative to some fixed direction. Although degrees are most commonly used to measure angles, they are not the SI unit. The unit used for angle (or angular displacement) is the radian, rad.

- The object's speed, v, with SI units m s^{-1}. Notice that the direction of motion is along the tangent to the circle.

- The **angular velocity**, ω (omega), measured in rad s^{-1}. The relationship between angular velocity and angular displacement is similar to that between linear velocity and linear displacement:

$$\omega = \frac{\Delta\theta}{\Delta t} \tag{22}$$

We will consider angular velocity in a little more detail. The rate at which many devices turn in a circle or spin is measured by counting how many revolutions they complete in a certain time. The most common everyday unit is 'revolutions per minute' (rpm). Most car engines rotate at between 2000 rpm and 3000 rpm in normal operation. A compact disc rotates at between 200 rpm and 500 rpm (depending on which part of the disc is being read).

In order to have consistent units, we do not measure angular velocity in rpm. Instead, ω is measured in radians per second. For example, if a wheel is rotating at a constant rate of 1 revolution per second, it turns through 2π rad (360°) every second. Its angular velocity is $\omega = 2\pi$ rad s^{-1}.

PRACTICAL SKILLS REFERENCE

Planning and experimental design
See Practical skills notes 2.1–2.4

Carrying out practical work
See Practical skills notes 3.1–3.2

Analysis and interpretation of data
See Practical skills notes 4.1–4.2

Conclusion and evaluation
See Practical skills notes 5.1–5.2

Figure 3.50 Fine-beam tube

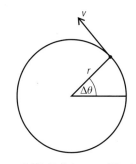

Figure 3.51 Defining quantities in circular motion

STUDY NOTE

You met radians in the chapter *The Sound of Music* and used them to describe the phase of waves.

MATHS REFERENCE

Degrees and radians
See Maths note 6.1

WORKED EXAMPLE

Q A compact disc is rotating at 300 rpm. How many revolutions does it complete every second? What is its angular velocity?

A The disc completes $\dfrac{300}{60}$ = 5 revs per second = $5\,\text{s}^{-1}$

In each revolution, the compact disc turns through $2\pi\,\text{rad}$.

So its angular velocity ω is:

$$\omega = 5\,\text{s}^{-1} \times 2\pi\,\text{rad} = 10\pi\,\text{rad}\,\text{s}^{-1} = 31.4\,\text{rad}\,\text{s}^{-1}$$

Another important quantity associated with circular motion is the period (symbol T), which is the time for one complete revolution. The angular velocity and the period are related:

$$\omega = \frac{2\pi}{T} \text{ or } T = \frac{2\pi}{\omega} \qquad (23)$$

For example, if a body in circular motion has a period $T = 0.5\,\text{s}$, it completes two revolutions every second, so the angle it turns through every second is $4\pi\,\text{rad}$. Thus, $\omega = 4\pi\,\text{rad}\,\text{s}^{-1}$.

Speed and angular velocity

Imagine two athletes jogging around a circular running track, side by side. The athlete in the outside lane has to run slightly faster than the athlete in the inside lane in order to complete a lap in the same time. Both athletes have the same angular velocity, ω, but the one whose track has a greater radius, r, has to run with a greater speed, v. How are these three quantities (v, ω and r) related?

In one complete circuit, the athlete runs a distance $2\pi r$. The time taken to run around 2π radians with angular velocity ω is $\dfrac{2\pi}{\omega}$. So the athlete's speed is given by:

$$v = \frac{\Delta s}{\Delta t} = 2\pi r \times \frac{\omega}{2\pi} = \omega r$$

Hence:

$$v = \omega r \qquad (24)$$

WORKED EXAMPLE

Q Turbine generators in a power station rotate at 3000 rpm. The longest blades in a turbine are about 1.5 m long. Calculate the linear speed of the rotor tip.

A First, calculate the angular velocity ω:

$$\omega = \frac{3000}{60}\,\text{rev}\,\text{s}^{-1} = 50\,\text{rev}\,\text{s}^{-1} = 50 \times 2\pi\,\text{rad}\,\text{s}^{-1} = 314\,\text{rad}\,\text{s}^{-1}$$

Now calculate the speed of the tip:

$$v = \omega r = 314\,\text{rad}\,\text{s}^{-1} \times 1.5\,\text{m} = 471\,\text{m}\,\text{s}^{-1}$$

(The rotor tip is moving faster than the speed of sound in air.)

QUESTIONS

Q 45 A car engine is rotating at 3000 rpm.

(a) How many revolutions does it complete in 1 second?

(b) What is its angular velocity?

(c) What is the period of its motion?

Q 46 A CD has an angular velocity of $30\,\text{rad}\,\text{s}^{-1}$.

(a) How many revolutions does it complete in 1 second?

(b) What is the period of its motion?

Q 47 A positron is travelling around the circular accelerator at CERN. Its speed is close to the speed of light ($c = 3.00 \times 10^8 \, \text{m s}^{-1}$). The accelerator's diameter is 8.6 km. Calculate the positron's angular velocity.

Q 48 A spinning roundabout of radius 2 m has a period $T = 1.25$ s.

(a) What is the angular velocity of the roundabout?

(b) (i) What is the speed of a boy sitting on the edge of the roundabout?
(ii) If the boy moves halfway to the centre of the roundabout, what is his speed there?
(iii) If he then moves to the centre of the roundabout, what is his speed there?

Q 49 A garden strimmer cuts grass with a fast rotating piece of cord. The cord is 15 cm long and the strimmer spins at 100 revs per second.

(a) What is the angular velocity of the cord?

(b) What is the speed of the tip of the cord as it spins?

Q 50 The head of a golf club has a speed of approximately $30 \, \text{m s}^{-1}$ when it strikes a ball. Estimate the angular velocity of a golf club at this point of a golf swing.

Centripetal acceleration

We will now return to the idea that an object moving in a circle at constant speed is accelerating. Imagine a car travelling at a steady speed around a bend in the road. Although its speed is constant, its direction is changing all the time, so its velocity is not constant. Figure 3.52 shows that the car's velocity is changing all the time as it moves around the bend. If its velocity changes, it must be accelerating, even though its speed is constant. Our car is moving at a constant speed, but it is not moving in a straight line. It is constantly changing direction.

In Figure 3.52, you can see that the change in velocity of the car, as it moves a short distance around the bend, is directed towards the centre of the circle. It is a centripetal acceleration, produced by the centripetal force that we identified with earlier. We can derive an equation for this acceleration, with the help of Figure 3.53.

The object shown in Figure 3.53(a) moves at a steady speed v from A to B in time Δt. It moves through angle $\Delta \theta$, so its angular velocity ω is:

$$\omega = \frac{\Delta \theta}{\Delta t} \qquad \text{(Equation 22)}$$

In this time, its velocity changes direction, as shown. Its velocity vector also moves through angle $\Delta \theta$.

Figure 3.53(b) shows how we work out the object's change in velocity. We draw a vector triangle; the short third side is the change in velocity Δv. From this triangle, *provided $\Delta \theta$ is small*, using the definition of an angle in radians, we can write:

$$\Delta v = \Delta \theta \times v$$

Dividing both sides by Δt gives:

$$\frac{\Delta v}{\Delta t} = \frac{\Delta \theta}{\Delta t} \times v$$

However:

$$\frac{\Delta v}{\Delta t} = a, \text{ and } \frac{\Delta \theta}{\Delta t} = \omega$$

So we have:

$$a = \omega v$$

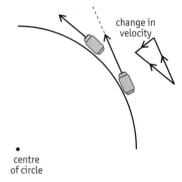

Figure 3.52 Car moving at constant speed around a bend

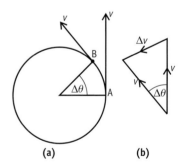

Figure 3.53 Analysing circular motion

> **MATHS REFERENCE**
> Degrees and radians
> See Maths notes 6.1 and 6.6

Now, if we substitute $\omega = \dfrac{v}{r}$, we get a useful expression for the centripetal acceleration:

$$a = \frac{v^2}{r} \tag{25}$$

and if we substitute $v = \omega r$, we get another useful expression:

$$a = r\omega^2 \tag{26}$$

QUESTION

Q 51 Go back to Questions 47–50 and calculate the centripetal acceleration in each case.

Centripetal force again

We can now work out what force is needed to make an object move in a circle, because we have two expressions for centripetal acceleration. Using the familiar relationship $F = ma$, we can simply multiply Equations 25 and 26 by the object's mass m to find the required centripetal force:

$$F = \frac{mv^2}{r} \tag{27}$$

and:

$$F = m \, r\omega^2 \tag{28}$$

QUESTIONS

Q 52 You are cycling along and decide to turn off to the left.

(a) Which requires a greater sideways force, turning sharply left or following a gentler path?

(b) If you were travelling faster, would you need a bigger or smaller sideways force?

Q 53 What force is required to keep a 5 kg object moving at $10 \, \text{m s}^{-1}$ in a circle of radius 2 m?

Q 54 What force is required to keep a 1 kg object moving with angular velocity $10 \, \text{rad s}^{-1}$ in a circle of radius 10 m?

Q 55 A man's competition hammer has a mass of 7.26 kg. The length of the hammer chain is 1.2 m and the length of the man's arms is 0.9 m. Before throwing, the man whirls the hammer round his head in a horizontal circle a few times. The average period for each revolution is 0.75 s.

(a) What is the average angular velocity of the hammer?

(b) What is the average tension required in the hammer chain if it is horizontal?

Q 56 A girl of mass 35 kg is swinging on a tyre of mass 10 kg attached to a tree branch by a rope of length 6 m.

(a) What is the weight of the tyre and the girl? (Use $g = 9.81 \, \text{N kg}^{-1}$)

(b) If her speed at the bottom of the swing is $4 \, \text{m s}^{-1}$, what centripetal force is required at this point by the tyre and the girl?

(c) Use your answers to **(a)** and **(b)** to calculate the tension in the rope at this point. Remember that the resultant force on the child and tyre must equal the value of the centripetal force and act in the right direction.

Q 57 A car rounding a bend requires a centripetal force of 2 kN. This is provided by friction between the tyres and the road.

(a) If the same car at the same speed rounds another bend of twice the radius, what centripetal force is required?

(b) If the same car rounds the original bend at twice the original speed, what centripetal force is required?

4.4 Steering and tracking charged particles

Round the bend

Now we can return to thinking about making charged particles move in circular orbits; you will recall that this is how we can avoid the problem of making very long linear accelerators. In Activity 29 you saw how a magnetic field can steer a beam of electrons into a circular path. In the chapter *The Medium is the Message*, you saw that a particle of charge q moving at speed v in a magnetic field of flux destiny B will experience a force of magnitude:

$$F = Bqv \sin\theta \tag{29}$$

where θ is the angle between the directions of the particle's motion and the magnetic field.

Fleming's left-hand rule (Figure 3.54) tells us how the directions of the magnetic field, the current and the force are related: the force always acts at right angles to the other two. But take care! Recall that for moving electrons, whose charge is negative, the conventional direction of current is in the opposite direction to the motion of the charges.

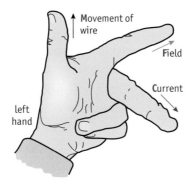

Figure 3.54 Fleming's left-hand rule

Figure 3.55 shows what happens when a positive charge $+q$ enters a region where there is a magnetic field; the crosses indicate that the flux is directed into the page. The direction of the current is left to right. From the left-hand rule, the force on the charge is directed upwards, at right angles to the motion of the charge. This force will alter the direction of motion of the charge, but will not affect its speed. Even when the direction of motion is changed, the force will still be at right angles to it. Because the force is always perpendicular to the motion, it will produce a circular path. The effect of the magnetic field on the charge is providing a centripetal force that enables the charge to move in a circle.

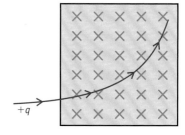

Figure 3.55 Positively charged particle moving in a magnetic field

In a circular particle accelerator or in a fine-beam tube, the magnetic force provides the centripetal force needed to keep the electrons moving in a circle. The field is applied at right-angles to the direction of motion ($\theta = 90°$, $\sin\theta = 1$), so from $ma = F$ we have:

$$\frac{mv^2}{r} = Bqv$$

Now we can rearrange this to deduce an expression for the radius of the particle's orbit:

$$r = \frac{mv}{Bq} \tag{30}$$

Substituting p for the momentum mv of the particle, we can write this expression as:

$$r = \frac{p}{Bq} \tag{30a}$$

So, the greater the momentum of a particle, the greater the radius of its path in a given field; put another way, the stronger the field needed to keep it in a given orbit.

> **STUDY NOTE**
>
> Equation 30(a) has the advantage that it applies equally to relativistic and non-relativistic particles.

STUDY NOTE

H⁻ ions are hydrogen atoms with an extra electron.

PRACTICAL SKILLS REFERENCE

Planning and experimental design

See Practical skills notes 2.1–2.4

Carrying out practical work

See Practical skills notes 3.1–3.2

Analysis and interpretation of data

See Practical skills notes 4.1–4.2

Conclusion and evaluation

See Practical skills notes 5.1–5.2

⚙ ACTIVITY 30   THE ISIS ACCELERATOR

ISIS, at the Rutherford Appleton Laboratory near Oxford, is an intense source of pulsed neutrons.

Via **www.pearsonhotlinks.co.uk**, go to the ISIS website's accelerator page.

Download the video animation that explains how the ISIS facility works. In particular, notice the use of both a linear accelerator to accelerate H⁻ ions and a synchrotron to accelerate the protons before they collide with a target and produce neutrons.

Look at the 'about ISIS' page to find out the advantages of using neutrons to probe the structure of materials.

⚙ ACTIVITY 31   THE FINE-BEAM TUBE

Use a fine-beam tube to explore how the accelerating voltage and the magnetic field affect the radius of the electrons' path.

QUESTIONS

Q 58   Redraw Figure 3.55 to show how the situation would change if the charge was negative.

Q 59   Beta particles, produced in radioactive decay, are fast-moving electrons. If a narrow beam of beta radiation from a source is passed into a magnetic field, it curves around and spreads out. What does this tell you about the energies of the beta particles?

Q 60   An electron is travelling at 1% of the speed of light ($c = 3.00 \times 10^8\,\mathrm{m\,s^{-1}}$). If a magnetic field, $B = 0.2\,\mathrm{T}$, is applied at right angles to its motion, what radius of path will result? ($m_e = 9.11 \times 10^{-31}\,\mathrm{kg}$; $h = 6.63 \times 10^{-34}\,\mathrm{J s}$)

Q 61   Electrons in a fine-beam tube are accelerated through a pd of 2.5 kV. A magnetic field of 1 T is applied at right angles to the electron beam.

(a) What is the speed of the electrons as they leave the electron gun? (Ignore relativistic effects.)

(b) What force is exerted on each electron because of the magnetic field? Describe the direction of the force.

(c) What acceleration will the electrons have as a result of this force? Does their speed increase?

(d) What will be the radius of the path that the electrons follow?

Making tracks

To finish this section, you will see how what you have learned about circular motion and about electric and magnetic fields is applied to another aspect of particle physics – that of studying the results of collisions between particles.

Figure 3.56 shows the layout of the LHC, which was switched on in 2008 and upgraded for 2015. There are four main experiments (ALICE, ATLAS, CMS and LHCb) located between 50 m and 150 m underground in huge caverns. An older accelerator called SPS (super proton synchrotron) is also shown; its function is to inject energetic particle beams into the LHC where they undergo their final acceleration before colliding.

In many particle-physics experiments, such as those carried out at the LHC, beams of particles are steered so that they collide at particular locations. In the LHC, a vast detector surrounds each of the four collision points, to record the resulting debris. Figure 3.57 shows the LHC detector known as ATLAS. Apart from LHCb, all the detectors have a similar layered structure.

The different layers are designed each to be sensitive to different types of particle, and each works in a slightly different way. But in each layer, passage of a charged particle is detected by the ionisation it produces in the low-density gas that fills the detector. To detect ionisation, most modern detectors use so-called multiwire chambers or drift chambers. A multiwire chamber consists of an array of fine wires, with potential differences applied between them.

Figure 3.56 LHC

Figure 3.57 Structure of the ATLAS detector

When ions are created in the detector, the electric fields between the wires sweep the ions onto the wires, and the resulting small pulse of current is recorded electronically (rather as a Geiger tube records the ionisation produced by the passage of an alpha or beta particle). The record of current pulses provides a record of a particle's passage through the chamber. A drift chamber works in a very similar way, except that it measures the time taken for ions to drift towards the nearest wire, enabling a more accurate record to be built up.

Figure 3.58 shows Peter Glassel, the technical coordinator for the original ALICE time projection chamber, sitting inside the detector. Thousands of wires are connected to read out electronic data produced as particles are created in collisions at the centre of the detector.

Figure 3.58 Inside the ALICE detector

Figure 3.59 Making detectors for ALICE

A detector such as ALICE or ATLAS incorporates some layers whose function is to measure the energy deposited – these are the calorimeters shown in Figure 3.57. Such measurements provide information about the energy of the particles produced in a collision, but this information on its own is not enough for particle physicists to unravel what went on. Figure 3.59 shows Louis Rose-Dulcina, a technician from the ATLAS collaboration, working on the ATLAS tile calorimeter in 2003. Special manufacturing techniques were developed to mass produce the thousands of elements in this detector. Tile detectors are made in a sandwich-like structure; scintillator tiles (which emit tiny flashes of light along the tracks of a particle) are placed between metal sheets.

As well as using electric fields to detect ionisation, detectors such as ALICE and ATLAS use magnetic fields to obtain information about particles. A strong magnetic field is applied along the axis of the detector. In order to make the field as strong as possible, superconducting magnets are used. The colliding particles travel in the same direction as the field, so their paths are not affected, but charged particles produced in a collision are likely to be travelling across the field and so they are forced into curved paths. The sign of their charge can be deduced from the direction of curvature of their tracks, and further measurements enable their momentum to be found.

The magnetic field produced by the ATLAS toroids (Figure 3.60) wraps around the beam path. It provides the field in the outer muon detectors. There is an axial field (parallel to the beam path) provided by a solenoid in the centre of ATLAS. This combination of magnetic fields improves resolution, as it is not possible to build a solenoid large enough to cover the whole area of the detector.

Figure 3.60 View of ATLAS during assembly, looking along the direction of the beam, showing huge toroidal magnets, muon chamber and the inner detector

In Activities 32 and 33 you will need to apply ideas from earlier in this chapter. There are several key points to keep in mind:

● only *charged* particles cause ionisation;

● therefore only charged particles make tracks in detectors;

● the direction of curvature of a particle's track in a magnetic field indicates the sign of its charge;

● the radius of curvature of a track provides information on the particle's charge and momentum;

- momentum is always conserved in any collision, interaction or decay;

- the presence of uncharged particles can be deduced by applying momentum conservation;

- energy is always conserved in collisions and interactions; there is interconversion between mass and energy;

- the equation $E = mc^2$ relates energy to mass;

- a particle's kinetic energy can be deduced from the amount of ionisation it produces along its track.

⚙ ACTIVITY 32 PARTICLE TRACKS

Many particle-physics websites include examples of particle tracks. To find out how such records are analysed, go to the Manchester Particle Physics Group's website (via **www.pearsonhotlinks.co.uk**) where there are lots of useful links.

You might also like to try the simulations in the Lancaster Particle Physics Package (also via **www. pearsonhotlinks.co.uk**). The animations aim to give you a feel for the real physics and experimental techniques involved in particle-physics research.

⚙ ACTIVITY 33 TRACK AND FIELD EVENTS

Write a short account of how electric and magnetic fields are used in particle detectors. Include an explanation of how momentum could be measured from a particle track (refer back to Section 4.3).

4.5 Summing up Part 4

In this part of the chapter, you have seen how an understanding of the way in which charged particles behave in electric and magnetic fields can be used to accelerate particles to high energies. High-energy particles are needed to recreate the conditions of the early Universe, on a small scale.

Use Activity 34 to help you look back over your work for this part of the chapter and see how the key ideas are being applied to the latest particle accelerators.

⚙ ACTIVITY 34 ACCELERATOR ENERGY

On 5 April 2015, after a two-year closure, the remodelled and upgraded LHC was restarted. Initially the two beams each had energy 450 GeV. While carrying out a series of trials, engineers increased this to 13 TeV, exceeding the previous record energy of 8 TeV. Before the upgrade, the LHC detectors analysed about 20 million collisions per second; this will rise to 40 million.

Find out the latest from CERN and other accelerator labs by checking their websites. (See **www.pearsonhotlinks.co.uk**)

Find out how the energy available from different particle accelerators has increased as the technology has developed. In doing this you may read about many types of accelerator: LINACs, cyclotrons, synchro-cyclotrons, etc. For each accelerator, state the type of particles it accelerates, how it accelerates them, how it directs the particle beam, and the energies it can achieve.

Quote all energies in electronvolts.

PRACTICAL SKILLS REFERENCE

Scientific questions and information research

See Practical skills notes 1.2, 1.3

LINK THE LEARNING

5 Link the Learning: On target

5.1 Big ideas

This chapter has looked at some of the most profound ideas in physics: the nature of the Universe, the fundamental particles of nature and the forces that act between them. Along the way, you have learned about circular motion, and the way charged particles behave in electric and magnetic fields. Use the following activity to remind yourself how these ideas fit together.

⚙ ACTIVITY 35 UNIFICATION

Copy the words below on to separate, small pieces of paper. With a partner, discuss the ideas that link groups of words. (Move the pieces of paper around to show how they can be grouped together.)

Now on a single sheet of paper, draw a concept map to show your thoughts. Write the words on the paper, and add arrows showing connections between them. Annotate the arrows to explain the connections.

fundamental particles	quarks	electrons
leptons	gravitation	electromagnetism
nuclear forces	magnetic fields	electric fields
electric charge	moving charges	circular motion
angular velocity	centripetal force	the Big Bang
particle accelerators	high-energy particles	the Universe
dark matter	Higgs boson	neutrinos

5.2 Questions on the whole chapter

You will need the following information when you answer these questions:

permittivity of free space $\varepsilon_0 = 8.85 \times 10^{-12}\,\mathrm{F\,m^{-1}}$

electron charge $e = 1.60 \times 10^{-19}\,\mathrm{C}$

electron mass $m_e = 9.11 \times 10^{-31}\,\mathrm{kg}$

proton mass $m_p = 1.67 \times 10^{-27}\,\mathrm{kg}$

speed of light in vacuum $c = 3.00 \times 10^8\,\mathrm{m\,s^{-1}}$

QUESTIONS

Q 62 We can picture an anti-atom of anti-hydrogen as a positron (an anti-electron) in a circular orbit around an anti-proton. The average separation of the two particles in the anti-atom is 0.037 nm.

(a) What force holds the positron in its orbit? Sketch a diagram of the anti-atom. Draw an arrow to show the force acting on the positron.

(b) Explain why a force is needed to keep the positron in its orbit.

(c) Calculate the force exerted on the positron by the anti-proton, and the force exerted on the anti-proton by the positron.

(d) How many times per second does the positron orbit the anti-proton?

Q 63   In an experiment to investigate the annihilation of particles, beams of high-energy electrons and positrons are produced, and then caused to collide with one another. In one collision, a 500 keV electron collides head-on with a 500 keV positron. The two particles annihilate one another, and two photons of electromagnetic energy are produced.

(a) What is the combined momentum of the two particles just before they collide?

(b) What is the energy of a 500 keV electron, in joules?

(c) What is the combined energy of the two photons, in joules?

Q 64   In 2017 it will be 120 years since J.J. Thomson's discovery of the electron. Thomson pictured the atom as being like a currant bun, with electron 'currants' embedded in a positively charged 'bun'. Describe at least three ways in which our model of matter on a sub-atomic scale has changed since 1897. For each change, identify the way the sub-atomic model has changed and the evidence that led to its acceptance. As well as drawing on your work on this chapter, look back at the earlier chapter *The Sound of Music* for other features of our sub-atomic model.

Q 65   This question is about the acceleration of protons in the Proton Synchrotron (PS) at CERN. See Figure 3.61. Table 3.13 lists some data about the PS and the Super Proton Synchrotron (SPS), also at CERN.

	PS	SPS
diameter of ring	200 m	2.2 km
circumference of ring	528 m	
number of accelerating points	14	
average accelerating voltage	4 kV	
proton energy at injection	50 MeV	
final energy of proton	28 GeV	
final momentum of proton	$1.7 \times 10^{-17}\,\text{kg m s}^{-1}$	
final speed of proton	almost $3 \times 10^8\,\text{m s}^{-1}$	almost $3 \times 10^8\,\text{m s}^{-1}$

Table 3.13 Data for Question 65

Figure 3.61 CERN Proton Synchrotron

Particle accelerators are used to increase the energy of charged particles such as protons and electrons. The accelerated particles are made to collide with other particles on a 'target' in order to investigate the structure of matter. Sub-atomic particles created in these collisions can be studied.

One type of particle accelerator is the synchrotron. In this machine a magnetic field causes charged particles to move in a circular path. The particles are accelerated to higher energies by an electric field. As their momentum increases, the magnetic flux density is increased to keep them travelling in a path of constant radius.

(a) Ignoring relativistic effects, show that:
    (i) the speed of a proton at injection into the PS is about $1 \times 10^8\,\text{m s}^{-1}$
    (ii) a proton takes about 6 μs to travel round the ring at this speed
    (iii) the momentum of a proton at injection is about $1.6 \times 10^{-19}\,\text{kg m s}^{-1}$.

(b) The accelerator ring is a pipe maintained at very low pressure. It is 'filled' with protons by injecting a proton current of 100 mA for 6 μs.
    (i) Calculate the number of protons injected.
    (ii) Explain why the ring must be maintained at very low pressure.

(c) Before the protons are accelerated, an electric field is used to group the protons in the ring into a number of bunches. The bunches are then accelerated as they pass through the acceleration points, which are spaced equally around the ring. An acceleration point is essentially a pair of electrodes between which an alternating voltage is applied.

(i) Explain why an *alternating* voltage is needed.

(ii) The proton bunches pass through an acceleration point when the potential difference between its electrodes is about 4 kV.
By how much does the energy of one proton increase in one revolution? (Give your answer in eV.)

(iii) Estimate the number of times a proton must travel round the ring in order to reach its final energy.

(iv) Explain briefly why linear accelerators are not used to accelerate protons to this final energy.

(d) (i) Show that the magnetic flux density, B, required to maintain a proton in a circular path of radius r is proportional to its momentum, p.

(ii) Estimate the magnetic flux density needed to maintain 50 MeV protons within the PS.

(iii) Explain why the frequency of the alternating voltage must change as the protons are accelerated to higher energies.

(e) Estimate:

(i) by what factor the magnetic field must be increased during the acceleration

(ii) by what factor the mass of the proton increases during acceleration.

(iii) Protons from the PS can be injected into the SPS for further acceleration. Explain why the magnetic field in the SPS is increased as protons are accelerated, but the frequency of the accelerating voltage is kept almost constant.

Figure 3.62 Bubble chamber tracks for Question 66

Q 66 The photograph in Figure 3.62 shows an interaction between an atomic electron and a high-energy photon recorded in a bubble chamber. The photon has entered from the left and collided with an electron already present in the bubble chamber. The photograph also captures a secondary event, but in this question we are focussing on the first event, shown to the left of the image.

This first interaction caused the release of the atomic electron and the creation of an electron-positron pair. The tracks of these three particles can be seen and have been labelled.

(a) What conservation laws apply in this type of interaction? For the interaction shown in the photograph, discuss the conservation of any *two* of these properties.

(b) Explain how you can deduce from the photograph that the initial particle enters *from the left* and carries *no charge*.

(c) What is the direction of the magnetic field? How can you tell?

(d) Explain which electron moves faster.

(e) The magnetic field strength is $B = 5.4 \times 10^{-3}$ T, and measurements from the photograph show the initial radius of curvature of the positron to be $r = 0.048$ m. Calculate the momentum of the positron. (Positron charge $= 1.60 \times 10^{-16}$ C.)

(f) Explain why the path of the positron is a spiral.

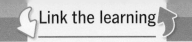

5.3 Achievements

Now you have studied this chapter you should be able to achieve the outcomes listed in Table 3.14.

TABLE 3.14 ACHIEVEMENTS FOR THE CHAPTER *PROBING THE HEART OF MATTER*

	Statement from Examination Specification	*Section(s) in this chapter*
1	know and understand the distinction between base and derived quantities and their SI units	3.2
2	demonstrate their knowledge of practical skills and techniques for both familiar and unfamiliar experiments	3.4
3	be able to estimate values for physical quantities and use their estimate to solve problems	3.5, 4.3
4	understand the limitations of physical measurement and apply these limitations to practical situations.	3.4
5	be able to communicate information and ideas in appropriate ways using appropriate terminology	2.2, 2.4
7	understand the role of the scientific community in validating new knowledge and ensuring integrity	4.1
99	understand how to apply conservation of linear momentum to problems in two dimensions	3.4
100	CORE PRACTICAL 10: Use ICT to analyse collisions between small spheres, e.g. ball bearings on a table top	3.4
102	be able to derive and use the equation $E_k = \dfrac{p^2}{2m}$ for the kinetic energy of a non-relativistic particle	3.4
103	be able to express angular displacement in radians and in degrees, and convert between these units	4.3
104	understand what is meant by *angular velocity* and be able to use the equations $v = \omega r$ and $T = \dfrac{2\pi}{\omega}$	4.3
105	be able to use vector diagrams to derive the equations for centripetal acceleration $a = \dfrac{v^2}{r} = r\omega^2$ and understand how to use these equations	4.3
106	understand that a resultant force (centripetal force) is required to produce and maintain circular motion	4.3
107	be able to use the equations for centripetal force $F = ma = \dfrac{mv^2}{r} = mr\omega^2$	4.3
110	be able to use the equation $F = \dfrac{Q_1 Q_2}{4\pi\varepsilon_0 r^2}$ for the force between two charges	3.2, 3.3
111	be able to use the equation $E = \dfrac{Q}{4\pi\varepsilon r^2}$ for the electric field due to a point charge	3.2, 3.3
114	be able to use $V = \dfrac{Q}{4\pi\varepsilon_0 r}$ for a radial field	3.2
122	be able to use the equation $F = Bqv \sin\theta$ and apply Fleming's left-hand rule to charged particles moving in a magnetic field	4.4
130	understand what is meant by nucleon number (mass number) and proton number (atomic number)	1.1, 3.1 (STA)
131	understand how large-angle alpha particle scattering gives evidence for a nuclear model of the atom and how our understanding of atomic structure has changed over time	3.1

continued

TABLE 3.14 ACHIEVEMENTS FOR THE CHAPTER PROBING *THE HEART OF MATTER* *continued*

	Statement from Examination Specification	*Section(s) in this chapter*
133	understand the role of electric and magnetic fields in particle accelerators (linac and cyclotron) and detectors (general principles of ionisation and deflection only)	3.5, 4.2, 4.4
134	be able to derive and use the equation $r = \dfrac{p}{BQ}$ for a charged particle in a magnetic field	4.4
135	be able to apply conservation of charge, energy and momentum to interactions between particles and interpret particle tracks	3.4
136	understand why high energies are required to investigate the structure of nucleons	4.2
137	be able to use the equation $\Delta E = c^2 \Delta m$ in situations involving the creation and annihilation of matter and anti-matter particles	2.4, 2.5
138	be able to use MeV and GeV (energy) and MeV/c^2, GeV/c^2 (mass) and convert between these and SI units	4.2
139	understand situations in which the relativistic increase in particle lifetime is significant (use of relativistic equations not required)	4.2
140	know that in the standard quark-lepton model, particles can be classified as: • baryons (e.g. neutrons and protons), which are made from three quarks • mesons (e.g. pions), which are made from a quark and an anti-quark • leptons (e.g. electrons and neutrinos), which are fundamental particles • photons and that the symmetry of the model predicted the top quark	2.2, 2.4
141	know that every particle has a corresponding anti-particle and be able to use the properties of a particle to deduce the properties of its anti-particle and vice versa	2.4
142	understand how to use laws of conservation of charge, baryon number and lepton number to determine whether a particle interaction is possible	2.5
143	be able to write and interpret particle equations given the relevant particle symbols	2.3
165	use the *atomic mass unit*, (u) to express small masses, and convert between this and SI units	4.2

Answers

Q 1 (a) strong; (b) gravity; (c) electromagnetic;
(d) strong and electromagnetic; (e) weak.

Q 2 $\mu^+ + \mu^- \rightarrow$ electromagnetic radiation

Q 3 A quark and its anti-quark have equal but opposite charges, so they cancel out.

Q 4 Creating the mass of the tau leptons requires much more than the energy equivalent of an electron–positron pair at rest; additional energy must be supplied in the form of kinetic energy of the colliding particles.

Q 5 $E = mc^2 = 1.67 \times 10^{-27}\,\text{kg} \times (3.00 \times 10^8\,\text{m s}^{-1})^2$

$= 1.50 \times 10^{-10}\,\text{J}$

Q 6 (a) $E = mc^2$

$= (2 \times 9.11 \times 10^{-31}\,\text{kg}) \times (3.00 \times 10^8\,\text{m s}^{-1})^2$

$= 1.64 \times 10^{-13}\,\text{J}$

(b) Energy 'left over' $= 2.00 \times 10^{-13}\,\text{J} - 1.64 \times 10^{-13}\,\text{J}$

$= 0.36 \times 10^{-13}\,\text{J}$

This is shared equally between the two particles, so the kinetic energy of each is $0.18 \times 10^{-13}\,\text{J}$ ($1.8 \times 10^{-14}\,\text{J}$).

Q 7 (a) electron and positron (anti-electron)

(b) They have opposite charges.

(c) They annihilate, producing electromagnetic radiation.

Q 8 hadrons: proton, neutron, pion
meson: pion
baryons: proton, neutron
quarks: charm, up
leptons: electron, anti-electron, electron-neutrino

Q 9 Notice how the closer the particle comes to the nucleus, the greater its change in direction.

Each particle is deflected in such a way that the path of the particle is most tightly curved at point of closest approach and instantaneously perpendicular to the line drawn to the centre of the nucleus.

This shows that the electrostatic force of repulsion increases as the separation of the charges decreases and acts along a line joining the centres of the charges. The greatest force occurs at the point of closest approach and this means that the centripetal force (required to deflect the particle into a curved path) is greatest at this point and so the direction changes most rapidly.

Q 10 To avoid the absorption of alpha particles: air is pumped out through tube T; source R and detector S are close to foil F; foil F is thin.

Q 11 Direct hits are more likely, so more alpha particles will be back-scattered.

Q 12 Slower alpha particles would not whiz straight past so easily, so more would be scattered through bigger angles.

Q 13 There is less chance of a direct hit, and the force is weaker, so fewer alpha particles will be back-scattered.

Q 14 See Figure 3.63

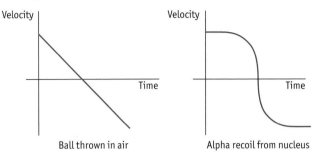

Figure 3.63 Answer to Question 14

Similarities:
Both show velocity changing sign, as ball and alpha change direction in the process.

Neglecting air resistance for the ball, the initial speed is the same as the final speed in each case.

Differences:
The acceleration of the ball is constant, because the force is constant. The acceleration of the alpha increases as it approaches the nucleus (because the field strength increases) and is a maximum when the velocity is zero.

The acceleration of the alpha at long distances is zero, whereas the ball never reaches a constant velocity.

Q 15 See Figure 3.64

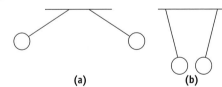

Figure 3.64 Answer to Question 15

Q 16 See Figure 3.65

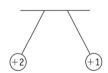

Figure 3.65 Answer to Question 16

Q 17 Figure 3.66 shows the forces acting on one ball.

Figure 3.66 Diagram for the answer to Question 17

Resolving horizontally: $F = T\sin30°$

Resolving vertically: $W = T\cos30°$

Dividing one equation by the other and cancelling T:

$$\frac{F}{W} = \tan30°$$

so $F = 0.10\,\text{N} \times \tan30° = 0.058\,\text{N}$

Q 18 $1\,\text{F} = 1\,\text{C}\,\text{V}^{-1}$, and $1\,\text{V} = 1\,\text{J}\,\text{C}^{-1}$, so $1\,\text{F} = 1\,\text{C}^2\,\text{J}^{-1}$

$1\,\text{J} = 1\,\text{N}\,\text{m}$, so $1\,\text{F} = 1\,\text{C}^2\,\text{N}^{-1}\,\text{m}^{-1}$

Therefore: $1\,\text{F}\,\text{m}^{-1} = 1\,\text{C}^2\,\text{N}^{-1}\,\text{m}^{-2} = (1\,\text{N}\,\text{m}^2\,\text{C}^{-2})^{-1}$ as required.

Q 19
$$F = \left(\frac{1}{4\pi\varepsilon_0}\right)\left(\frac{Q_1 Q_2}{r^2}\right)$$

$$= \frac{9.0 \times 10^9\,\text{N}\,\text{m}^2\,\text{C}^{-2} \times (10^{-8}\,\text{C})^2}{(0.05\,\text{m})^2}$$

$$= 3.6 \times 10^{-4}\,\text{N}$$

Q 20 **(a)**
$$F = \left(\frac{1}{4\pi\varepsilon_0}\right)\left(\frac{Q_1 Q_2}{r^2}\right)$$

$$= \frac{9.0 \times 10^9\,\text{N}\,\text{m}^2\,\text{C}^{-2} \times (1.60 \times 10^{-19}\,\text{C})^2}{(0.037 \times 10^{-19}\,\text{m})^2}$$

$$= 1.68 \times 10^{-7}\,\text{N}$$

(b) See Figure 3.67

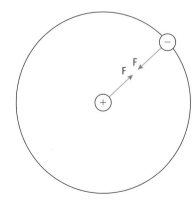

Figure 3.67 Answer to Question 20(b)

Q 21 **(a)** The nuclear charge is smaller, so the force is weaker at a given distance (about one-third of that with gold). As shown in Figure 3.68, the graph plotted on a linear scale is scaled down by a factor of about 3, and the log graph is *shifted* downwards.

(b) $V = \dfrac{kQ}{r}$, so the graphs of V against r are as shown in Figure 3.68. At any given radius, V for copper is about one-third the value for gold.

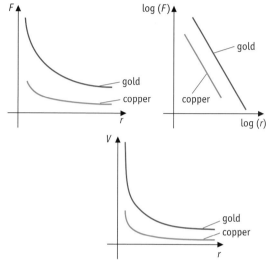

Figure 3.68 Answers to Question 21

Q 22 **(a)** The student has multiplied the mass in g by 9.8 mN/g. This is the appropriate value since the data has a minimum of 2 significant figures.

(b) The log–log graph yields a straight line with gradient −2.06 (Figure 3.69). The gradient of a log–log graph reveals the power law that connects the two variables. So this value suggests an inverse square relationship within the uncertainty of the experiment.

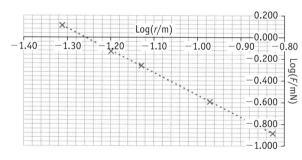

Figure 3.69 Answer to Question 22

Q 23 Initial $E_k = \Delta E = \dfrac{kQ_1Q_2}{r}$ (Equation 10), so:

$$r = \dfrac{kQ_1Q_2}{E_k}$$

$$= \dfrac{9.0 \times 10^9 \, \mathrm{N\,m^2\,C^{-2}} \times 79 \times 1.60 \times 10^{-19}\,\mathrm{C} \times 2 \times 1.60 \times 10^{-19}\,\mathrm{C}}{8.0 \times 10^{-13}\,\mathrm{J}}$$

$$= 4.6 \times 10^{-14}\,\mathrm{m}$$

Q 24 Lines of force go *towards* a negatively-charged sphere. See Figure 3.70

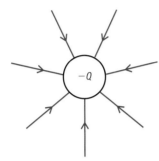

Figure 3.70 Answer to Question 24

Q 25 See Figure 3.71

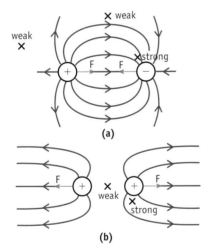

Figure 3.71 Answers to Question 25

Q 26 $E = \dfrac{F}{Q} = \dfrac{10\,\mathrm{N}}{20 \times 10^{-3}\,\mathrm{C}} = 500\,\mathrm{N\,C^{-1}}$

Q 27 $F = Eq$

$= 2000\,\mathrm{N\,C^{-1}} \times 1.60 \times 10^{-19}\,\mathrm{C}$

$= 3.2 \times 10^{-16}\,\mathrm{N}$

Acceleration $a = \dfrac{F}{m} = \dfrac{3.2 \times 10^{-16}\,\mathrm{N}}{9.11 \times 10^{-31}\,\mathrm{kg}}$

$= 3.5 \times 10^{14}\,\mathrm{m\,s^{-2}}$

(Note that this is more than a million million times the acceleration due to gravity.)

Q 28 $E = \dfrac{kQ}{r^2}$

$$= \dfrac{9.0 \times 10^9\,\mathrm{N\,m^2\,C^{-2}} \times 79 \times 1.60 \times 10^{-19}}{(1.0 \times 10^{-13}\,\mathrm{m})^2}$$

$$= 1.1 \times 10^{19}\,\mathrm{N\,C^{-1}}$$

Q 29 For someone of mass 60 kg capable of running at $10\,\mathrm{m\,s^{-1}}$ (close to the world record for the 100 m sprint!):

$p = mv = 60\,\mathrm{kg} \times 10\,\mathrm{m\,s^{-1}} = 600\,\mathrm{kg\,m\,s^{-1}}$

$E_k = \frac{1}{2}mv^2 = \frac{1}{2} \times 60\,\mathrm{kg} \times (10\,\mathrm{m\,s^{-1}})^2 = 3\,\mathrm{kJ}$

$\dfrac{p^2}{2m} = \dfrac{(600\,\mathrm{kg\,m\,s^{-1}})^2}{2 \times 60\,\mathrm{kg}} = 3\,\mathrm{kJ}$

Q 30 $E_k = \dfrac{p^2}{2m}$

$$= \dfrac{(5.0 \times 10^{-20}\,\mathrm{kg\,m\,s^{-1}})^2}{2 \times 6.7 \times 10^{-27}\,\mathrm{kg}}$$

$$= 1.9 \times 10^{-13}\,\mathrm{J}$$

Q 31 **(a)** Momentum would be conserved – it is *always* conserved in any interaction.

(b) It is *not* elastic. Some of the initial kinetic energy would provide the additional mass of the muons (see Section 2.4 of this chapter) so there would be an overall decrease in the kinetic energy.

Q 32 They would measure the momentum before and after the collision (you will see in Part 4 how this can be done). Any discrepancy can be accounted for in terms of the momentum of 'unseen' particles.

Q 33 $E_k = eV = 1.60 \times 10^{-19}\,\mathrm{C} \times 500\,\mathrm{V} = 8.00 \times 10^{-17}\,\mathrm{J}$

$p = \sqrt{2meV} = \sqrt{2 \times 9.11 \times 10^{-31} \times 8.00 \times 10^{-17}}$

$= 1.21 \times 10^{-23}\,\mathrm{kg\,m\,s^{-1}}$

$v = \dfrac{p}{m} = \dfrac{1.21 \times 10^{-23}\,\mathrm{kg\,m\,s^{-1}}}{9.11 \times 10^{-31}\,\mathrm{kg}}$

$= 1.33 \times 10^7\,\mathrm{m\,s^{-1}}$

Q 34 $E_k = \dfrac{1}{2}mv^2 = \dfrac{1}{2} \times 9.11 \times 10^{-31}\,\mathrm{kg} \times (1.0 \times 10^6\,\mathrm{m\,s^{-1}})^2$

$= 4.55 \times 10^{-19}\,\mathrm{J}$

$V = \dfrac{E_k}{e} = \dfrac{4.55 \times 10^{-19}\,\mathrm{J}}{1.60 \times 10^{-19}\,\mathrm{C}} = 2.84\,\mathrm{V}$

Q 35 Greater kinetic energy means greater momentum and therefore shorter wavelength.

Q 36 Greater voltage means greater kinetic energy and momentum, so shorter wavelength.

Q 37 De Broglie wavelength $\lambda = \dfrac{h}{p}$

$$= \frac{6.63 \times 10^{-34}\,\text{J}\,\text{s}}{9.11 \times 10^{-31}\,\text{kg} \times 1.0 \times 10^{6}\,\text{m}\,\text{s}^{-1}}$$

$$= 7.3 \times 10^{-10}\,\text{m}$$

(This wavelength is comparable to the spacing of atoms in a solid.)

Q 38 $p = \dfrac{h}{\lambda} = \dfrac{6.63 \times 10^{-34}\,\text{J}\,\text{s}}{3.0 \times 10^{-10}\,\text{m}}$

$$= 2.21 \times 10^{-24}\,\text{kg}\,\text{m}\,\text{s}^{-1}$$

Q 39 For a person of mass 60 kg walking at $1\,\text{m}\,\text{s}^{-1}$:

$p = 60\,\text{kg}\,\text{m}\,\text{s}^{-1}$

$\lambda = \dfrac{h}{p} = \dfrac{6.63 \times 10^{-34}\,\text{J}\,\text{s}}{60\,\text{kg}\,\text{m}\,\text{s}^{-1}} \approx 10^{-35}\,\text{m}$

This wavelength is *extremely* small. You would only notice diffraction effects when passing through an aperture of comparable size – which is about 35 orders of magnitude smaller than your own physical size. A doorway is some 35 orders of magnitude too large, so any diffraction effects are unobservable.

Q 40 (a) (i) $E_0 = m_0 c^{-2}$

$\qquad = 1.67 \times 10^{-27}\,\text{kg} \times (3.00 \times 10^{8}\,\text{m}\,\text{s}^{-1})^2$

$\qquad = 1.50 \times 10^{-10}\,\text{J}$

(ii) $E_0 = m_0 c^{-2}$

$\qquad = \dfrac{1.50 \times 10^{-10}\,\text{J}}{1.60 \times 10^{-19}\,\text{J}\,\text{eV}^{-1}}$

$\qquad = 9.39 \times 10^{8}\,\text{eV} = 0.939\,\text{GeV}$

(b) $m = 0.939\,\text{GeV}/c^2$

Q 41 Total energy available = 2 GeV. Energy needed to create a single pi-zero at rest is 0.14 GeV.

$$\frac{2\,\text{GeV}}{0.14\,\text{GeV}} = 14.28$$

So at most 14 pi-zeros could be created. (This assumes the pi-zeros are created at rest, and that no other particles are created. In practice, neither of those assumptions is likely to be correct.)

Q 42 $m = 20\,\text{GeV}/c^2$

Q 43 $1\,\text{u} = 1.66 \times 10^{-27}\,\text{kg}$

$E = mc^2 = 1.66 \times 10^{-27}\,\text{kg} \times (3.00 \times 10^{8}\,\text{m}\,\text{s}^{-1})^2$

$\qquad = 1.49 \times 10^{-10}\,\text{J}$

$\qquad = \dfrac{1.49 \times 10^{-10}\,\text{J}}{1.60 \times 10^{-19}\,\text{J}\,\text{eV}^{-1}}$

$\qquad = 9.31 \times 10^{8}\,\text{eV} = 931\,\text{MeV}$

Q 44 $m_e = 9.11 \times 10^{-31}\,\text{kg}$

$$= \frac{9.11 \times 10^{-31}\,\text{kg}}{1.66 \times 10^{-27}\,\text{kg}\,\text{u}^{-1}}$$

$$= 5.49 \times 10^{-4}\,\text{u}$$

Q 45 (a) $\dfrac{3000\,\text{rpm}}{60\,\text{s}} = 50\,\text{s}^{-1}$

(b) Angular velocity, $\omega = 2\pi \times 50\,\text{s}^{-1} = 314\,\text{rad}\,\text{s}^{-1}$

(c) Period, $T = \dfrac{1}{50\,\text{s}^{-1}} = 0.02\,\text{s}$

Q 46 (a) Number of revolutions per second $= \dfrac{30}{2\pi} = 4.8$

(b) Period, $T = \dfrac{1}{4.8\,\text{s}^{-1}} = 0.21\,\text{s}$

Q 47 Radius of accelerator $r = \dfrac{8600\,\text{m}}{2} = 4300\,\text{m}$

Angular velocity $\omega = \dfrac{v}{r} = \dfrac{3.00 \times 10^{8}\,\text{m}\,\text{s}^{-1}}{4300\,\text{m}}$

$\qquad = 6.98 \times 10^{4}\,\text{rad}\,\text{s}^{-1}$

Q 48 (a) Angular velocity $\omega = \dfrac{2\pi}{T} = \dfrac{2\pi}{1.25\,\text{s}} = 5.0\,\text{rad}\,\text{s}^{-1}$

(b) (i) Speed $v = \omega r = 5.0\,\text{rad}\,\text{s}^{-1} \times 2\,\text{m} = 10\,\text{m}\,\text{s}^{-1}$

(ii) Speed is halved, i.e. $5\,\text{m}\,\text{s}^{-1}$

(iii) Speed $= 0\,\text{m}\,\text{s}^{-1}$

Q 49 (a) Angular velocity $\omega = 100 \times 2\pi\,\text{rad}\,\text{s}^{-1}$

$\qquad = 628\,\text{rad}\,\text{s}^{-1}$

(b) Speed $v = \omega r$

$\qquad = 628\,\text{rad}\,\text{s}^{-1} \times 0.15\,\text{m}$

$\qquad = 94.2\,\text{m}\,\text{s}^{-1}$

Q 50 Radius $r \approx$ distance from shoulder to ground

$\qquad \approx 1.5\,\text{m}$, say.

Angular velocity $\omega = \dfrac{v}{r} = \dfrac{30\,\text{m}\,\text{s}^{-1}}{1.5\,\text{m}} = 20\,\text{rad}\,\text{s}^{-1}$

Q 51 From Q47: $a = \dfrac{v^2}{r} = \dfrac{(3.00 \times 10^{8}\,\text{m}\,\text{s}^{-1})^2}{4300\,\text{m}}$

$\qquad = 2.1 \times 10^{13}\,\text{m}\,\text{s}^{-2}$

From Q48: (i) $a = \dfrac{v^2}{r} = \dfrac{(10\,\text{m}\,\text{s}^{-1})^2}{2\,\text{m}} = 50\,\text{m}\,\text{s}^{-2}$

(ii) $a = \dfrac{v^2}{r} = \dfrac{(5\,\text{m}\,\text{s}^{-1})^2}{1\,\text{m}} = 25\,\text{m}\,\text{s}^{-2}$

(iii) $v = 0$ so $a = 0$

From Q49: $a = \dfrac{v^2}{r} = \dfrac{(94.2\,\text{m s}^{-1})^2}{0.15\,\text{m}} = 5.9 \times 10^4\,\text{m s}^{-2}$

From Q50: $a = \dfrac{v^2}{r} = \dfrac{(30\,\text{m s}^{-1})^2}{1.5\,\text{m}} = 600\,\text{m s}^{-2}$

Q 52   **(a)** turning sharply left

      (b) A bigger force is needed.

Q 53 $F = \dfrac{mv^2}{r} = \dfrac{5\,\text{kg} \times (30\,\text{m s}^{-1})^2}{2\,\text{m}} = 2250\,\text{m s}^{-2}$

Q 54 $F = mr\omega^2 = 1\,\text{kg} \times (10\,\text{rad s}^{-1})^2 \times 10\,\text{m} = 1000\,\text{N}$

Q 55   **(a)** $\omega = \dfrac{2\pi}{T} = \dfrac{2\pi}{0.75\,\text{s}} = 84\,\text{rad s}^{-1}$

      (b) $F = mr\omega^2 = 7.26\,\text{kg} \times (8.4\,\text{rad s}^{-1})^2 \times 2.1\,\text{m}$

           $= 1.1 \times 10^3\,\text{N}$

Q 56   **(a)** $W = mg = 45\,\text{kg} \times 9.81\,\text{N kg}^{-1} = 441\,\text{N}$

      (b) $F = \dfrac{mv^2}{r} = \dfrac{45\,\text{kg} \times (4\,\text{m s}^{-1})^2}{6\,\text{m}} = 120\,\text{N}$

      (c) Tension in rope = 441 N + 120 N = 561 N

      (The rope must support the weight of the girl + tyre, *and* supply the centripetal force.)

Q 57   **(a)** $F \propto \dfrac{1}{r}$, so twice the radius requires half the force, i.e. 1 kN.

      (b) $F \propto v^2$, so twice the speed requires four times the force, i.e. 8 kN.

Q 58   See Figure 3.72

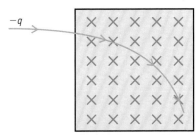

Figure 3.72 Answer to Question 58

Q 59   The electrons do not all have the same momentum (or, therefore, the same kinetic energy). The most energetic electrons will curve least.

Q 60 $r = \dfrac{mv}{Bq}$

      $= \dfrac{9.11 \times 10^{-31}\,\text{kg} \times 0.01 \times 3.00 \times 10^8\,\text{m s}^{-1}}{0.2\,\text{T} \times 1.60 \times 10^{-19}\,\text{C}}$

      $= 8.54 \times 10^{-5}\,\text{m}$

Q 61   **(a)** $v = \sqrt{2eV/m}$

      $= \sqrt{\dfrac{2 \times 1.60 \times 10^{-19}\,\text{C} \times 2500\,\text{V}}{9.11 \times 10^{-31}\,\text{kg}}}$

      $= 2.96 \times 10^7\,\text{m s}^{-1}$

      (b) $F = Bqv$

      $= 1\,\text{T} \times 1.60 \times 10^{-19}\,\text{C} \times 2.96 \times 10^7\,\text{m s}^{-1}$

      $= 4.74 \times 10^{-12}\,\text{N}$

      (c) $a = \dfrac{F}{m} = \dfrac{4.74 \times 10^{-12}\,\text{N}}{9.11 \times 10^{-31}\,\text{kg}}$

      $= 5.20 \times 10^{18}\,\text{m s}^{-2}$

      The electrons change direction but not speed.

      (d) $r = \dfrac{mv}{Bq} = \dfrac{9.11 \times 10^{-31}\,\text{kg} \times 2.96 \times 10^7\,\text{m s}^{-1}}{1\,\text{T} \times 1.60 \times 10^{-19}\,\text{C}}$

           $= 1.70 \times 10^{-4}\,\text{m}$

      or use $a = \dfrac{v^2}{r}$

      $r = \dfrac{v^2}{a} = \dfrac{(2.96 \times 10^7\,\text{m s}^{-1})^2}{5.20 \times 10^{18}\,\text{m s}^{-2}} = 1.70 \times 10^{-4}\,\text{m}$

BUILD OR BUST?

Why a chapter called *Build or Bust?*

This chapter focuses on two aspects of building design: earthquake resistance and temperature control.

In 2015 a major earthquake devastated the country of Nepal. Official estimates put the death toll at over 8000, with another 20 000 people injured. An estimated 250 000 were left homeless as buildings collapsed in the quake or were destroyed by the landslides that followed (Figure 4.1a).

An earthquake consists of shock waves travelling through the Earth. Physicists can study these waves, allowing them to discover a great deal about the structure and origin of the Earth and how it is evolving.

In April and May 2011 two small earthquakes struck near the town of Blackpool. No damage or injuries were reported, but suspicion immediately fell on the process of fracking (hydraulic fracturing) that had been taking place nearby. The epicentre was less than 500 metres from the drilling site. An independent report confirmed the link with fracking. Fracking is a method of extracting natural gas from surrounding rock. It involves drilling down into the rock and pumping in high pressure water and chemicals to rupture the rock, allowing the gas to escape. Fracking has only recently started in the UK and its use is controversial. It has been going on in other countries for much longer and there have been numerous studies investigating fracking and possible increases in seismic activity. It is not yet clear whether fracking could cause a significant earthquake.

Figure 4.1 (a) A major earthquake caused widespread devastation in Nepal in 2015

Heatwaves kill more people than other extreme weather events. The summer of 2003 was the hottest in central Europe since 1540, maybe longer. More than 20 000 people died in the heatwave – including about 15 000 in France. (Compare these figures with Hurricane Katrina (2005), in which about 1800 people died.) Due to global warming we can expect more hot summers, so we need to design modern buildings that will stay cool in summer and warm in winter (Figure 4.1b).

Figure 4.1 (b) Many modern buildings are designed with effective temperature control in mind

An understanding of waves enables us to design buildings and create materials that will withstand most earthquakes – and to locate valuable deposits of oil. An understanding of material properties is critical to this.

An understanding of thermal properties of materials, and how energy is stored and transferred by heating and cooling, is essential in designing buildings that make the best use of resources and are also pleasant to live and work in.

OVERVIEW OF PHYSICS PRINCIPLES AND TECHNIQUES

In this chapter you will continue your study of waves, energy and materials. Much of the work will build on and extend ideas you have met in previous chapters. You will also study the type of oscillation known as simple harmonic motion, consider why objects vibrate most at certain frequencies (resonance) and what that means for building design. Simple harmonic motion underlies most studies of waves and vibrations, and is the most important topic in this chapter. In addition, you will extend your understanding of how materials absorb and transmit heat, and learn about phase changes such as melting and boiling. You will see how that understanding is useful in building design.

In this chapter you will extend your knowledge of the following from earlier chapters:

- travelling waves from *The Sound of Music* and *Spare Part Surgery*
- energy and its conservation from *Higher, Faster, Stronger*
- refraction and reflection from *Good Enough to Eat* and *Spare Part Surgery*
- bulk properties of solids from *Good Enough to Eat* and *Spare Part Surgery*.

You will do more work on:

- waves in *Reach for the Stars*
- energy and temperature in *Reach for the Stars*.

1 Earthquakes

1.1 Shaking the Earth

Earthquakes – the wrath of the gods?

Earthquakes can be killers. In 2015, an earthquake devastated the country of Nepal. Official estimates put the death toll at over 8000, with more than 20 000 people injured. Over 250 000 were made homeless as buildings collapsed in the quake or were destroyed by the landslides that followed.

In 2011, a major earthquake and tsunami hit the Japanese island of Honshu. The damage was extensive and official estimates put the death toll at nearly 20 000. The tsunami also caused meltdowns in nuclear reactors at the Fukushima power plant.

Between 2010 and 2012, multiple major earthquakes struck near Christchurch, one of the largest urban areas in New Zealand. The most devastating of the earthquakes was in 2011 when 185 people were killed and many buildings either collapsed or had to be demolished due to damage sustained in the earthquakes.

Modern cities in earthquake zones represent a particular hazard; in cities that are at high risk of earthquakes, the major concerns are to predict accurately when an earthquake will occur so that people can leave, and to build so that the buildings, bridges and roads will not collapse. A lot of effort has been put into predicting earthquakes, but with little success. Building earthquake-resistant cities makes prediction less important and has been more successful. If we specify that we want a building to be 'earthquake-proof', how do engineers meet that specification?

In Bronze Age Crete, the Minoans revered the bull (Figure 4.2) as the incarnation of Poseidon, in his role as god of earthquakes. There is even evidence of human sacrifice, with the remains of a youth found on an altar, his body having been drained of blood to offer to the gods. This failed to avert the earthquake: the bodies of the priest, the sacrificial knife and the vessel for containing the victim's blood were all found where they were crushed by the falling masonry as the earthquake shook the Earth.

We now know that the Earth is not solid and it is constantly moving and changing. Figure 4.3 shows the Earth's structure. The central solid core is made of iron and nickel and is held at an immensely high temperature and pressure. The liquid core of molten iron and nickel surrounding this is continuously moving in convection currents, and there are also convection currents in the mantle, which is soft and liquid in places. The Earth's crust is relatively thin (only a few kilometres in places) and is made of a number of continental and oceanic plates that float on the mantle. It is the movement of these plates that causes earthquakes.

As the plates are forced slowly towards, or past, one another, the forces increase and the strain gradually builds up until the plates suddenly move: this produces an earthquake. Seismic waves (shock waves) spread out and travel all round the world. As rocks settle after the earthquake there are smaller aftershocks. Seismologists study how earthquake waves travel in order to establish more about the Earth; some seismologists use this information to locate oil and gas deposits.

Figure 4.4 shows the edges (margins) of the plates. At so-called constructive margins, the plates are spreading or diverging and new oceanic crusts form mid-ocean ridges with volcanoes. There are two types of destructive margin: subduction zones, where the oceanic crust moves towards the light continental crust, sinks and is destroyed; and collision zones, where two continental crusts collide and are forced up into mountains. At conservative margins, two plates move sideways past each other, and land is neither formed nor destroyed.

Figure 4.2 Minoan bull statue

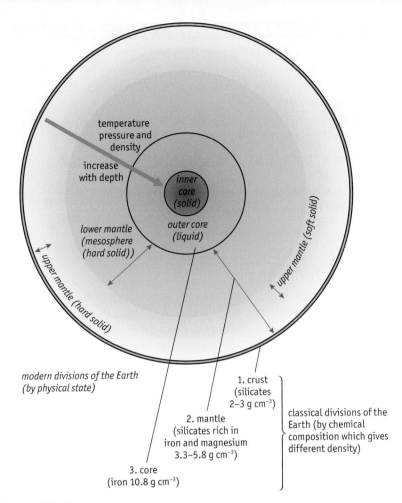

Figure 4.3 Structure of the Earth

constructive plate margin   conservative plate margin

destructive plate margin   - - - - - plate boundary uncertain

Figure 4.4 Plate margins of the Earth

PRACTICAL SKILLS REFERENCE

Scientific questions and information research

See Practical skills notes 1.2–1.3

⚙ ACTIVITY 1 EARTHQUAKES ON THE INTERNET

Use the internet to find out about one or more of the following topics. Some useful links can be found at **www.pearsonhotlinks.co.uk**.

● Where was the most recent earthquake recorded in the UK? In the world? What can you say about the sizes of these earthquakes?

● What was the largest earthquake recorded in the UK? Roughly how many earthquakes occur in the UK each year?

● How are instruments designed to detect and measure earthquakes? What principles of physics are involved?

● How do engineers and architects choose materials, and use principles of physics, when designing earthquake-resistant buildings?

1.2 Seismology

Did the Earth move for you?

The study of earthquakes – seismology – increases our knowledge of the Earth and enables engineers to design earthquake-resistant cities. Earthquakes originate below the surface of the Earth at a point called the focus, or hypocentre; the point on the Earth's surface directly above the earthquake is known as the epicentre. The seismic waves spread out from the focus in all directions (see Figure 4.5).

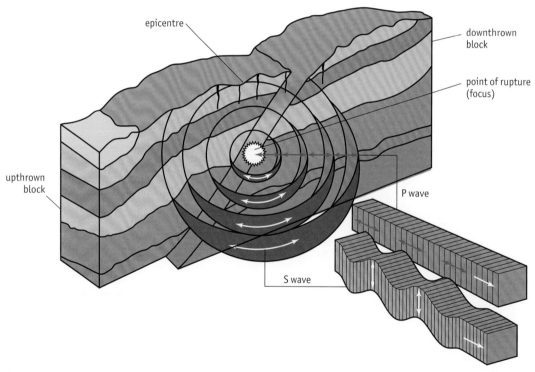

Figure 4.5 Anatomy of an earthquake

Seismic waves are categorised as body waves, which travel through the whole body of the Earth, and surface waves, which travel only through the crust. It is the surface waves that cause the most damage.

There are two types of body waves (Figure 4.6), called primary waves (P waves) and secondary waves (S waves). They typically have frequencies of a few hertz.

P waves are **longitudinal waves** and can travel through the whole Earth, including the molten core. They travel through the rocks as a series of compressions and rarefactions. They are called primary waves because they travel faster than the S waves and arrive first at seismic monitoring stations.

STUDY NOTE

You have met longitudinal and transverse waves before, notably in the chapter *The Sound of Music*. Animations of these waves may be viewed via **www.pearsonhotlinks.co.uk**.

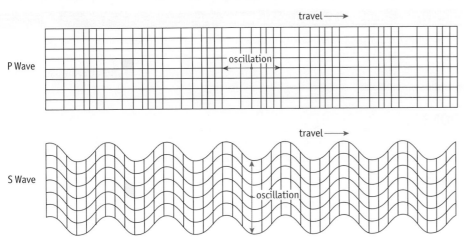

Figure 4.6 Body waves

S waves (sometimes also known as shear waves or shake waves) are **transverse waves** and cannot travel through the liquid parts of the Earth. The particles oscillate in a direction that is perpendicular to the direction of wave travel, so there is a shearing movement in the rock. The delay between the arrival of the P and S waves can be used to calculate the distance of the monitoring station from the earthquake.

There are also two types of surface waves (see Figure 4.7), named Rayleigh waves and Love waves after the scientists who studied them. Surface waves occur only in the Earth's crust, when one boundary of the material in which the wave is travelling is a free boundary. They have longer wavelengths (hundreds of kilometres) and lower frequencies than body waves (typically 0.1 Hz or less), and the displacement is largest at the surface, decreasing exponentially with the depth.

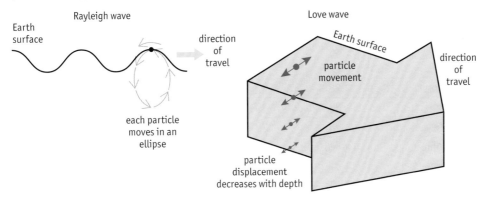

Figure 4.7 Surface waves

Love waves are transverse (or shear) waves where there is a strong horizontal displacement at right angles to the direction of travel. These waves are like water waves that have been turned on their side, through 90°.

Rayleigh waves are transverse waves, similar to water waves, where there is a strong vertical movement at right angles to the direction of travel. The particles that are being displaced have what is known as retrograde elliptical motion. This means that they are moving in the opposite direction to the wave at the crest. The path of each particle moved by the wave is an ellipse. Rayleigh waves travel more slowly than S or P waves.

⚙ ACTIVITY 2 REVISION OF WAVES

On a set of plain postcards, generate a set of 'key facts' revision cards that summarise what you have previously learned about waves. Include definitions of key terms such as longitudinal, wavelength, phase and so on.

STUDY NOTE

You learned about the Young modulus in the chapter *Spare Part Surgery*.

⚙ ACTIVITY 3 REVISION OF MATERIAL PROPERTIES

Earthquakes result from the exposure of rocks to compressive stresses in three dimensions. Generate a concept map that summarises everything you need to know about Young modulus, including the definitions of terms such as stress and strain. Find out what is meant by **bulk modulus** and **shear modulus** and include these terms in your concept map.

QUESTIONS

Q 1 If the speed of waves in the Earth's crust is about $3\,\text{km s}^{-1}$, estimate how long it would take for seismic waves to travel through the crust to reach the opposite side of the world.
(Radius of Earth = $6.378 \times 10^6\,\text{m}$)

Q 2 A quartz wire 10 cm long and 6 μm in diameter is used to suspend a small metal sphere of mass 2 g.

(a) Calculate the extension of the wire.

(b) What is the maximum mass that can be suspended from the wire?

(Use $g = 9.81\,\text{N kg}^{-1}$. Data for quartz: Young modulus $E = 73.1\,\text{GN m}^{-2}$; tensile strength = $1.00\,\text{GN m}^{-2}$.)

Q 3 Table 4.1 shows the speeds of seismic waves in materials commonly found in and on the Earth.

Material	Elastic modulus/$10^9\,\text{N m}^{-2}$		Density/ $10^3\,\text{kg m}^{-3}$	Seismic wave speed/$10^3\,\text{m s}^{-1}$	
	Bulk	Shear		P waves	S waves
air at 273 K	1.0×10^{-4}	0.0	1.0×10^{-3}	0.3	0.0
water at 298 K	2.2	0.0	1.0	1.4	0.0
ice	3.0	4.9	0.92	3.2	2.3
shale	8.8	17	2.4	3.6	2.6
sandstone	24	17	2.5	4.3	2.6
salt	24	18	2.2	4.7	2.9
limestone	38	22	2.7	5.0	2.9
quartz	33	39	2.7	5.7	3.8
granite	88	22	2.6	6.7	2.9
peridotite (compressed)	1.4×10^2	58	3.3	8.1	4.2

Table 4.1 Waves speeds, densities and elastic moduli for various materials

(a) From the data in Table 4.1, suggest, qualitatively, how elastic modulus affects wave speed. (Does increasing the modulus increase or decrease the wave speed for a given density?)

(b) Suggest how density affects wave speed.

Journey to the centre of the Earth

In recent decades the study of earthquakes has provided us with a great deal of information about the Earth. It is now apparent that the Earth is made of many layers that have different characteristics (see Figure 4.3), so that seismic waves have different speeds when travelling through these layers. Figure 4.8 is based on the Preliminary Earth Reference Model, which was developed in the 1980s using seismic data. It shows how density and wave speed vary with depth.

The surface of the Earth is made of rigid plates that consist of three layers: the crust, the Moho and the upper mantle. The crust has P wave speeds of about $2\text{–}5\,\text{km s}^{-1}$ for sedimentary rocks and about $6\,\text{km s}^{-1}$ for igneous rocks. The Moho (short for Mohorovičić discontinuity) separates the crust from the upper mantle, and is about 10 km down under the oceans and 20–70 km down under continents; here P wave speeds increase to about $8\,\text{km s}^{-1}$.

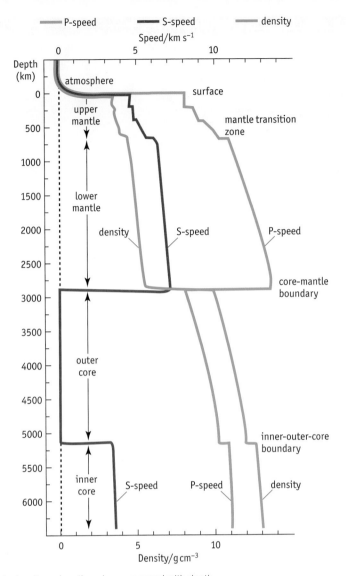

Figure 4.8 Variation in density and earthquake wave speed with depth

Below the plates, the lower part of the mantle is a soft plastic solid; some of it may be partially melted. The P and S waves have speeds of about 7–8 km s^{-1}, gradually increasing to about 10 km s^{-1} with depth.

Pressure increases with depth, so the lower mantle is more rigid. Then there is a sudden decrease in P and S wave speeds, marking the boundary with the core.

Wave speeds suddenly decrease at the boundary and then slowly increase further with depth. The outer core transmits no S waves, which results in a shadow zone (see Figure 4.9a). This tells us that the outer core is a fluid. There is also a P wave shadow zone (Figure 4.9b), due to the sudden change in speed. The inner core is assumed to be solid since some weak P waves arrive on the other side of the Earth sooner than expected.

The layered structure of the Earth has been deduced from measurements of seismic waves together with knowledge of the waves' speeds in different types of rock. Yet what properties of a material will determine the speed of waves within it? Question 3 showed that density cannot be the only factor and Activity 3 gave a hint concerning elastic moduli. As Activity 4 demonstrates, the density and elastic modulus together determine the wave speed.

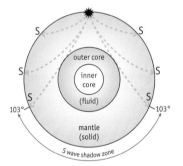

(a) No S arrival beyond 103°

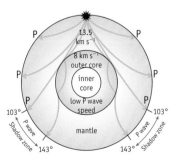

(b) No P between 103° and 143°

Figure 4.9 Shadow zones (a) for S waves and (b) for P waves

⚙ ACTIVITY 4 MODELLING EARTHQUAKE WAVES

Use the apparatus shown in Figure 4.10 to explore transverse and longitudinal waves. Investigate how the wave speed is affected by (i) adding mass to the trolleys (increasing the density), and (ii) adding more springs in parallel (increasing the stiffness).

(Notice that you are modelling waves in one dimension, whereas earthquake waves travel in three dimensions.)

Figure 4.10 Modelling earthquake waves

STUDY NOTE

For shear waves, E is replaced by the shear modulus.

As you will see in Section 2.3 of this chapter, the speed, v, of longitudinal waves is given by:

$$v = \sqrt{\frac{E}{\rho}} \tag{1}$$

where E is the elastic modulus (Young modulus) and ρ the density.

At a boundary between different materials, the wave speed changes – the waves are refracted. If the waves do not meet the boundary head-on, their direction also changes. Snell's law relates the speeds and directions:

$$n_1 \sin\theta_1 = n_2 \sin\theta_2 \tag{2}$$

STUDY NOTE

You have met examples of wave refraction before, in the chapters *The Sound of Music*, *Good Enough to Eat*, *Spare Part Surgery* and *The Medium is the Message*. You may wish to look back at this earlier work as you tackle Questions 4–7.

where n_1 and n_2 are the refractive indices of the two materials, and θ_1 and θ_2 are the angles between the normal to the boundary in materials 1 and 2 respectively (Figure 4.11).

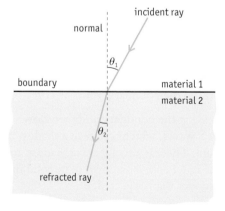

Figure 4.11 Refraction of waves at a boundary

QUESTIONS

Q 4 At the Moho there is a sudden increase in the velocity of seismic waves. Draw a ray diagram to show what happens to a wave when it crosses a boundary and its velocity increases. What is this phenomenon called? Then draw a wavefront diagram to show how the change in wavelength results in the change in direction of travel.

Q 5 Figure 4.12 shows the paths of several waves that have been formed by a controlled explosion at $X = 0$. Several monitoring stations detect the waves.

(a) Describe what has happened to each wave to cause it to travel in the path shown.

(b) Explain how information from the monitoring stations could be used to find out about the structure of the Earth below the stations.

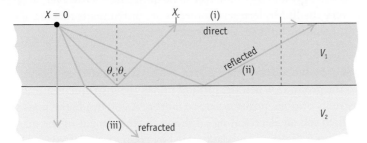

Figure 4.12 Waves from a controlled explosion

Q 6 A P wave crosses the boundary from the lower mantle into the outer core at an angle of incidence, $\theta_1 = 30°$. Using information from Figure 4.8, calculate a value for the refractive index between the two layers and in this way calculate the angle of refraction.

Q 7 **(a)** The waves' paths shown in Figure 4.9 are curved. What does this imply for the way that wave speed changes with depth in a given material?

(b) Deep within the Earth, the pressure is very high so materials are highly compressed. What effect will this have on the density of a given material?

(c) In view of your answers to **(a)** and **(b)**, what can you deduce about the way elastic modulus varies with depth (and therefore with pressure)?

Seismology in action

Mike Evans is an oil and gas exploration geophysicist who graduated with a degree in geological geophysics from the University of Reading and went on to manage an E&P (Exploration and Prospection) consultancy, Focus Exploration.

It can be very exciting as it's a high-risk game with companies committing millions of pounds to sink a well, often based primarily on your interpretation of seismic data.

Geophysics plays an important role in locating oil and gas deposits and deciding the best place to sink a well. The most important geophysical tool is the seismic method, which is a three-stage process requiring plenty of communication between each stage. The first stage is the seismic acquisition where acoustic waves are sent into the Earth. As the waves travel through the different rock types they are variably reflected, refracted and transmitted at each rock boundary. Hydrophones and geophones are used to pick up the reflected waves, building up a mass of data covering the whole area. Data processing is the next stage, where analysts use computers to produce a representation of the sub-surface – seismic profiles (2D seismic) or a seismic volume (3D seismic).

Mike was talking about his work at the third stage, as a seismic interpreter. He had to use his interpretive experience and knowledge of geology to develop an understanding of the present-day geometries of rock in the sub-surface, and their structural history. He was looking for evidence of mature hydrocarbon source rocks and structural or stratigraphic features where oil or gas may be trapped. Mike said:

It's satisfying when all the bits of the jigsaw come together, and we're fortunate that our search for oil takes us all over the world from the remotest jungles of the Far East to downtown Los Angeles.

Figure 4.13 Early Chinese seismometer

STUDY NOTE

For information about using seismometer traces to locate an earthquake's epicentre, see BLD Additional Sheet 1.

PRACTICAL SKILLS REFERENCE

Scientific questions and information research

See Practical skills notes 1.2–1.3

Seismometers

Earthquakes have long been the subject of human study; Figure 4.13 shows an early Chinese seismometer. Inside the large jar is a pendulum. An earthquake starts the pendulum swinging, and rods inside move and open a dragon's mouth so that the ball falls into the mouth of one of the frogs, in this way indicating the direction of the earthquake.

Close to the epicentre, ground movements can be quite large. However, at large distances the movements are very small. Modern seismometers are instruments that are able to detect very small movements from earthquakes thousands of kilometres away.

In the 1960s, seismometers such as the one in Figure 4.14 began to be used. This one records vertical movements. It has a heavy mass inside, which is mounted on springs. The instrument is bolted firmly to the ground. When the ground moves the instrument moves with it, but the mass will 'try' to stay in its original position due to inertia. A coil inside the mass will register changes in the magnetic field and an electrical signal shows the movement of the mass and the instrument. This electrical output is digitised and datalogged by computer. Networks of seismic stations are linked in real time via satellites, to enable global monitoring of earthquakes. The output can be displayed visually in a record known as a seismogram. Seismic monitoring stations usually have three seismometers to measure oscillations in three directions at right angles to each other.

⚙ ACTIVITY 5 SEISMIC WAVES

Explore animations and recordings of earthquake waves that can be found on the internet. Make a summary of the key points shown by each animation. If you have the opportunity, download the Seismic Waves Software via **www.pearsonhotlinks.co.uk**.

Figure 4.14 Modern seismometer

Seismic alert

Since 2010, multiple earthquakes and aftershocks have hit Christchurch, New Zealand. Christchurch is the largest city on New Zealand's South Island and is home to over 300 000 people. The most devastating earthquake occurred on February 2011 and caused 185 deaths. It had magnitude 6.3 on the Richter scale, and occurred at a depth of around 5 km. The epicentre was just a few kilometres from the city centre and caused major damage in and around the central business district of the city (Figure 4.15). Six months earlier, an earthquake with magnitude 7.1 had struck the city. Although this earthquake did not directly cause any fatalities, it caused widespread damage to the city's buildings and infrastructure.

The lifestyle in Christchurch is very similar to that in a small UK city. Imagine what it would be like if buildings and services we rely on were destroyed overnight. There is a video available from Pearson hotlinks that shows the damage to streets, as well as the unsafe areas and abandoned buildings. One of the common problems was the destruction of effective sanitation and sewerage facilities. People were asked to make their own toilets in their gardens, and there was a competition for the best one.

Figure 4.15 Damage caused by the 2010 and 2011 earthquakes in Christchurch, New Zealand

ACTIVITY 6 SEISMIC ALERT

Use the Global Earthquake Explorer Software to explore seismograph data from a recent earthquake. Find data for two different recording stations near to the earthquake's epicentre on **www.pearsonhotlinks.co.uk**.

Write a paragraph to explain the differences between the traces you observe. What happens to the traces on the seismograph as the recording station gets further away from the epicentre?

Write a second paragraph describing how both traces would look for an earthquake of larger magnitude. What would stay the same and what would change?

Figure 4.16 Earthquake monitoring stations in the UK

Earthquake scales

There are two scales commonly used to characterise the strength of an earthquake: the Richter scale and the Mercalli scale. The Richter scale describes the earthquake itself, whereas the Mercalli scale describes its observable effects at locations where it is detected.

Richter scale

The Richter scale is summarised in Table 4.2.

Magnitude on Richter scale	Effects
0.0 to 1.9	Recorded by instruments
2.0 to 2.9	Felt by very sensitive people; suspended objects swing slightly
3.0 to 3.9	Felt by some people; vibration similar to that of a heavy vehicle
4.0 to 4.9	Felt by most people; hanging objects swing; dishes and windows rattle and may break
5.0 to 5.9	People frightened; chimneys topple, furniture moves
6.0 to 6.9	Buildings may suffer substantial damage
7.0 to 7.9	Few buildings remain standing; large landslides and fissures
8.0 to 8.9	Complete devastation; ground waves

Table 4.2 The Richter scale for earthquakes

MATHS REFERENCE

Logarithmic scales

See Maths notes 8.6 and 8.7

STUDY NOTE

You might like to look back at your work on logarithmic scales in *The Medium is the Message*.

Ground movements in an earthquake can be as large as several metres (Figure 4.17), but movements of only a few micrometres can be detected with modern instruments. With such a wide range of values, a linear scale would be inconvenient. The Richter scale is a logarithmic scale that is based on the amplitude of the P waves. On this scale, earthquakes are given a magnitude, defined such that a magnitude 3 earthquake has movements ten times that of a magnitude 2 earthquake, and so on. The greater the distance from the epicentre the less the ground will move, so the scale must take into account the distance from the epicentre.

Charles Richter (1900–1985) devised the scale that bears his name in the 1930s. He defined an earthquake of zero magnitude to be one that produced a zero-to-peak deflection on his seismometer of 0.001 mm at a distance of 100 km from the epicentre. A magnitude 1 earthquake would cause a disturbance with 10 times this amplitude, and a magnitude 6 earthquake would cause a disturbance a million times as big.

Figure 4.17 Ground subsidence in Alaska

As different seismometers produce different deflections for a given disturbance of the ground, each instrument has to be calibrated so that its output corresponds to the Richter scale.

Seismometers are very sensitive instruments, and the deflection of the instrument is generally much greater than the actual ground movement. For this reason, it is very difficult to measure a major earthquake close to its epicentre, and observations from very distant monitoring stations must be used.

Once the distance has been allowed for, each earthquake can be described in terms of a single magnitude. The largest recorded earthquake, with a magnitude of 9.5, occurred in Chile in 1960.

The Mercalli scale

The Mercalli scale uses the effects at the surface, as seen by people there at the time, to define the 'intensity'. Like the Beaufort scale for wind conditions, it is empirical – in other words, it is defined initially in terms of observable effects. Intensity will vary according to distance from the epicentre and the type of ground you are standing on. The Mercalli scale ranges from 1 – which is detected by instruments, but not felt by people except in special circumstances – to 12 – where there is almost total destruction, objects are thrown in the air, and waves are seen on the Earth's surface.

> ### ⚙ ACTIVITY 7 EARTHQUAKE DATA
>
> Use a resource such as the US Geological Survey's Earthquake Center, available via **www.pearsonhotlinks.co.uk**, to research a number of earthquakes and plot them on the Richter scale (log scale). Plot them according to the magnitude of the Earth movement resulting from each one (mm scale). You will find that the linear scale has to be so big that it will not fit around the room; a logarithmic scale is more convenient.

1.3 Summing up Part 1

This part of the chapter has been mainly revision of earlier work. You have reviewed some key ideas about waves, and have seen how the layered structure of the Earth affects the passage of seismic waves. You have also seen how logarithmic scales are used to describe earthquakes. Use the following activities to check your knowledge and understanding of waves in materials and of log scales.

> ### ⚙ ACTIVITY 8 EXAM HOWLERS
>
> In an exam, the question 'Why do P and S waves follow curved paths?' produced the following answers:
>
> **(a)** ...because the Earth spins and causes the waves to spin to the surface.
>
> **(b)** ...because the waves are affected by the gravity of the core.
>
> **(c)** ...because the magnetic core repels the seismic waves because they are of opposite charge.
>
> **(d)** ...because the waves cannot travel through a vacuum.
>
> For each answer, discuss and explain why it is incorrect. Then give the correct explanation in answer to the question.

> ### ⚙ ACTIVITY 9 VIRTUAL EARTHQUAKE
>
> Visit the Virtual Courseware for Earth and Environmental Science website via **www.pearsonhotlinks.co.uk**.
>
> Set off a virtual earthquake. Locate your earthquake and find its magnitude on the Richter scale.

STUDY NOTE

The software mentioned in Activity 5 includes information about the magnitudes of earthquakes measured on the Richter scale.

2 Shaken not stirred

Earthquakes are not the only cause of major building collapse. In 1940 the famous Tacoma Narrows Bridge in the USA (Figure 4.18) finally shook itself to bits. For many years previously, the fact that it would sway and twist violently when the wind blew up the river drew people from miles around to drive up and down this natural 'fairground' ride. You can see the collapse of the bridge for yourself on YouTube – search for 'Tacoma Narrows Bridge collapse'.

Figure 4.18 Tacoma Narrows Bridge

In fact all buildings will sway if the wind is blowing at the right speed, but taller ones are more impressive – the Empire State Building sways with a frequency of 1/8 Hz when the wind blows, while the old Severn Bridge vibrates at 1/7 Hz. Exactly how these wind-induced oscillations get started is quite complicated and depends on the complex pattern of air flow round a shaped beam.

In this part of the chapter, you will see how oscillations affect buildings and other structures; earthquakes may produce the oscillations, but there are other possible sources too. You will also see how one common type of oscillation can be analysed mathematically.

2.1 Free and forced oscillations

Free oscillations

Figure 4.19 shows a mass attached to one end of a spring whose other end is fixed in place. If you extend or compress the spring, the mass oscillates. If the mass rests on a smooth frictionless surface, these oscillations will persist indefinitely with constant amplitude as long as no other forces act. These **free oscillations** have a constant period known as the **natural period** of oscillation, T_0. For a mass m attached to a spring with stiffness k:

Figure 4.19 A mass atttached to a spring

$$T_0 = 2\pi\sqrt{\frac{m}{k}} \tag{3}$$

The **natural frequency** f_o is the frequency with which the system would oscillate if perturbed and no subsequent forces were to act. As frequency, f, is related to period:

$$f = \frac{1}{T} \tag{4}$$

We can obtain a formula for f_0:

$$f_0 = \frac{1}{2\pi}\sqrt{\frac{k}{m}} \tag{5}$$

The greater the mass or the less stiff the spring, the slower the oscillations – that is, the lower the natural frequency.

You may have seen that a pendulum also has a natural period of oscillation which depends upon its length l:

$$T_0 = 2\pi\sqrt{\frac{l}{g}} \tag{6}$$

and thus:

$$f_0 = \frac{1}{2\pi}\sqrt{\frac{g}{l}} \tag{7}$$

The longer the pendulum or the weaker the gravitational field (lower g), the lower the natural frequency will be. Note that the natural frequency of a pendulum is not dependent upon the mass of the bob.

Resonance rings a bell

Every structure has at least one **natural frequency** where it will go into violent oscillation with large amplitude when driven at this frequency. This phenomenon is called **resonance**, and engineers have to consider it very seriously when designing everything from skyscrapers to cars. A seated human body resonates at around 5 Hz. Car designers must ensure that the car body does not vibrate at this frequency, otherwise passengers would be shaken up quite badly by the end of the journey – regardless of the driving abilities of the driver. Examples of resonance include: a wineglass 'humming' and then shattering when vibrated by a pure note (although this is very difficult to do with modern glasses, as they are made to be much stronger); a flag flapping audibly; and parts of vehicles that suddenly start vibrating as the revs increase and then stop as the revs climb higher.

Resonance occurs when a structure is driven or forced to vibrate at its natural frequency and consequently absorbs maximum energy from the driving source. It is possible to drive structures to vibrate at other frequencies, but their response will be much less impressive (they vibrate with much smaller amplitude as much less energy is transferred).

The University of Salford's A level *Sounds Amazing* website (available via **www.pearsonhotlinks.co.uk**) is an excellent resource on the topic of resonance and includes a video showing the shattering of a wineglass by a pure note. You can use this website to explore the characteristics of simple harmonic motion and the causes and effects of resonance in different objects. Activities 10 and 11 study resonance in more detail.

⚙ ACTIVITY 10 WIND-INDUCED OSCILLATIONS

Explore the oscillations of a simple structure such as that shown in Figure 4.20. The structure is a horizontal beam of semicircular cross-section suspended between four springs. (A rectangular strip would do, provided it is fairly thin compared with its width.) The air from a fan or blower is directed on to it, and as the air speed is varied the beam leaps into violent vertical oscillations at a particular air speed. (Are there any other ways – 'modes' – in which you can see the structure vibrating, apart from those in a vertical plane?)

As an extension, devise some means of measuring air speed then measure the amplitude of vibrations produced at different speeds. Explore the effect of using different beams.

STUDY NOTE
You have used the idea of stiffness of an object in *Higher, Faster, Stronger, Good Enough to Eat* and *Spare Part Surgery*.

You met the relationship between period and frequency in *The Sound of Music* and *The Medium is the Message*.

STUDY NOTE
You will see how to derive Equations 3 and 5 in Section 2.2 of this chapter.

STUDY NOTE
You met gravitational field strength in *Higher, Faster, Stronger*.

Figure 4.20 Exploring wind-induced oscillations

PRACTICAL SKILLS REFERENCE

Scientific questions and information research

See Practical skills notes 1.1–1.3

Planning and experimental design

See Practical skills notes 2.1–2.4

Carrying out practical work

See Practical skills notes 3.1–3.2

Analysis and interpretation of data

See Practical skills notes 4.1–4.2

Conclusion and evaluation

See Practical skills notes 5.1–5.2

STUDY NOTE

You learned about phase difference in the chapter *The Sound of Music*.

⚙ ACTIVITY 11 BARTON'S PENDULUMS

Barton's pendulums (see Figure 4.21) are a set of pendulums of different lengths all attached to the same horizontal piece of string. One (the driver) is heavier than the rest. Predict and then observe what happens when you set the driver pendulum in motion.

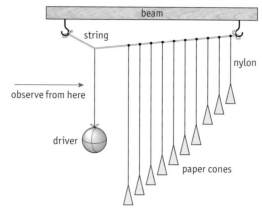

Figure 4.21 Barton's pendulums

Activities 10 and 11 demonstrate just how easy it is to set something oscillating at its natural frequency compared with other frequencies. In Activity 11 you will have seen that the pendulums first begin to oscillate with what appear to be their natural frequencies, but these **transient** oscillations die away and the pendulums settle down to oscillating with the same frequency as the driver, though with different amplitudes and phases. Activity 11 also shows how the phases of the driver and the driven oscillators are related. Perhaps surprisingly, at resonance, there is a phase difference of a quarter of a cycle – the two oscillators are said to be **in quadrature**.

Marching soldiers will be told to 'break step' when going over a bridge, just in case the rhythm of their marching matches the natural frequency of the bridge and sets it vibrating (see Figure 4.22). Anyone can set a rope bridge swinging by timing their bounce just right.

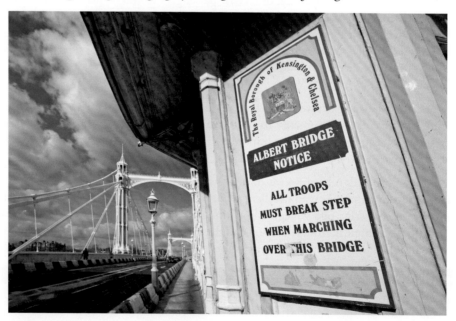

Figure 4.22 Break step when walking over a suspension bridge!

Resonance and civil engineering

Figure 4.23 shows a ruler held on to a table with one end overhanging the edge. If you push down the ruler's free end and release it, the overhanging length oscillates at its natural frequency. If you repeat the process with a shorter length overhanging, you will be able to hear that the natural frequency of oscillation is now higher.

Imagine that, instead of pushing and releasing the rule, you were able to shake the table. This is an example of a **forced oscillation** – you force the ruler to oscillate at the frequency of the shake. Now suppose you could you could shake the table at the ruler's natural frequency. The free end of the ruler would **resonate** – it would undergo very large amplitude oscillations. The greater the difference between the table oscillation frequency and the ruler's natural frequency, the smaller the amplitude of the ruler's oscillations, as shown in Figure 4.24. This graph shows a typical resonant peak. Applied to the example of the ruler, it shows that the amplitude of oscillation is greatest when the **driving frequency** (i.e. the frequency with which the table is shaken) is equal to the natural frequency of the free length of ruler.

Figure 4.23 Oscillating a ruler

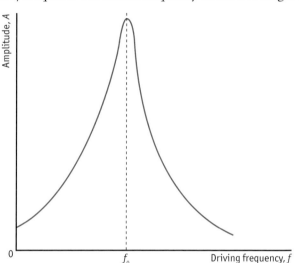

Figure 4.24 Amplitude of forced oscillations depends on frequency

In a similar way, a building has a natural frequency that depends upon its height above the ground and on other features of the structure (Figure 4.25 and Table 4.3). If an earthquake shakes the ground at a frequency that happens to be equal to the building's natural frequency, it will respond with very large amplitude oscillations and damage may occur as a result.

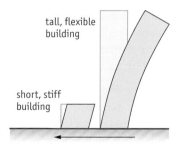

Figure 4.25 A building's response to an earthquake depends on its height

Number of storeys	Typical natural period/s
2	0.2
5	0.5
10	1.0
20	2.0
30	3.0
50	5.0

Table 4.3 A building's natural period of oscillation depends on its height

STUDY NOTE

A building may be able to oscillate in several different ways (known as modes of oscillation). The fundamental mode, which involves all parts of the building oscillating in phase, is the one with the lowest frequency. You met this idea when you studied stationary waves in *The Sound of Music*. In our discussion of buildings, we are only considering the fundamental mode.

In 2011 an earthquake measuring 9.0 on the Richter scale struck 70 km off the coast of Japan. As well as the direct damage caused by the earthquake, it gave rise to a devastating tsunami, which led to many deaths and damaged the Fukushima nuclear power plant, causing the release of radioactive material. This was the most powerful earthquake ever recorded in the region (the fourth largest globally) and is considered to be the costliest; almost 16 000 people lost their lives and the estimated costs of repairs in Japan and nearby countries reached over £100 billion. Surprisingly, despite the energy released during the earthquake, the Taipei 101 building in Taiwan (Figure 4.26) barely moved, despite other nearby buildings being affected. The 500 m tall Taipei 101 skyscraper has a natural frequency of approximately 0.15 Hz. Research into the phenomenon noted that the frequencies of seismic waves recorded at the building were in the range 0.015–0.1 Hz. The building design incorporated 'tuned mass dampers' that reduced the effect of the vibrations – as you will see in Section 3.2.

Many major engineering disasters, including those involving bridges, buildings and turbine blades in marine and aero engines, have been attributed to resonance. What is interesting, though, is that – with the notable exception of London's Millennium Bridge, which, until it was re-regineered, wobbled in time to the steps of those walking on it – it is difficult to find recent examples of such failures. So, are we getting better at design? It does seem that experiment, together with theoretical and computer analysis, is increasing engineers' understanding of oscillations and resonance, and that this understanding is feeding into better design practice. However, complacency could be disastrous; the problems with the Millennium Bridge occurred despite thorough testing and computer modelling. Engineers have to establish the natural frequencies of any structure they propose to build to ensure that it will not be set into violent oscillation, especially if the structure is in an earthquake zone. Studying how structures oscillate is an important part of this.

During earthquakes in cities, some buildings are destroyed, while others in the same district seem barely affected. One of the reasons for this effect is resonance.

Figure 4.26 The 500 m tall Taipei 101 building barely moved in the powerful 2011 earthquake

⚙ ACTIVITY 12 EARTHQUAKES AND RESONANCE IN BUILDINGS

Make and study some simple models that illustrate the effect of resonance in buildings.

Seismometers

In Activity 13 you will explore resonance further and see how it comes into play when designing a seismometer.

A seismometer is an example of a **forced oscillator**, whereby a system capable of undergoing free oscillations is driven from the outside by some external force that itself oscillates at a frequency called the **driver frequency**, f_d. For a seismometer, the earthquake itself constitutes the driver. There is further slight complication in that an earthquake is unlikely to have a single frequency; there will be a spectrum of frequencies. However, in the simple model that you are going to use, it will be assumed that an earthquake can be modelled by a single frequency.

A simple seismometer consists of a mass hanging vertically on a spring (Figure 4.27). A vibrator at fixed amplitude drives the top of the spring from a variable-frequency generator. The generator–vibrator combination simulates a variable frequency earthquake. The design has the two essential ingredients for an oscillation to occur: mass and stiffness. How likely is it in practice that the mass will stay still and not oscillate? One way of re-stating that the mass should be large in a 'good' seismometer is to say that the natural frequency should be low; but what does 'low' mean? What can we compare it with?

⚙ ACTIVITY 13 MAKE AND MARKET A SEISMOMETER

Make and test a simple seismometer as shown in Figure 4.27a.

When you have made and tested your instrument, imagine that you represent a company that markets seismometers. Prepare a sales presentation to convince a client that your instrument has features that will make it work well.

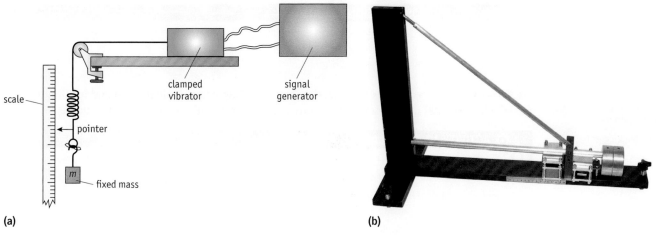

(a)   **(b)**

Figure 4.27 (a) Simple seismometer (b) the SEP (science enhancement programme) vibration detector

A system that oscillates in response to external vibrations is the beginning of a seismometer, but to record those vibrations electronically requires the use of electromagnetic induction.

⚙ ACTIVITY 14 THE EARTH *DID* MOVE

Study the output of a simple vibration detector, such as the SEP (science enhancement programme) vibration detector, and use it to extend your understanding of real seismometers.

⚙ ACTIVITY 15 FURTHER EXAMPLES OF RESONANCE

Discuss, or make notes on, the following questions.

- Machinery is normally operated at speeds well above any possible resonances. What problems might be presented, though, when the machinery is started up or shut down? (This can be a particular problem for ships with a very long transmission shaft from engine to propeller.) What instructions might you give to an operator?

- What does a garage always do – and charge for – when a tyre is changed and why?

- It is popularly supposed that a singer can shatter a wineglass by singing loudly at a certain frequency. Has this ever actually been done? What is the physics involved?

- Extension: Electrical systems can also resonate when driven by an alternatiing voltage of the right frequency. Find out how a radio receiver makes use of this.

2.2 A useful type of oscillation

To ensure that the effects of resonance are controlled, engineers must understand the fundamentals of vibration. We describe vibration, and any kind of motion that repeats itself, such as a simple pendulum or a mass bouncing on a spring, as a **harmonic motion**. There is a particular form of harmonic motion, called **simple harmonic motion** (SHM), which has the following very important properties:

- the period (or frequency) is independent of the **amplitude** (maximum displacement) of the motion

- the force is always directed towards the central point of the oscillation: if displacement is positive the force is negative, and vice versa. Mathematically this can be expressed as:

$$F = -kx \qquad\qquad (8)$$

where F and x are the magnitudes of the force and displacement, respectively, and k is a constant.

The second property means that the force – and thus the acceleration – must be zero as the system passes through its central position. This is therefore the **equilibrium position**: the one to which the system gradually returns as the oscillations die away. Because of this change in direction of force either side of equilibrium, it is usually called the **restoring force** and is related to the 'stiffness', k, of the system. At the equilibrium position there is no resultant force, so the system continues to move at constant velocity.

PRACTICAL SKILLS REFERENCE

Planning and experimental design

See Practical skills notes 2.1–2.4

Carrying out practical work

See Practical skills notes 3.1–3.2

ACTIVITY 16 OSCILLATION CIRCUS

Observe a variety of oscillating systems and decide whether each exhibits simple harmonic motion. Try to identify what provides the inertia and what provides the stiffness of the system. Some systems require detailed measurements, while others simply require observation.

As Activity 17 shows, there is a connection between simple harmonic motion and motion in a circle. This provides us with a useful way to analyse SHM mathematically.

ACTIVITY 17 CIRCULAR MOTION AND SHM

Observe the shadow of a mass undergoing SHM on the end of a spring alongside the shadow of a ball rotating in a vertical circle, as in Figure 4.28.

Hold your arms out in front of you with your index fingers pointing forwards and level with each other. Set your left hand moving anticlockwise at a steady speed and move your right hand vertically so that your index fingers stay at the same horizontal level. While your left hand executes a circle, the right is executing SHM.

You have just demonstrated that SHM and circular motion are closely related. Make sure you are familiar with the terms angular velocity, period, centripetal force and acceleration, and the relationships between them. Also make sure you are familiar with angular displacements expressed in radians.

A quick Internet search reveals many websites that go over this in more detail and provide animations to back up your work. For example, see the University of New South Wales website (available via **www.pearsonhotlinks.co.uk**).

STUDY NOTE

Look back at your work on circular motion in *Probing the Heart of Matter*.

Figure 4.28 Shadows of SHM and circular motion

Analysing SHM

Figure 4.29 shows an object, mass m, moving with angular velocity ω (measured in radians per second) in a circle of radius A. As was demonstrated in Activity 17, its 'shadow' cast on the x-axis performs simple harmonic motion; to describe its motion we need expressions for the x component of displacement, the velocity and so on.

MATHS REFERENCE

Angular measurements

See Math note 6.1

MATHS REFERENCE

Sine, cosine and tangent of an angle

See Maths note 6.2

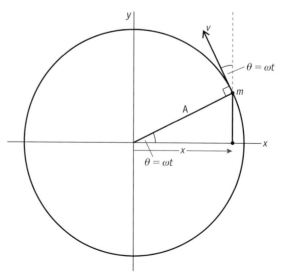

Figure 4.29 SHM as a projection of circular motion

The circular motion also enables us to deduce the period, T, of the SHM. The time to make one complete oscillation (there and back) is the same as the period of the circular motion; that is:

$$T = \frac{2\pi}{\omega} \tag{9}$$

Notice that the period, T, is independent of the amplitude, A; this is the other key property of SHM. Using Equation 4, we can also write an expression for frequency:

$$f = \frac{1}{T} = \frac{\omega}{2\pi} \tag{10}$$

Suppose the amplitude of the oscillating motion is A. For circular motion with an angular velocity ω, the angular displacement after time t is ωt, and so from Figure 4.29 the horizontal displacement is:

$$x = A\cos\omega t \tag{11}$$

The maximum displacement is when $\cos\omega t = \pm 1$; in other words, $x_{max} = \pm A$.

In circular motion, velocity is given by:

$$v = A\omega \tag{12}$$

at an angle ωt to the y-axis in Figure 4.29. Here we need the horizontal component:

$$v_x = -A\omega\sin\omega t \tag{13}$$

The negative sign shows that the mass is moving in the negative direction; that is, back towards its equilibrium position. The x-component of velocity (the speed in the x direction) is greatest when $\sin\omega t = \pm 1$. Therefore:

$$v_{max} = A\omega \tag{14}$$

For the circular motion, the force acting on m always has magnitude:

$$F = \frac{mv^2}{A} = m\omega^2 A \tag{15}$$

and is directed towards the centre (it is a *centripetal* force). The x component is thus:

$$F_x = -m\omega^2 A\cos\omega t \tag{16}$$

Comparing Equations 11 and 16, we have:

$$F_x = -m\omega^2 x \tag{17}$$

Since m and ω are both constants, Equation 17 is equivalent to Equation 8 and describes a key property of SHM – the restoring force is proportional to displacement. Equation 17 is often expressed in terms of the acceleration:

$$a_x = \frac{F_x}{m} = -\omega^2 A\cos\omega t \tag{18}$$

Thus:

$$a_x = -\omega^2 x \tag{19}$$

The quantitiy ω is important and useful when describing all types of oscillations, not just mechanical oscillations that can be analysed in terms of a related circular motion. It is usually referred to as the **angular frequency** of an oscillation.

STUDY NOTE

For a derivation of the equations of SHM using calculus, see BLD Additional Sheet 2.

STUDY NOTE

You met Equation 9 in connection with circular motion in *Probing the Heart of Matter*.

Figure 4.30 shows graphs of the variation of displacement x, velocity v and acceleration a with time for a simple harmonic oscillator. We have dropped the subscript x because we are now dealing with motion in just one dimension.

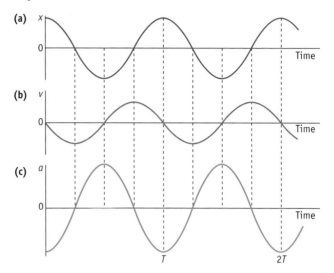

Figure 4.30 Graphs of (a) x versus t, (b) v versus t, and (c) a versus t for an object executing SHM

From the displacement graph (a) we can see that:

● the displacement starts at its maximum positive value;

● a quarter of the way through the oscillation, the displacement is zero;

● halfway through the cycle, the displacement reaches its maximum negative value;

● the displacement reaches zero again when we are three-quarters of the way through the cycle;

● after one full cycle, the displacement is back to its original maximum positive value.

The velocity graph (b) shows that:

● maximum negative velocity occurs one-quarter of the way through the cycle;

● maximum positive velocity occurs three-quarters of the way through the cycle.

These are the times when the displacement is zero.

● the velocity is zero at the beginning of the cycle, halfway through the cycle and at the end of the cycle.

These are the times when the displacement is a maximum.

The acceleration graph (c) shows that:

● maximum positive acceleration occurs halfway through the cycle;

● maximum negative acceleration occurs at the beginning and end of the cycle;

● acceleration is zero one-quarter and three-quarters of the way through the cycle.

Notice that the acceleration is zero when displacement is zero, and that maximum *positive* acceleration occurs at the same times as maximum *negative* displacement (and vice versa), as described by Equation 19.

⚙ ACTIVITY 18 SUMMARISING SHM

Complete a copy of Table 4.4 to summarise the key features of SHM. In each cell of the table, say whether the relevant quantitiy is zero or takes its maximum positive or negative value.

	Displacement $x = A\cos\omega t$	Velocity $v = -A\omega \sin\omega t$	Acceleration $a = -\omega^2 A\cos\omega t$
$t = 0$	maximum positive	0	maximum negative
$t = \dfrac{T}{4}$	0		
$t = \dfrac{T}{3}$			
$t = \dfrac{3T}{4}$			
$t = T$			

Table 4.4 Summarising SHM

⚙ ACTIVITY 19 SPRINGY OSCILLATIONS

Use Equations 18 and 19, and the worked example below, to predict the frequency of a mass oscillating on various combinations of springs (Figure 4.31a). Then compare your predictions with experimental measurements of frequency.

Produce a time-trace for the apparatus shown in Figure 4.31b (or similar) and compare the results with the theoretical Equations 9, 11 and and 13.

PRACTICAL SKILLS REFERENCE

Carrying out practical work

See Practical skills notes 3.1–3.2

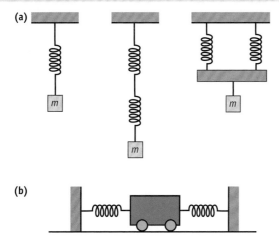

Figure 4.31 Apparatus for Activity 19

A useful recipe

The above theoretical treatment of SHM provides a very useful 'recipe' for analysing the motion of an oscillator. Simply carry out the following steps.

1   Is the restoring force proportional to displacement? If so, then the motion is simple harmonic. (If it is not, the motion is not SHM and you cannot proceed with this recipe.)

2   Since restoring force is proportional to displacement, then acceleration must also be proportional to displacement. Find the constant that relates displacement to acceleration. Sometimes it is easy to find this constant directly; sometimes it is easier to find the relationship between displacement and restoring force and then divide by mass.

3   The constant you have found in Step 2 is equal to ω^2 in Equation 19. Find its square root. You now know the value of ω for your oscillator.

4   Using your value of ω, you can now work out the period of the oscillation using Equation 9. If you know the amplitude, you can also find the displacement, velocity and acceleration at any given time using Equations 11, 13 and 18.

WORKED EXAMPLE

Q A mass, m, of $0.1\,kg$ oscillates on the end of a spring that obeys Hooke's law and has a stiffness, k, of $50\,N\,m^{-1}$. Is the motion simple harmonic? If so, what is its period?

A Since the spring obeys Hooke's law, $F = -kx$, and so the motion must be SHM:

$$a = \frac{-kx}{m} = -\left(\frac{k}{m}\right)x$$

so:

$$\omega^2 = \frac{k}{m}$$

and:

$$\omega = \sqrt{\frac{k}{m}} = \sqrt{\frac{50\,N\,m^{-1}}{0.1\,kg}} = 22.36\,s^{-1}$$

$$T = \frac{2\pi}{\omega} = 0.28\,s \text{ (to 2 significant figures)}$$

Q Suppose the mass in the previous question is displaced by $3.0\,cm$ from its equilibrium position, then released.

(a) What is its maximum speed?

(b) What are its displacement and velocity after $0.05\,s$?

A We need to use Equations 11 and 13, and so we need to know values of A and ωt. Since $3.0\,cm$ is the maximum displacement, the amplitude, $A = 3.0\,cm$.

(a) By inspection of Equation 13, speed is maximum when $\sin\omega t$ is greatest; that is, when $\sin\omega t = 1$, therefore:

$$v_{max} = A\omega = 3.0\,cm \times 22.36\,s^{-1} = 67\,cm\,s^{-1} \text{ (to 2 significant figures)}$$

(b) After $0.05\,s$, $\omega t = 0.05\,s \times 22.36\,s^{-1} = 1.118$

Since ω can be interpreted as an angular velocity in radians per second, ωt can be interpreted as an angle expressed in radians:

$$\cos\omega t = \cos 1.118 = 0.1834$$

From Equation 11:

$$x = A\cos\omega t = 3.0\,cm \times 0.1834 = 0.55\,cm \text{ (to 2 significant figures)}$$

$$\sin\omega t = \sin 1.118 = 0.8992$$

From Equation 13:

$$v = -A\omega\sin\omega t = -3.0\,cm \times 22.36\,s^{-1} \times 0.8992$$

$$= -60\,cm\,s^{-1} \text{ (to 2 significant figures)}$$

QUESTIONS

Q 8 **(a)** Express the units of the quantities in Equation 3 in terms of their SI base units and in this way show that the right-hand side has SI units of seconds.

An object supported by a 'Hooke's law' spring oscillates with SHM and has $A = 0.30\,m$ and $\omega = 5.0\,s^{-1}$.

(b) What are:
  (i) its maximum speed?
  (ii) the frequency of the oscillations?

(c) The object has mass $0.40\,kg$. What is the stiffness of the spring?

(d) After 0.20 seconds, what are:
  (i) the displacement?
  (ii) the acceleration?

Q 9 (a) An ultrasonic oscillator has a frequency of 50 kHz and amplitude of 2×10^{-6} m. What are the maximum values of **(i)** speed and **(ii)** acceleration?

(b) Expressing the acceleration as a multiple of g (= $9.81\ \mathrm{m\,s^{-2}}$), do you think the value has any implications for the mechanical strength of the oscillating part?

(c) By irradiating gallstones in the body with ultrasonic vibrations, they can often be broken up. Why is increasing the frequency more effective than increasing the amplitude in this operation?

Simple pendulum

A simple pendulum (Figure 4.32) formed by a mass hanging from a string oscillating through a *small* angle obeys simple harmonic motion. To show that this is the case we need to find the restoring force acting on the mass and relate that to the displacement.

When a pendulum of length l is at an angle θ to the vertical, the horizontal displacement, x, of the bob is:

$$x = l \sin\theta \tag{20}$$

There are two forces acting upon the mass as it moves: its weight acting down with a magnitude of mg, and the tension in the string. When the pendulum is at an angle θ to the vertical, the vertical component of the tension is $T\cos\theta$, and by equating the vertical forces we can write:

$$T\cos\theta = mg \tag{21}$$

The horizontal restoring force, F, is the horizontal component of the tension:

$$F = -T\sin\theta \tag{22}$$

The minus sign signifies that the force is directed back towards the equilibrium position.

Substituting for T from Equation 21:

$$F = -\frac{mg\sin\theta}{\cos\theta}$$

$$= mg\tan\theta \tag{23}$$

If the angle θ is small, then with θ in radians:

$$\sin\theta \approx \tan\theta \approx \theta \tag{24}$$

So, at small enough values of θ, Equations 20 and 23 become:

$$x = l\theta \tag{25}$$

and:

$$F = -mg\theta \tag{26}$$

We can combine Equations 25 and 26 to produce:

$$F = -\frac{mg}{l}x \tag{27}$$

So we have shown that, *at small angles*, the force is proportional to the displacement from equilibrium. Comparing Equation 27 with Equation 19, we can see that:

$$-\frac{mg}{l}x = -m\omega^2 x \tag{28}$$

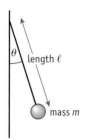

Figure 4.32 Simple pendulum

MATHS REFERENCE

Degrees and radians
See Maths note 6.1

Sine, cosine and tangent of an angle
See Maths note 6.2

The small angle approximations
See Maths note 6.6

Therefore we can identify a value for the angular frequency and write:

$$\omega^2 = \frac{g}{l}$$

$$\omega^2 = \sqrt{\frac{g}{l}} \tag{29}$$

The 'useful recipe' set out above then leads to the expressions for the pendulum's natural period and frequency that we stated earlier:

$$T_0 = 2\pi\sqrt{\frac{l}{g}} \tag{Equation 6}$$

$$f_0 = \frac{1}{2\pi}\sqrt{\frac{g}{l}} \tag{Equation 7}$$

QUESTION

Q 10 A simple pendulum has a length of 0.60 m and oscillates with SHM with amplitude of 5.0 cm.

(a) What is:

    (i) the angular frequency, ω, of the oscillations?

    (ii) the period?

(b) What is the maximum speed of the pendulum bob?

(c) The same pendulum is now pulled to an angle of 60° from the vertical and released. The period is carefully measured and found to be different from your answer to **(a) (ii)**. Explain why this would be expected.

⚙ ACTIVITY 20 A SIMPLE AND A LESS-SIMPLE PENDULUM

Explore how the horizontal restoring force acting on the bob of a simple pendulum depends on angle. In this way explore the limitations of Equations 6 and 7.

Use a computer simulation to explore the factors affecting the period of a simple pendulum. Then apply your understanding as you investigate a less-simple pendulum.

Energy in SHM

The previous analysis can be extended to consider the energy of a simple harmonic oscillator. The general principles can be illustrated by considering a mass oscillating horizontally on the end of a light Hooke's law spring, resting on a frictionless surface (Figure 4.33) – the only energies that change are the kinetic energy of the mass and the potential energy due to the stretched spring.

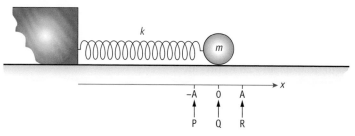

Figure 4.33 Simple harmonic oscillator

Suppose the mass is held at point P, where energy E is stored in the compressed spring, and then released. As the mass passes through its equilibrium position, Q, the spring is neither compressed nor stretched, so stores no energy. At this point the mass has kinetic energy E. At Q, the mass has its greatest speed and therefore its greatest kinetic energy, but the potential energy is momentarily zero. As the mass moves from Q to R it slows down, then at R it is again momentarily at rest, and energy E is stored in the stretched spring. At all positions other than P, Q and R, the system has both kinetic and potential energy.

There is a clear relationship between the kinetic and the potential energy of the system: their sum total is constant, as required by energy conservation. Thus any increase in one must correspond to a decrease of the same amount in the other.

The kinetic energy, E_k, of the mass, m, is related to its speed, v, by the familiar equation:

$$E_k = \frac{1}{2}mv^2 \tag{30}$$

By squaring Equation 13 to obtain an expression for v^2 for the oscillator, we can get an expression showing that kinetic energy of the oscillating mass varies with time:

$$E_k = \frac{1}{2}mA^2\omega^2\sin^2\omega t \tag{31}$$

The (elastic) potential energy, E_p, stored in the stretched spring with stiffness k, is given by:

$$E_p = \frac{1}{2}kx^2 \tag{32}$$

STUDY NOTE

You met and derived Equation 32 in the chapter *Higher, Faster, Stronger*.

By squaring Equation 11 to get an expression for x^2, we can see how the potential energy of the spring changes with time:

$$E_p = \frac{1}{2}kA^2\cos^2\omega t \tag{33}$$

The relationship between Equations 31 and 33 becomes more apparent if we note that $\omega^2 = \frac{k}{m}$, and so Equation 31 becomes:

$$E_k = \frac{1}{2}kA^2\sin^2\omega t \tag{34}$$

The total energy, E_{tot}, of the system is found by adding Equations 33 and 34:

$$E_{tot} = E_p + E_k$$

$$E_{tot} = \frac{1}{2}kA^2(\cos^2\omega t + \sin^2\omega t)$$

$$E_{tot} = \frac{1}{2}kA^2 \tag{35}$$

because $\cos^2\omega t + \sin^2\omega t = 1$, irrespective of the value of ωt. In other words, the **total energy of a simple harmonic oscillator** is constant. This is true for *all* SHM systems, not just for the simple one we have chosen to study.

MATHS REFERENCE

Sine, cosine and tangent of an angle

See Maths note 6.2

The energies can also be expressed in terms of m and ω rather than k and A, in which case Equation 35 becomes:

$$E_{tot} = \frac{1}{2}m\omega^2A^2 \tag{36}$$

STUDY NOTE

For further discussion of energy in SHM, see BLD Additional Sheet 3.

Figure 4.34 shows graphs of kinetic and potential energy plotted against time and against displacement.

(a)

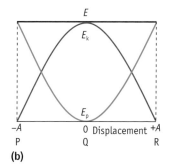

(b)

Figure 4.34 Variation of E_k and E_p for a mass oscillating on a spring as a function of (a) time and (b) displacement

QUESTIONS

Q 11 Calculate the total energies of **(a)** the oscillator considered in the previous worked example and **(b)** the oscillator in Question 8. In each case, show that Equations 35 and 36 give the same answer.

Q 12 **(a)** A spring of stiffness $1.0 \times 10^2 \, \text{N m}^{-1}$ is stretched by 0.15 m. What is its elastic potential energy?

(b) The spring is used (at this stretch) to catapult a small missile of mass 2.0 g. At what speed does the missile leave the catapult? What assumption did you have to make?

(c) Using appropriate symbols k, x and m (mass of missile), turn the reasoning of **(b)** into a general theoretical argument to produce an algebraic expression for the launch speed v. Show that the launch speed is proportional to the initial stretch.

(d) Without completely re-working the problem, find the launch speed of:
   (i) the same missile for an initial extension of 0.05 m
   (ii) a missile twice as massive launched using the original extension of 0.15 m.

2.3 Earthquake waves revisited

Knowing something about SHM, we can now return to the question of waves in a solid material. As you saw in Activity 4, a longitudinal wave in a solid can be modelled with a row of trolleys, as in Figure 4.35. Each trolley (or atom in a solid) is held to its neighbours by springy connections, so you can consider each one as an oscillator similar to that in Activity 19. We can use the equations of SHM to deduce the speed of a compression wave along the row of oscillators.

Figure 4.35 Compression pulse travelling along a row of oscillators

Imagine pushing the oscillator at the end of the row. The spring joining it to its neighbour becomes compressed, exerting a force on the next oscillator and setting it in motion so that it passes on the force to its neighbour – and so a compression pulse travels along the row with a time lag between one oscillator and the next. If s is the distance from one oscillator to the next, and τ is the time interval, then the wave travels at speed v where:

$$v = \frac{2}{\tau} \tag{37}$$

If you assume that each successive oscillator is only affected once its neighbour has reached its maximum displacement, then the time taken for the pulse to be 'handed on' will be one-quarter of a cycle. In fact, the time interval is rather shorter – only $\frac{1}{2\pi}$ of a cycle:

$$\tau = \frac{T}{2\pi} \tag{38}$$

STUDY NOTE

Look back at Activity 19 and the worked example in Section 2.2.

We know that for an oscillator mass m held by springs of stiffness k:

$$T = 2\pi \sqrt{\frac{m}{k}} \tag{Equation 3}$$

Therefore we now have an expression for the speed at which the pulse travels along the row:

$$\tau = \sqrt{\frac{m}{k}}$$

So:

$$v = s \sqrt{\frac{k}{m}} \tag{39}$$

Waves in a solid

The arrangement in Figure 4.35 can be thought of as modelling a row of atoms, and we can use Equation 39 to deduce the speed of longitudinal waves in a solid. Figure 4.36 shows a three-dimensional model of a solid in which atoms are held by spring bonds that, for small extensions and compressions, obey Hooke's law.

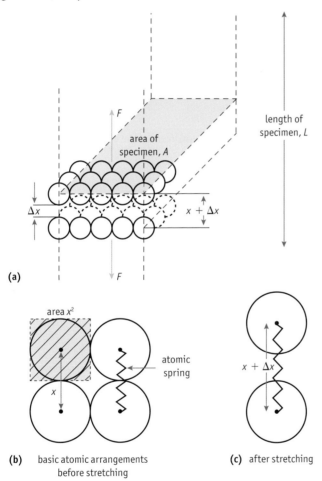

Figure 4.36 Three-dimensional model of a solid

For each bond we can write:

$$\Delta f = k\Delta x \qquad (40)$$

where Δf is the force and Δx is the resultant extension. k is the stiffness of each bond. We can relate this inter-atomic k to the Young modulus of the metal.

Each atom in Figure 4.36 occupies an area x^2. If the cross-sectional area of the sample is A then the number of atoms in the layer is $\dfrac{A}{x^2}$ and this is the number of springs being stretched.

Thus the total force F is given by:

$$F = \frac{Ak\Delta x}{x^2} \tag{41}$$

which leads to an expression for stress:

$$\sigma = \frac{k\Delta x}{x^2} \tag{42}$$

We can also write down an expression for strain:

$$\varepsilon = \frac{\Delta x}{x} \tag{43}$$

STUDY NOTE

Look back at your work on the chapter *Spare Part Surgery* for a reminder about stress and strain.

Combining Equations 29 and 30 leads to an expression for the Young modulus:

$$E = \frac{k}{x} \tag{44}$$

Putting $s = x$ and $k = Ex$ in Equation 32 we get a new expression for the wave speed:

$$v = x\sqrt{\frac{Ex}{m}} = \sqrt{\frac{Ex^3}{m}} \tag{45}$$

But as x^3 is the volume each atom occupies, then the density ρ is:

$$\rho = \frac{m}{x^3}$$

Therefore we can write the equation that you met in Part 1 of this chapter:

$$v = \sqrt{\frac{E}{\rho}} \tag{Equation 1}$$

Although we assumed a square arrangement of atoms, Equation 1 applies to longitudinal waves in any solid, regardless of the way its atoms are arranged.

⚙ ACTIVITY 22 WAVE SPEED IN A SOLID

Measure the speed of a longitudinal wave in a steel or aluminium rod. The principle is the same as that used in Activity 21 – the rod behaves like the row of trolleys and so the contact time between rod and hammer is equal to the there-and-back travel time for a longitudinal pulse.

Compare your result with the speed calculated using Equation 1 and the data given below.

Aluminium: $E = 7.0 \times 10^{10}\,\text{N m}^{-2}$, $\rho = 2.7 \times 10^3\,\text{kg m}^{-3}$

Steel: $E = 2.0 \times 10^{11}\,\text{N m}^{-2}$, $\rho = 7.8 \times 10^3\,\text{kg m}^{-3}$

PRACTICAL SKILLS REFERENCE

Carrying out practical work

See Practical skills notes 3.1–3.2

Table 4.1 (in Part 1 of this chapter) lists data for density and the speed of P waves in various materials. Seismologists use such data, derived from calculations and laboratory measurements, to help them interpret observations of seismic waves and in this way make deductions about the internal structure of the Earth. As you will see in Part 3 of this chapter, a knowledge of wave speeds in materials also helps engineers to design buildings that are safe and comfortable for the occupants, and machines that are safe to operate.

QUESTIONS

Q 13 (a) Figure 4.37 shows some results obtained in Activity 22. On the screen trace shown in Figure 4.36, the contact time between a metal bar and the hammer that hit it was estimated at 512 µs. The bar was 0.70 m long. What was the speed of longitudinal waves in the bar?

(b) The trace after 512 µs appears to be exponential. What does that suggest about the circuitry within the computer interface?

x = –165 μs, o = 347 μs, xo = 512 μs

24Apr1999 11:56

Figure 4.37 Results from Activity 22

MATHS REFERENCE

SI units

See Maths notes 2.1, 2.2 and 2.3

Q 14 **(a)** Use SI base units to show that the right-hand side of Equation 1 has unit $m\,s^{-1}$.

Table 4.5 lists the speed of longitudinal waves in various common materials along with their densities.

(b) Without doing any calculations, what can you say about the Young modulus of oak compared with that of pine, and the Young moduli of metals compared with those of wood?

(c) Calculate the Young moduli of **(i)** pine and **(ii)** copper.

Material	Density, $\rho/kg\ m^{-3}$	Wave speed, $v/m\ s^{-1}$
pine	500	3313
oak	700	3837
aluminium	2698	5100
steel	7800	5060
copper	8933	3650
lead	11343	1230

Table 4.5 Densities and longitudinal wave speeds in various materials

2.4 Build it better – Summing up Part 2

The physics that you have studied in this part of the chapter is the physics that enables an engineer to design a building for an earthquake zone. You have seen how an oscillator can be driven to oscillate by an external force, and that if the frequency of the driving force matches the oscillator's natural frequency, then resonance occurs and oscillations build up to large amplitude. You then went on to study and analyse simple harmonic motion – a type of motion that occurs in many natural and man-made systems, and which underlies many more complex oscillatory motions.

Figure 4.38 shows a model structure being tested on an earthquake table in an engineering laboratory. As the 'Earth' shakes, the building vibrates. It is the engineer's task to ensure that the amplitude of the vibrations remains fairly small, and also to choose materials that will withstand deformation. In Activity 23, you will apply your knowledge of resonance and simple harmonic motion to the matter of limiting the vibrations. (In Part 3 you will return to the question of deformation.)

Figure 4.38 Testing a model structure on an earthquake table

⚙ ACTIVITY 23 EARTHQUAKES IN THE LABORATORY

The behaviour of buildings in earthquakes was explored initially in Activity 12. The models used in this activity will exhibit more complex modes of vibration.

As you work through this activity, take the opportunity to look back through Part 2 of the chapter and make sure you understand the meaning of the key terms printed in **bold**.

Make some simple model building structures (Figure 4.39). By assuming that each model vibrates like a mass on a spring, calculate its natural frequency of vibration. Test each model by vibrating it on an earthquake table.

Figure 4.39 Typical models for use in Activity 23

⚙ ACTIVITY 24 RESONANCE OF A VIBRATING MASS

Explore the resonance of a mass-and-spring system driven by a vibration generator (Figure 4.40). Explore how resonant frequency depends on mass, and in this way determine the value of an unknown mass from its resonant frequency.

Figure 4.40 Vibrating a mass-and-spring system

QUESTIONS

Q 15 The natural vibration frequency of a building can be dramatically reduced (to remove it from typical earthquake frequencies) by mounting the whole building on rubber bearings. The building then oscillates as a single mass without bending.

(a) A 40 tonne model had a natural frequency of horizontal oscillations of 0.60 Hz when mounted on rubber. What is the horizontal stiffness of the bearings? (1 tonne = 1000 kg)

(b) The same model has a vertical natural frequency of 10 Hz.
   (i) Using the fact that frequency squared is proportional to stiffness, use your answer to (a) to find the vertical stiffness of the bearings.
   (ii) Through what distance will the weight of the model compress the bearings? (Use $g = 9.81$ N kg^{-1}.)

(c) On the model *without* rubber bearings, the maximum acceleration at third floor level was measured to be 12 m s^{-2} at a frequency of 3.0 Hz.
   (i) What is the displacement amplitude?
   (ii) The building can be thought of as bending on a fixed base (like your model in Activity 24). If the model is one-third linear scale, what would be the amplitude of oscillation of the real building? Would you find this alarming?

Q 16 The model in Question 15 is built to a one-third linear scale using the materials that would be used in the actual building.

(a) What would be the mass of the actual building?

(b) What would have to happen to the stiffness of the real building rubber bearings compared to that of the model ones if the natural frequency of horizontal oscillation were to stay the same?

(c) Why are vertical earthquake oscillations likely to cause less damage than horizontal ones?

Q 17 Figure 4.41 shows a pendulum made up of a spring of natural length $l = 0.40$ m attached to a string (which doesn't significantly stretch) of length $L = 0.20$ m. A mass is hung on the spring, which stretches it by 0.20 m.

(In your answer use $g = 9.81$ m s^{-2}.)

(a) What is the frequency of the 'springy' oscillation?

(b) What is the frequency of the pendulum oscillations – the sideways swinging of the whole arrangement?

(c) What is the ratio of these two frequencies? How could you have written down the ratio straight away?

(d) When this arrangement is set oscillating as a pendulum, it starts to 'go wild'. Suggest how your answer to (c) might help to explain this. If you have a moment it is worth setting this up and playing with it – it has some interesting properties. The actual values are not important; just adjust L for any mass and spring until you get the ratio of lengths you need (you will know when you're there).

Figure 4.41 Complex pendulum

Q 18 The top of a car's wheel arch is 50 cm from the ground when the car is unloaded. A person of weight 700 N sits on the bonnet above the wheel, and the distance from the ground to the top of the wheel arch is now measured to be 47 cm.

(a) Calculate the spring constant of the suspension system.

(b) Given that the mass supported by each wheel is 400 kg, calculate the system's natural frequency of oscillation.

(c) The driver notices that at a speed of 70 km h^{-1} the car oscillates significantly as it passes over humps in the road. Calculate how far apart the undulations in the road are likely to have been if the effect noticed was due to resonance of the car suspension system.

3 Design for living

In this part of the chapter you will use and extend what you have learned about vibrations and resonance, and see how buildings (and machines) can be designed to withstand vibrations. Whether the vibrations are the results of earthquakes or something else, the principles are the same.

3.1 Damping the motion

'The wheels on the bus go round and round…', but they also go up and down every time they hit a bump in the road. Passengers expect a smooth ride so shock absorbers perform an essential task: they absorb the energy of the vibration and stop the car body from bouncing up and down long after the bump has been passed.

The amplitude of oscillations in a system that is not subject to an external driving force will decrease over time due to resistive forces within the system; this **damping** is a consequence of friction and/or viscous drag in mechanical systems. All real oscillating systems are damped; for example, a pendulum swinging in the air stops after a while because its energy is dissipated into the air (warming it slightly).

When designing vehicles and other systems that undergo oscillations, engineers go to great lengths to ensure an optimum degree of damping, which allows the system to function as required without undergoing large amplitude oscillations for extended periods. One method of deliberately damping back-and-forth motion is to use a dashpot. This uses the viscous drag of a fluid to provide a force that opposes the motion. The principle can be seen in a coffee pot (Figure 4.42) where motion of the liquid (coffee) through the holes in the plunger provides a resistive force. The size of the force depends on the size and number of the holes, the area of the plunger and the viscosity of the fluid. Dashpots can be used to damp the oscillations of spring–mass systems (Figure 4.43); they are used in vehicle suspension systems where the dashpot is positioned within the spring (Figure 4.44).

Figure 4.42 A coffee plunger illustrates the principle of a dashpot

Figure 4.43 A dashpot can be used to damp a mass–spring system

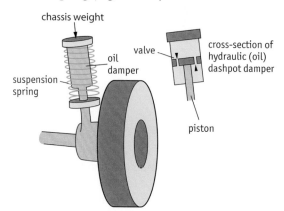

Figure 4.44 Dashpots are used in vehicle suspension

The damping affecting a system (Figure 4.45) is classified as follows.

● Heavy damping: when perturbed, the system takes a long time to return to its equilibrium position, then stays there without oscillating. (Imagine a swing door with with a very stiff hinge. If you open it and let go, it closes very slowly.)

● Light damping: when perturbed, the system oscillates several times before coming to rest. (Imagine opening and releasing a swing door that oscillates several times before coming to rest.)

● Critical damping: the system comes to rest at its equilibrium position with little (almost no) overshoot in the shortest possible time.

Often oscillating systems are designed to be critically damped, so that they dissipate their energy in the shortest possible time.

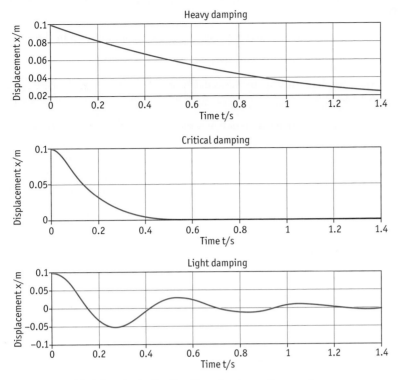

Figure 4.45 Heavy, light and critical damping (notice the different scales on the y-axes)

⚙ ACTIVITY 25 DAMPED OSCILLATIONS

Use a motion sensor to produce a displacement–time graph of a damped oscillator. Examine the results to see whether the amplitude decays exponentially with time.

Figure 4.46 shows typical results from Activity 25, demonstrating how the amplitude of a damped oscillation dies away exponentially with time.

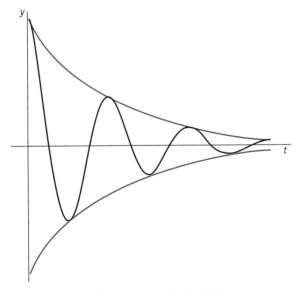

Figure 4.46 Amplitude of damped oscillations decreases exponentially with time

PRACTICAL SKILLS REFERENCE

Carrying out practical work

See Practical skills notes 3.1–3.2

MATHS REFERENCE

Exponential changes; exponential functions; exponentials and logs.

See Maths notes 9.1, 9.2, 9.3

STUDY NOTE

It is a common mistake to describe all growths and decays as exponential, but they're not.

STUDY NOTE

For further discussion of damping, energy and resonance, see BLD Additional Sheet 4.

QUESTIONS

Q 19 What would it feel like if you were in a car that went over a bump in the road when:

(a) the suspension was heavily damped?

(b) the suspension had only springs, no dashpot dampers (as was the case for very early vehicles)?

Q 20 A tall tower-like model structure (such as a radio mast) of mass 5.5×10^3 kg is being tested for its vibration characteristics to see whether there is any risk of the actual tower being affected by earthquakes. It is initially given a static test by being pulled sideways by a steady force applied near the top. It is found that for the very small displacement possible, the force obeys Hooke's law, with a force of 1.0×10^5 N producing a deflection of 50 mm.

(a) At this deflection, how much elastic potential energy is stored in the structure?

(b) The load is suddenly removed and the structure 'twangs' like a gigantic tuning fork. As an engineer involved in the test, you decide to make two drastic assumptions:

- damping may be ignored

- the kinetic energy of the tower as it passes through the central position can be represented by the motion of a point mass (equal to the tower mass) at the centre of mass of the tower moving with a speed v.

Using these assumptions estimate the speed v.

(c) The dynamic test (the oscillation) now continues and the amplitude of the point of attachment is measured over five complete cycles. The results are listed in Table 4.6.

What evidence is there that the damping is exponential (that is, the amplitude decays exponentially with the number of cycles)? State as many tests as you can for an exponential variation and apply one of these accurately.

Cycle number, n	Amplitude, $A/10^{-3}$ m
0	50
1	36
2	26
3	19
4	13
5	10

Table 4.6 Data for Question 20

(d) On a copy of Table 4.6, add a third column headed 'stored energy at end of cycle' and fill in the values. The first entry should be your answer to (a).

(e) Compare the fractional decay in energy per cycle with the fractional decay in amplitude per cycle. Why do you think the energy decay rate is considerably greater?

(f) (i) What is the actual loss in energy in joules over the first cycle?
   (ii) Assuming (again not quite correctly) that this damping loss takes place uniformly over the cycle, how much will have been lost over the first quarter cycle?
   (iii) Use this value to re-calculate an answer to part (b) which takes account of damping.

(g) Assuming that the speed found in (b) is not too different from the speed of the point where the displacement was measured, find:
   (i) the average speed over the first quarter cycle (remember it starts from rest)
   (ii) the time taken for this quarter cycle
   (iii) the period, T
   (iv) the frequency, f. (Does the frequency seem a sensible value? Does it relate to anything you know about earthquakes?)

Damping and resonance

Not only can damping make an oscillating system come to rest more quickly, it can also prevent it resonating quite so violently when driven at its natural frequency. A damped system will absorb energy and vibrate when driven (forced vibration), but, as you saw in Activity 10, the resonant vibrations will have a smaller 'spread' over a wider frequency range, as shown in Figure 4.47.

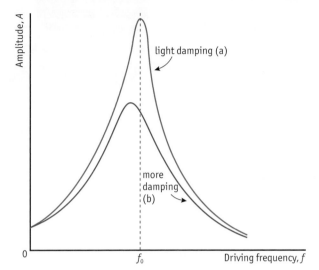

Figure 4.47 Two frequency response curves: (a) light damping, and (b) more damping

Damping is therefore one way in which a building can be prevented from shaking itself to bits; even if the earthquake frequency happens to match the natural frequency of the building, it will not vibrate with such large amplitudes.

A very effective way to reduce resonance within buildings is to install a tuned-mass damper (TMD). This involves adding either a massive pendulum or spring-mass system to the building that is designed to have the same natural frequency as the building. If the building is driven to oscillate by an external periodic force of the same frequency as that of the structure, both the structure and the TMD will resonate; however, the TMD will oscillate in antiphase with the structure and thus the amplitude of the resultant oscillations will be reduced, as shown in Figure 4.48.

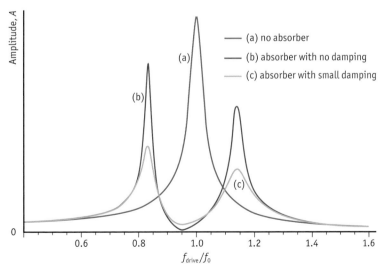

Figure 4.48 Introducing a TMD effectively substitutes one pronounced resonant peak (a) at the natural frequency of the original structure for two smaller peaks (b)

The Taipei 101 skyscraper incorporates a TMD in the form of a 660 tonne pendulum (Figure 4.49). Dashpot dampers are attached to the bottom of the pendulum (Figure 4.49) and dissipate energy by heating the liquid of the hydraulic dampers. This reduces the kinetic energy of the building's oscillations, and so reduces the amplitude of oscillation. The systems effectively reduce the size of the double resonant peaks from red line to yellow line in Figure 4.48.

91st Floor [390.60 m]
(Outdoor Observation Deck)

89th Floor [382.20 m]
(Indoor Observation Deck)

88th Floor

87th Floor

Figure 4.49 The Taipei 101 skyscraper's TMD pendulum

QUESTION

Q 21 Calculate the length of the pendulum used as the TMD within Taipei 101, given that its purpose is to reduce the effect of resonance on the building, which has a natural frequency of 0.15 Hz.

PRACTICAL SKILLS REFERENCE

Scientific questions and information research

See Practical skills notes 1.2–1.3

⚙ ACTIVITY 26 RESISTANT BUILDINGS

Go to the University of Bristol's IDEERs website (available via **www.pearsonhotlinks.co.uk**) and explore the section 'Resistant buildings', to understand how structures are made to withstand earthquakes.

The University of Bristol's Earthquake Engineering Research Centre's website (also available via **www.pearsonhotlinks.co.uk**) may also be of interest.

3.2 Other ways to reduce earthquake damage

As well as avoiding resonance and incorporating some damping, there are other methods used by engineers to protect buildings and their occupants from vibrations. While the following information relates mainly to buildings and earthquakes, the same principles are used to prevent damage caused by vibrating machinery in factories and in vehicles.

Structural methods

Isolation

In this method the building is separated from the Earth in some way that prevents the vibrations being transmitted. In other words, the structure is 'decoupled' from the ground. A very good example of such a structure is a ship. As water cannot transmit the horizontal components of the earthquake waves, the ship will be unaffected.

Other possibilities include putting the building on rollers, or using a layer of sand, which would allow the building to slide. One type of construction uses sliders. If the structure can slide on the foundations, because of a low-friction interface, the building will move and so will not suffer a shear force. However, there can be problems with high winds causing movement. Also, in an earthquake the building may move suddenly so that new vibrations caused by the sudden movement are set up in the structure. Connection of services such as gas, water and electricity would also be a problem. The movement could be in any direction, so the foundations on which the building moves would have to be large and extend in all directions. This could be reduced if the building were anchored with a weak link that would break when the force was large enough that the building needed to move.

Another type of isolation involves using bearings made mainly of rubber placed between the foundations and the building (Figure 4.50). For example, the Parliament buildings and the Te Papa Museum in Wellington, New Zealand, are protected from large earthquakes by rubber bearings.

Modern designs use steel and rubber laminated bearings that are strong enough to take the weight of the building (rubber alone would bulge outwards). The stiffness of the bearings is less than that of the ground and the building, so the fundamental frequency of the whole structure is lower than that of the building on the ground, ensuring that an earthquake cannot set up vibrations at a resonance frequency.

Figure 4.50 Building mounted on bearings

Choice of materials

Every precaution must be taken to protect people from building collapse, and the development of better building materials is one more weapon in the armoury.

Reinforced concrete design

Cracking is the main cause of structural failure. If important structural elements crack, the building may collapse. Reinforced concrete (which is strong in compression) contains steel bars (which are strong in tension). If a concrete column has bars, as shown in Figure 4.51a, it will be strong provided it is not moved; however, in an earthquake the concrete will crack outside the steel links because there is nothing to hold it together. As it falls away the steel reinforcement is no longer held in position, so it can move and the column loses its strength – often it will collapse. Therefore the links must be anchored to the centre of the column by concrete that is held in place by the links themselves (see Figure 4.51b).

Figure 4.51 Reinforced concrete (a) for normal use, and (b) for use in earthquake zones

Figure 4.52 Braced steel frame

Steel frames

Steel is a ductile material and will absorb energy while it is being deformed; earthquake-resistant buildings can make effective use of this property. The design must ensure that floors fail before columns, so that a whole building will not collapse. There should be extra beams included so that the building will stand if some fail. Figure 4.52 shows a braced steel-frame design. Buckling should occur first in the beams, and not in the diagonals or columns. The Transamerica building in San Francisco (Figure 4.53) is protected by cross-bracing at its base.

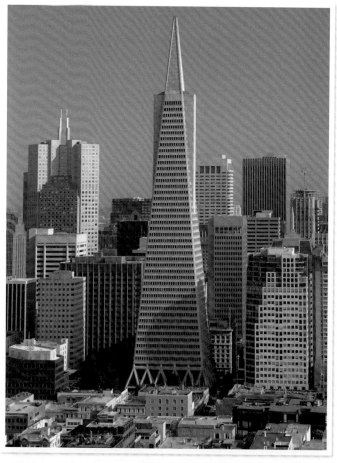

Figure 4.53 Transamerica building

⚙ ACTIVITY 27 MATERIALS

In the chapters *Good Enough to Eat* and *Spare Part Surgery*, you explored ductile and brittle materials. In a small group, discuss the following questions.

- How would a structure behave in an earthquake if it were made entirely from materials that were **(i)** elastic, **(ii)** brittle and **(iii)** ductile?
- What property is needed for earthquake-resistant building materials? How do you think reinforced concrete could be made so that it had this property?

3.3 Summing up Part 3

This part of the chapter has been mainly revision. You have revisited the notion of damping and seen how damping affects resonance. You have considered some of the various ways in which engineers design buildings to withstand earthquakes, and you have related building design to your previous knowledge of materials.

⚙ ACTIVITY 28 SUMMING UP PART 3

Look back through your work on this part of the chapter and make sure you have noted and understood the key points about damping.

4 Keeping buildings warm or cool

According to the UK Building Research Establishment (BRE), the most frequent complaint that people make about where they work involves temperature – that it is either too hot or too cold. To design buildings that are pleasant to live and work in, we must know how energy is transferred thermally, and how much heat is required to change the temperature of something. We will look at each of these in turn. A thermal transfer of energy to or from a material can also cause a phase change (between solid and liquid or between liquid and gas), and we will look at how this effect is being used in the design of modern buildings.

4.1 Energy and temperature change

When an object is heated, the resulting rise in temperature, $\Delta\theta$, depends on the amount of energy transferred to it, ΔE or ΔQ, its mass, m, and the material(s) from which it is made.

$$\Delta E = mc\Delta\theta \tag{46}$$

The constant c is called the **specific heat capacity** of the material. With energy change in joules (J), mass in kg and temperature change in degrees Celsius or kelvin, K, the units of c are $\mathrm{J\,kg^{-1}\,°C^{-1}}$ or $\mathrm{J\,kg^{-1}\,K^{-1}}$. For many materials, specific heat capacity varies over different temperature ranges, so a mean value has to be used.

> **ACTIVITY 29 HEATING WATER**
> Use an electric kettle to estimate the specific heat capacity of water.

> **ACTIVITY 30 COMPARING MATERIALS**
> Use a low voltage heater to heat two different metal blocks by about 10 °C, and compare their specific heat capacities.

> **ACTIVITY 31 A MODEL RADIATOR**
> Use a model radiator system (Figure 4.54) to find the energy transferred to the surroundings as hot water flows through it at a known rate. Alternatively, use a continuous flow calorimeter to model the radiator.

Figure 4.54 Model radiator

The following example based on a domestic 'radiator' shows how to deal with flowing fluids.

> **WORKED EXAMPLE**
>
> **Q** A domestic 'radiator' has water pumped through it at a rate of 6 kg per minute. Before entering, the water temperature is 60 °C; it leaves at a temperature of 50 °C. Calculate:
>
> **(i)** the energy transferred each minute by this heater
>
> **(ii)** its power output.
>
> **A** Consider a time of 1 minute. In this time, 6 kg of water cools from 60 °C to 50 °C, so $\Delta\theta = 10\,°C$:
>
> $\Delta Q = 6\,\mathrm{kg} \times 4.2 \times 10^3\,\mathrm{J\,kg^{-1}\,°C^{-1}} \times 10\,°C$
>
> $\quad = 252\,000\,\mathrm{J}\ (= 2.52 \times 10^5\,\mathrm{J})$
>
> $\text{Power transferred} = \dfrac{\text{energy lost}}{\text{time taken}}$
>
> $\quad = \dfrac{252\,000\,\mathrm{J}}{60\,\mathrm{s}} = 4200\,\mathrm{W}$

STUDY NOTE

The kelvin and the degree Celsius are the same size, so $1\,\mathrm{J\,kg^{-1}\,°C^{-1}} = 1\,\mathrm{J\,kg^{-1}\,K^{-1}}$.

SAFETY NOTE

Take care to avoid steam burns.

PRACTICAL SKILLS REFERENCE

Planning and experimental design
See Practical skills notes 2.1–2.4

Carrying out practical work
See Practical skills notes 3.1–3.2

Analysis and interpretation of data
See Practical skills notes 4.1–4.2

Conclusion and evaluation
See Practical skills notes 5.1–5.2

STUDY NOTE

A domestic radiator transfers most energy by convection, so the name is misleading.

QUESTIONS

Q 22 A kilogram of water cools from 100 °C to 80 °C in two minutes. Calculate the average power transfer from the water. (In this temperature range, water has $c = 4.2 \times 10^3 \, \text{J kg}^{-1} \, °\text{C}^{-1}$.)

Q 23 A room is heated during daytime by a 2 kW radiant heater that is on for 12 hours. It is replaced by a storage heater that contains a concrete block, which is heated at night to 70 °C and cools during the day to 25 °C. Both heaters transfer the same amount of energy to the room during the day.

Use this information to calculate the volume of the concrete block:

Density of concrete = $2700 \, \text{kg m}^{-3}$. Specific heat capacity of concrete = $0.85 \, \text{kJ kg}^{-1} \, °\text{C}^{-1}$.

Q 24 A 1.6 kW radiator has a temperature difference between the input and output water flow of 12 °C. Calculate the water flow rate in kg s^{-1}.

(Specific heat capacity of water = $4.2 \, \text{kJ kg}^{-1} \, °\text{C}^{-1}$.)

Testing radiators

The performance of a radiator is related to its temperature, its surface area, the nature of its surface and the temperature of the surroundings. Figure 4.55 shows data obtained by researchers working at ESTEC in the Netherlands (the European Space Research and Technology Centre, run by the European Space Agency – as in Earth-based buildings, temperature control in spacecraft is important). In these tests, the same type of surface was used throughout, and the input power to the radiator was kept constant. The radiator was allowed to reach a 'steady state'; that is, a steady temperature at which the power radiated balanced the input power. Figure 4.55 shows how this steady-state temperature depends on the surface area and on the temperature of the surroundings (called the sink temperature). Note that the radiator area axis has a logarithmic scale rising in factors of 10.

MATHS REFERENCE

Using log scales

See Maths note 8.6

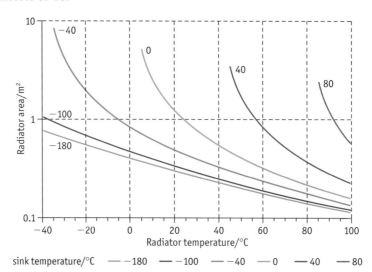

Figure 4.55 Radiator performance data

QUESTIONS

Q 25 Figure 4.55 gives some radiator performance data. Use this information to answer the following questions.

(a) Write a short paragraph (of approximately two sentences) that states how the temperature of the radiator is affected by **(i)** its area and **(ii)** the temperature of the environment or sink.

(b) If the environment or sink temperature was 0 °C, what radiator area would be needed to maintain its temperature at 24 °C?

Q 26 In the continuous casting process of steel production, the molten steel has to be cooled before it is straightened, cut and shaped. The steel is poured into a water-cooled mould and further cooling takes place in the curved cooling chamber (Figure 4.56). This question illustrates the large amounts of energy involved in the process.

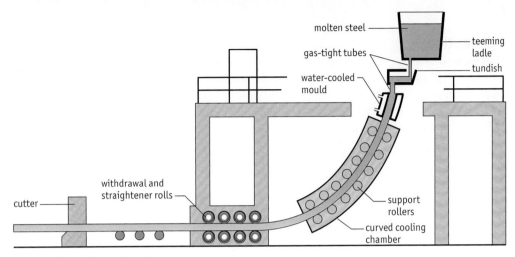

Figure 4.56 Cooling molten steel

Water entering the mould passes through each of two wide and two narrow faces. Flow to each of the wide faces is 2400 kg (2400 litres) per minute and to each of the narrow faces 240 kg per minute. The water enters at a temperature of 40 °C and the outlet temperatures are 47 °C on each wide face and 48 °C on each narrow face. The average specific heat capacity of water over this temperature range is 4179 J kg^{-1} °C^{-1}.

(a) How much energy is transferred to the water in each minute?

(b) What is the total power transfer to the water?

4.2 Thermal energy transfer processes

The thermal transfer of energy, giving rise to heating and cooling, takes place in three basic ways: conduction, convection and radiation. Each of these has its place in a building.

Conduction

Conduction involves energy 'spreading' through a material by virtue of atomic and molecular vibrations, and by the motion of free electrons. Generally, materials that are good electrical conductors are also good thermal conductors, as they have a large number of free electrons.

Thermal conduction can be described in a similar way to electrical conduction, but conventionally (rather than using resistance) the thermal **conductance** of building materials is used. The properties of building materials are compared using U-values. The U-value of a building material is defined as the rate of energy transfer through a metre square of material for a 1 K (or 1 °C) temperature difference. U-value thus depends on the thickness of a sample, as well as on the material from which it is made. U-value has SI units of W m^{-2} K^{-1} or W m^{-2} °C^{-1}. If there is a temperature difference $\Delta\theta$ across a sample of area A, then the rate of energy transfer (i.e. the power), P, is given by:

$$P = UA\Delta\theta \qquad (47)$$

QUESTION

Q 27 **(a)** Express the units of U in terms of SI base units.

(b) For a single-thickness brick wall, $U = 2.2$ W m^{-2} °C^{-1}. The temperature indoors is 24 °C and outdoors is 14 °C. What is the rate of thermal energy transfer through such a wall 3 m wide and 2 m high?

STUDY NOTE

For further discussion of thermal conductance, and a comparison between thermal and electrical conduction, see BLD Additional Sheet 5.

MATHS REFERENCE

SI units

See Maths notes 2.1, 2.2 and 2.3

You might have studied methods of insulating houses for GCSE. Activity 32, which is about double-glazing, will help you to revise these ideas.

ACTIVITY 32 DOUBLE-GLAZING

Explore how effective double-glazing can be for heat insulation.

Convection

Convection involves the heating of a fluid (i.e. a liquid or gas) that expands when heated. Expansion reduces its density below that of the surrounding cooler fluid, and so it experiences an upthrust and rises. As cooler, denser, fluid sinks to take its place, this results in a convection current. So-called 'radiators' transfer most energy by convection; they heat the air around them and set up convection currents.

A so-called gravity-fed heating and hot-water system relies on convection to move heated water from a boiler to a water tank and radiators (Figure 4.57). In most buildings, however, a pump assists this movement.

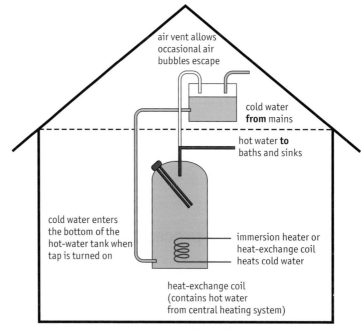

Figure 4.57 Gravity-fed hot-water system

ACTIVITY 33 COOLING BY CONVECTION

Use the internet to research how buildings are designed to use convection in air, and sometimes water, to maintain a pleasant temperature while minimising energy input. Present your findings on a poster.

Radiation

All objects, even cool ones, lose energy by emitting electromagnetic **radiation**. The hotter the object, the shorter the wavelength of the radiation it emits, and the greater its radiated power. For objects with temperatures typical of buildings and their surroundings, the radiation is mainly in the infrared part of the electromagnetic spectrum.

Radiation from buildings can be monitored using infrared-sensitive cameras. Images such as Figure 4.58 show which parts of the building's exterior are warmest and so indicate where energy is being transferred from the interior (often by conduction and convection).

Buildings can be heated by absorbing solar radiation. Although this 'solar gain' might be welcome in a cold climate, many workplace windows in warm places are fitted with infrared reflective glass to prevent overheating on sunny days.

ACTIVITY 34 RADIATION

Design and carry out an investigation to compare the effectiveness of different building materials at absorbing and transmitting infrared radiation.

PRACTICAL SKILLS REFERENCE

Planning and experimental design
See Practical skills notes 2.1–2.4

Carrying out practical work
See Practical skills notes 3.1–3.2

Analysis and interpretation of data
See Practical skills notes 4.1–4.2

Conclusion and evaluation
See Practical skills notes 5.1–5.2

Figure 4.58 False colour image showing infrared radiation

STUDY NOTE

For Activity 35 you will need to review your work on thermistors in the chapter *Technology in Space* and potential divider circuits in the chapter *Digging Up the Past*.

Temperature control

In buildings, temperature is often controlled by a combination of *passive* control (buildings cool at night, the Sun warms them during the day), and *active* control (where there is some energy input to cause a required change). Sometimes temperatures are monitored electronically and controlled by automatic thermostat systems, or the system may be manual – for example, opening the window or closing the blinds in your classroom.

ACTIVITY 35 THERMOSTAT

Calibrate a thermistor in a potential divider circuit as a thermostat.

ACTIVITY 36 ENERGY-EFFICIENT BUILDING DESIGN

Use the internet to find out about methods of temperature control in buildings. Write a short report to summarise and explain what you find. There are links to start with at **www.pearsonhotlinks.co.uk**.

PRACTICAL SKILLS REFERENCE

Scientific questions and information research
See Practical skills notes 1.1–1.4

Planning and experimental design
See Practical skills notes 2.1–2.4

Carrying out practical work
See Practical skills notes 3.1–3.2

Analysis and interpretation of data
See Practical skills notes 4.1–4.2

Conclusion and evaluation
See Practical skills notes 5.1–5.2

4.3 Energy and phase change

If you have used a cool-box for your lunch or for a picnic, you may have added a block, or bag, of liquid, frozen in the freezer, to keep the contents cool for longer. By doing so, you are making use of the fact that energy is needed to produce a **phase change**: the energy transferred to the contents of the box will cause the solid to melt, rather than increase the temperature inside the box. This is not a new idea; people have been using ice to keep food and drink at 0 °C for thousands of years. What is new is that phase change materials (PCMs) are being developed that change state at other useful temperatures.

PCMs (Figure 4.59a and b) are used to keep pharmaceuticals, food and even the body at the correct temperature. When the body is too hot, the PCMs absorb heat from the body and melt. When the body is too cold they freeze, releasing the heat. This process can be repeated indefinitely. Later in this section you will look at how PCMs can be used in buildings to keep them at a comfortable temperature.

Figure 4.59a A scanning electron micrograph of a PCM (blue microcapsules) coating fabric fibres

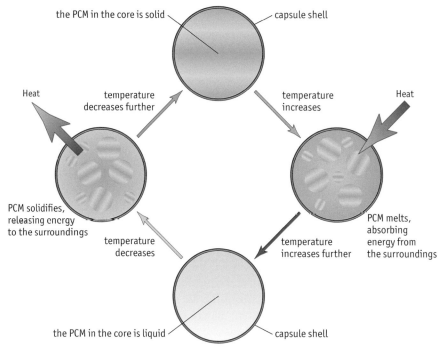

the PCM in the core is solid

capsule shell

Heat

temperature decreases further

temperature increases

Heat

PCM solidifies, releasing energy to the surroundings

temperature decreases

temperature increases further

PCM melts, absorbing energy from the surroundings

the PCM in the core is liquid

capsule shell

Figure 4.59b PCMs absorb and release heat

PCMs in buildings

In 2004 the city of Melbourne in Australia needed a new council building. They built CH2 (Council House 2), a building designed to use energy as efficiently as possible (Figure 4.60). To keep the building cool, chilled water flows through the ceiling panels. The water comes from tanks in the basement that contain thousands of small stainless steel spheres filled with a salt suspension – a PCM that freezes at 16 °C. The chilled water, which is kept at about 16 °C by the spheres, is pumped to the ceiling panels when cooling is needed, then back to the tanks. The chilled water is a few degrees warmer when it returns and is cooled by the PCM-filled spheres. When all the PCM has melted, the system is switched off and a backup chilling system is used. At night, when it is cool outside, the water is pumped to the roof, cooled, and returns to the tanks below 16 °C so that it freezes the PCM. A backup system is needed on hotter days, but this system reduces the electricity needed for cooling.

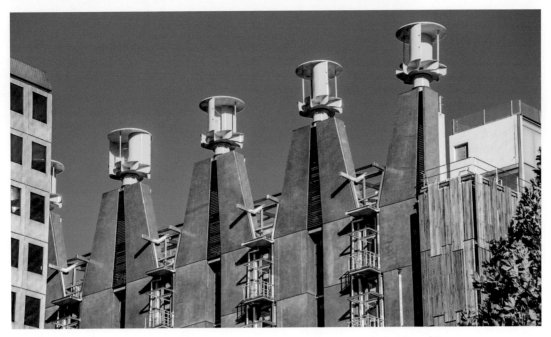

Figure 4.60 The CH2 council building in Melbourne, Australia uses PCMs to keep the building cool

The other method of using PCMs to cool buildings is to include small spheres containing the PCM in the wallboard that is being used to make the wall. The board is usually made of gypsum or clay. If small capsules are included in the board, containing a PCM that changes phase at the maximum comfortable temperature, then the temperature of the inside of the wall will not increase until all the PCM has melted. Cold outside temperatures at night will freeze the PCM again – keeping the room warmer, as the room temperature will not drop until all the PCM is frozen.

Latent heat

Phase change is central to understanding how PCMs work. It was Joseph Black in the 18th century who discovered that when ice at its melting point (0 °C) was heated its temperature did not increase until all the ice had melted, and that when water reached its boiling point (100 °C) it required a lot more heating to turn it all to steam at 100 °C. He called the 'heat' required **latent heat** (latent means hidden).

Nowadays we think in terms of energy. When the temperature is increased, the particles move faster so they have more kinetic energy – and this is true for a solid, liquid or gas. However, in the solid phase there are forces holding the particles in a structure, and they have less potential energy than they do in the liquid phase. In the liquid phase, the particles are much closer together than in the gas phase, so they have less potential energy than in the gas phase.

For a collection of particles, the sum of the kinetic energy and the potential energy is called the **internal energy**. When a material experiences a phase change, its internal energy changes, because the potential energy of the particles changes; however, the temperature remains constant because the average kinetic energy of the particles stays unchanged.

Latent heat is the energy involved in a phase change. When a material melts, evaporates, boils or sublimes (changes from solid to gas), energy is needed to increase the potential energy of the particles. When a material condenses or solidifies, energy must be transferred to its surroundings.

When a sample of material of mass Δm experiences a phase change, the energy transferred to, or from, the material ΔE is:

$$\Delta E = L\,\Delta m \tag{48}$$

The constant L is called the specific latent heat of the phase change and has SI units $J\,kg^{-1}$. For a change between solid and liquid, L is called the **specific latent heat of fusion**, and for a change between liquid and gas it is called the **specific latent heat of vaporisation**.

STUDY NOTE

The particles' potential energy depends on their separation, because in order for them to move apart, work needs to be done against the attractive forces that the particles exert on one another. You met this idea in *The Medium is the Message* when dealing with potential energy in an electric field, and in *Higher, Faster, Stronger* when dealing with potential energy in a gravitational field. The forces that atoms and molecules exert on one another are electrostatic in origin.

STUDY NOTE

You will learn more about the kinetic and potential energy of particles and about internal energy in the chapter *Reach for the Stars*.

WORKED EXAMPLE

Q Use the following information to calculate a value for the specific latent heat of fusion for water.

Some students are measuring the specific latent heat of ice. They use a copper can and stirrer of mass 70 g containing 200 g of water, starting at a room temperature of 19 °C. They heat the water until it has risen in temperature to 24 °C. They then add small pieces of dried ice at 0 °C, one at a time, stir until each has completely melted, and measure the temperature. When the temperature reaches 14 °C they stop the experiment and weigh the calorimeter of water and melted ice. The total mass is 292 g.

Specific heat capacity of copper, $c_{copper} = 385 \, J \, kg^{-1} \, °C^{-1}$

Specific heat capacity of water, $c_{water} = 4180 \, J \, kg^{-1} \, °C^{-1}$

A We have to assume that no energy is transferred to or from the surroundings as the ice is added and melts. Then we can write:

$$
\begin{array}{ccccccc}
\text{energy gained by} & + & \text{energy gained by water} & = & \text{energy lost by water} & + & \text{energy lost by} \\
\text{ice as it melts} & & \text{(melted ice) as it warms} & & \text{(originally in can) as it cools} & & \text{copper as it cools} \\
\Delta E_1 & + & \Delta E_2 & = & \Delta E_3 & + & \Delta E_4
\end{array}
$$

We can define these symbols:

m_{ice} is the mass of the ice added (22 g)

m_{water} is the mass of water originally in the can (200 g)

m_{copper} is the mass of copper can and stirrer (70 g)

$\Delta\theta_{warm}$ is the temperature change of the melted ice (14 °C)

$\Delta\theta_{cool}$ is the temperature change of the can and the stirrer and the water originally in the can (10 °C)

Then we can write:

$\Delta E_1 = L m_{ice}$

$\Delta E_2 = m_{ice} \, c_{water} \, \Delta\theta_{warm}$

$\qquad = 22 \times 10^{-3} \, kg \times 4180 \, J \, kg^{-1} \, C^{-1} \times 14 \, °C$

$\qquad = 1287 \, J$

$\Delta E_3 = m_{water} \, c_{water} \, \Delta\theta_{cool}$

$\qquad = 200 \times 10^{-3} \, kg \times 4180 \, J \, kg^{-1} \, C^{-1} \times 10 \, °C$

$\qquad = 8360 \, J$

$\Delta E_4 = m_{copper} \, c_{copper} \, \Delta\theta_{cool}$

$\qquad = 70 \times 10^{-3} \, kg \times 385 \, J \, kg^{-1} \, C^{-1} \, 10 \, °C$

$\qquad = 269.5 \, J$

And so:

$L m_{ice} = 8360 \, J + 269.5 \, J - 1287 \, J$

$\qquad = 7342.5 \, J$

$L \quad = \dfrac{7342.5 \, J}{22 \times 10^{-3} \, kg}$

$\qquad = 3.3 \times 10^5 \, J \, kg^{-1}$

Notice that we kept all the 'calculator' answers until the final step, then rounded to 2 significant figures to match the least precise values used in the calculation.

MATHS REFERENCE

Significant figures: a useful rule of thumb

See Maths note 7.3

PRACTICAL SKILLS REFERENCE

Carrying out practical work

See Practical skills notes 3.1–3.2

⚙ ACTIVITY 37 PHASE CHANGE

Heat a sample of octodecanoic acid until it has all melted, then let it cool until it has all solidified. Plot a graph of temperature against time for the sample as it is heated and cooled.

⚙ ACTIVITY 38 THE SPECIFIC LATENT HEAT OF FUSION OF WATER

Use a low voltage heater to melt some ice and in this way determine the specific latent heat of fusion of water.

PRACTICAL SKILLS REFERENCE

Planning and experimental design
See Practical skills notes 2.1–2.4

Carrying out practical work
See Practical skills notes 3.1–3.2

Analysis and interpretation of data
See Practical skills notes 4.1–4.2

Conclusion and evaluation
See Practical skills notes 5.1–5.2

Measuring the specific latent heat of vaporisation of a liquid generally involves collecting the liquid that has been turned to vapour and condensing it in order to measure the mass. Question 29 is about using the equipment shown in Figure 4.61. The double-walled glass container contains the liquid, which is heated by a heater immersed in the liquid. When the liquid boils, the vapour goes through the holes into the outer 'jacket' and passes down into the condenser. Cold water flowing around the condenser cools the vapour and condenses it. The vapour in the 'jacket' insulates the liquid and prevents heat losses. Once the liquid is boiling, and dripping steadily from the condenser, it is collected in a beaker for a measured time.

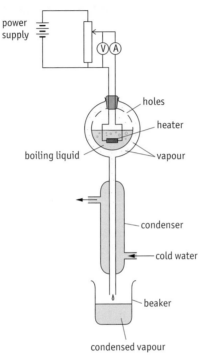

Figure 4.61 Apparatus for determining the latent heat of vaporisation of a liquid

QUESTIONS

Q 28 In February 2015, heavy snow in Massachusetts, USA, led to dangerous conditions in which over 100 roofs collapsed or partly collapsed. Wet snow has a density of about $200\,\text{kg}\,\text{m}^{-3}$ and a depth of 1.6 m could cause a roof to collapse. If the area of snow covering a roof is $55\,\text{m}^2$, how much energy is needed to melt this amount of snow?

(Specific latent heat of fusion of ice, $L = 3.34 \times 10^5\,\text{J}\,\text{kg}^{-1}$.)

Q 29 Some students use the equipment shown in Figure 4.61 to find the specific latent heat of vaporisation of water. Their results are listed in Table 4.7.

Current in heater/A	2.9
Voltage across heater/V	11.9
Time for which water collected/minutes	20
Mass of water collected/g	18

Table 4.7 Data for Question 29

Use the data in Table 4.7 to calculate the specific latent heat of vaporisation of water.

PRACTICAL SKILLS REFERENCE

Planning and experimental design

See Practical skills notes 2.1–2.4

Carrying out practical work

See Practical skills notes 3.1–3.2

Analysis and interpretation of data

See Practical skills notes 4.1–4.2

Conclusion and evaluation

See Practical skills notes 5.1–5.2

ACTIVITY 39 PHASE CHANGE MATERIALS

Compare the thermal properties of a commercially available PCM and an icepack.

4.4 Summing up Part 4

In this part of the chapter you have considered how temperature control is achieved in modern buildings. You have reviewed the mechanisms of heat transfer and learned how to calculate the energy transfer involved when an object is warmed or cooled. You have learned about the latent heat involved in phase changes.

Activity 40 is designed to help you review your progress, and Questions 30 to 33 are designed to help you put into practice what you have been learning.

ACTIVITY 40 HEATING AND COOLING IN THE LABORATORY

First, check through Part 4 and ensure you know the meaning of each of the terms printed in **bold**.

Consider the heating and cooling systems in your school or college laboratory. With the aid of two labelled sketches (or photographs), explain how conduction, convection and radiation are involved in (a) cold and (b) warm weather. Make some recommendations for improvements, or suggestions for a new laboratory that would reduce the fuel needed for heating and cooling the room.

QUESTIONS

STUDY NOTE

For some of these questions you will need to refer to work that you did in the chapter *Technology in Space*.

Q 30 A storage heater has a 3.4 kW electric heating element that heats a brick block enclosed in a casing. The specific heat capacity of the brick is 950 J kg °C^{-1}. The *U*-value of the casing when the flaps are opened is 11.2 W m^{-2} °C^{-1} and its surface area is 0.23 m^2. The mass of the brick block is 152 kg.

At night the brick block is heated electrically, then flaps are opened during the day allowing it to allow warmed air to circulate through the casing.

(a) Calculate the average temperature rise of the brick block if the heating element is switched on for three hours. Assume the block emits no heat during this time.

(b) Explain why the value calculated in **(a)** is an average.

(c) The ventilation flaps are then opened and the heater heats the room. Assuming that the brick block has reached a uniform temperature, calculate the initial power output of the heater.

Q 31 The passage below refers to Figure 4.62. When you have read it and studied the diagram, answer the questions that follow.

Sun-catching venetian blinds could help to store the energy transferred from solar radiation and help heat buildings. Each slat has a front reflector made of aluminised plastic. Nested in a bed of insulation below the front reflector is a glass tube containing water. This nests in a second reflector and is protected by a long, narrow window of special glass.

Each front curved reflector bounces the Sun's rays up to the slat above. Here the rays are focused through the narrow glass window and their energy transferred to the water. These water tubes are linked by pipes to the building's hot water system. The slats are re-angled occasionally, but do not need to track the Sun precisely.

During the day the Sun-warmed water passes along the pipes and provides some heating for the room as well as adding hot water to the building's heating system. At night, with the blinds shut, the heated water is circulated back through the water tubes to heat the room. The original prototype transferred 70% of the available solar energy.

(a) Why are both the large front reflector and the small rear reflector needed?

(b) Why do the blind slats need insulation behind the reflectors? Suggest a suitable insulating material and explain what makes it a good insulator.

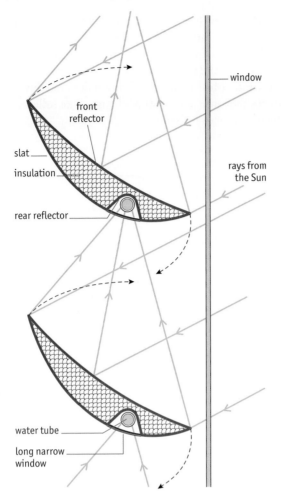

Figure 4.62 Sun-catching venetian blinds

(c) If the hot-water tank is positioned above the level of the blind, explain how the heated water **(i)** reaches the tank during the day and **(ii)** circulates through the blind at night.

(d) By what thermal energy transfer process(es) is the room heated by the warm water passing through the blind?

(e) Taking the various energy transfer processes into account, what properties would be ideal for the material making the water tubes? It is suggested that copper tubes might be better than the glass ones. What advantages and disadvantages might the copper tubes have?

(f) In a test of the prototype, the solar radiation intensity measured square-on to the blind was $15\,\mathrm{MJ\,m^{-2}}$ per day, the blind measured $2\,\mathrm{m} \times 2\,\mathrm{m}$ and the water circulated at a rate of $6000\,\mathrm{kg}$ per day.

Calculate:
(i) the energy transferred to the water
(ii) the resulting temperature rise of the water.

Assume that the blind is 70% efficient at transferring energy to the water and that energy transfers from the heated water to the surroundings can be ignored. The average specific heat capacity of water over this temperature range is $4200\,\mathrm{J\,kg^{-1}\,{}^{\circ}C^{-1}}$.

Q 32 This question is about a technique used in the oil and gas industries to measure rates of fluid flow. It involves measuring the temperature upstream and downstream of a small heater placed in the path of the fluid, as shown in Figure 4.63. The apparatus is first calibrated by measuring the temperature difference produced by a known heater power for a known gas flow rate. Measuring an unknown flow rate involves readjusting the flow rate so that it produces the same temperature change. The following example illustrates how this enables the flow rate to be deduced.

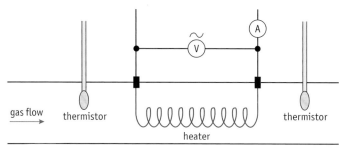

Figure 4.63 Technique for measuring fluid flow rate

(a) If the thermistors are the NTC type, how will their resistance depend on the temperature of the gas?

(b) (i) The voltmeter and ammeter connected to the heater read 240 V and 15 A respectively. Calculate the power transferred by the heater.

(ii) At a flow rate of $0.10 \, \text{kg s}^{-1}$, gas passing the heater rose in temperature by 13 °C. Taking the specific heat capacity of the gas to be $2200 \, \text{J kg}^{-1}\,°\text{C}^{-1}$ over this temperature range, calculate the power transferred to the gas.

(iii) Your answer to **(i)** should not be the same as your answer to **(ii)**. Explain why this is the case, and calculate the power that is *not* being transferred to the gas. Which conservation law are you using in this calculation?

(c) On another occasion, with the gas at the same initial temperature flowing at a different rate, the power transferred by the heater had to be reduced to 2500 W in order to keep the temperature rise from the same starting temperature to 13 °C and thus to keep the power transfer to the surroundings the same as in **(b)**.

(i) Using your answer to **(b)(iii)**, calculate the power transferred to the gas.

(ii) Calculate the flow rate of the gas on this occasion.

Q 33 Table 4.8 lists data for four commercially available PCMs.

The graph in Figure 4.64 shows the outside temperature and the temperature inside two test rooms over 24 hours. One room has a PCM in the walls and the other is a control; it does not contain PCM but is otherwise identical.

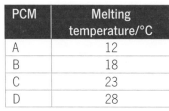

PCM	Melting temperature/°C
A	12
B	18
C	23
D	28

Table 4.8 Data for Question 33

Figure 4.64 Graphs for Question 33

Use Figure 4.64 to answer parts **(a)** to **(c)**.

(a) Which PCM from Table 4.8 is in the walls of the test room?

(b) State the difference this PCM makes, in °C, to:
 (i) the minimum temperature inside the room during the 24 hours
 (ii) the maximum temperature inside the room during the 24 hours.

(c) State:
 (i) the time at which this PCM starts to melt
 (ii) the time at which has all melted
 (iii) the time at which it starts to freeze.

Parts **(d)**, **(e)** and **(f)** refer to a different test.

In this new test, the control room is built of material with total mass 8400 kg and specific heat capacity $1080\,\mathrm{J\,kg^{-1}\,°C^{-1}}$.

The test room is built using the same material mixed with PCM B. 8400 kg of the combined material is used, and the specific heat capacity of the combined material is $1160\,\mathrm{J\,kg^{1}\,°C^{-1}}$.

The combined material used in the test room contains 300 kg of PCM B, whose specific latent heat of fusion is $189\,\mathrm{kJ\,kg^{-1}}$.

Both rooms have an initial temperature of 16 °C. The control room is heated so that its walls reach a temperature of 24 °C. The same amount of energy is used to heat the test room.

(d) Calculate the energy used to heat the walls of the control room during the test.

(e) Calculate:
 (i) the energy needed to raise the temperature of the test room to the melting point of the PCM
 (ii) the energy needed to melt the PCM.

(f) Using your answers to **(d)** and **(e)**, deduce the resulting temperature of the test room walls.

LINK THE LEARNING

5 Link the Learning: Rebuilding

5.1 Building up knowledge

Having finished this chapter, you are now quite close to the end of your course. In this chapter, you have revisited some earlier work on waves and material properties, and have also used ideas about forces, energy transfer and motion. Use Activities 41–44 to help you review your work.

ACTIVITY 41 BUILDING ON FIRM FOUNDATIONS

Take time to have a thorough look at all the work you have done in this chapter, and at all the related work in other parts of the course.

Go back to Activities 2 and 3 in Part 1 of this chapter and extend your summaries so that they include the new things you have learned.

ACTIVITY 42 WHY SHM?

Imagine you are designing a course of study for engineers. Write the prospectus entry that will persuade them that studying SHM will be of the value to their future careers. Outline briefly what they will cover in each section and explain:

- why it is important to understand the fundamentals of SHM
- why it is important to know a building's natural frequency
- whether a large structure is likely to have a high or low natural frequency
- what could excite a building at low frequencies
- why it is useful to know about damping.

ACTIVITY 43 BOUNCY CASTLE

Many principles of physics can be illustrated through everyday situations and events. In a 'bouncy castle', a large pillow is kept inflated by means of an air pump while small children bounce around on it; it illustrates many aspects of the physics of oscillations and materials that you have covered in this chapter, as well as principles from earlier in the course (such as forces and energy).

Imagine you have been asked to write an illustrated article for a newspaper or magazine about the physics of a bouncy castle. Draft a set of headings and notes, with diagrams, equations and order-of-magnitude calculations, showing some of the things that you could include in the article. Include clear statements of principles of physics and show how they are relevant to a bouncy castle.

ACTIVITY 44 BUILDING A LEISURE CENTRE

A city in a semi-tropical earthquake-prone area requires a leisure centre that includes a cinema, bowling alley, swimming pool, gym and a supermarket.

- How would you design the centre to be earthquake resistant up to magnitude 6 on the Richter scale?
- How would you design the building so as to maintain the interior at a comfortable temperature?
- How would you try to balance any conflicting demands that might arise from considering these two requirements?

5.2 Questions on the whole chapter

Q 34 An earthquake occurs at a point X on the surface of the Earth, shown in Figure 4.65. Draw the ray paths to show how the P waves and the S waves would travel within the Earth if:

(a) the Earth were completely liquid

(b) the Earth were completely solid

(c) the Earth were solid with a liquid core starting halfway to the centre.

Figure 4.65 Diagram for Question 34

Q 35 Domestic washing machines often incorporate washing, rinsing, spinning and drying. This question is about the spinning.

(a) The inner drum of the machine into which the clothes are placed has quite large holes in it. Explain carefully how, when the clothes are being spun-dried, the water gets from the clothes and out through the holes.

(b) One of the spin speeds in a certain model of washing machine was listed as 1000 rpm (revolutions per minute).
  (i) What is this spin speed in radians per second?
  (ii) If the radius of the spinning drum is 12.5 cm, what would be the highest centripetal force that could be exerted on a wet sweatshirt of mass $m = 0.5$ kg?

(c) If clothes are unevenly distributed in the machine, it vibrates slightly as it rotates. The outer drum within which the spinning drum rotates is attached to the rest of the framework of the washing machine by springs (Figure 4.66). What is the purpose of these springs?

(d) For each spring, the spring constant $k \approx 200$ Nm^{-1}. In use, the loading on each spring is effectively 5 kg. Explain, with the aid of a calculation, what is likely to happen when an unevenly loaded machine begins to spin the clothes.

Figure 4.66 Schematic diagram of a washing machine

Q 36 A very accurate method for monitoring the density of a liquid in many industrial processes involves filling a hollow U-tube with the liquid and then measuring the period of natural oscillations by making it resonate to an external variable frequency oscillator. (The arrangement is rather like a hollow tuning fork.) The period of oscillation, T, is given by:

$$T = 2\pi\sqrt{\frac{(m + \rho V)}{c}}$$

where m is the mass of the empty tube of volume V, c is the elastic stiffness of the tube and ρ is the density of the liquid filling the tube.

For a particular tube, $m = 15 \times 10^{-3}\,\mathrm{kg}$ and $V = 2.2 \times 10^{-6}\,\mathrm{m^3}$.

(a) The evacuated tube has a natural frequency of $800\,\mathrm{Hz}$. What is the value of c, the elastic stiffness of the tube?

(b) The expression for T can be re-written as:

$$T^2 = A\rho + B$$

What are:
(i) the algebraic expressions for A and B?
(ii) their numerical values and SI units?

(c) A manufacturer of an instrument using this technique quotes only numerical values of A and B. Assuming frequencies are measured in Hz, why do you not need to ask the manufacturer what units are being used for B, but you do for A?

(d) What is the frequency of oscillation when the tube is filled with a liquid of density $1.2 \times 10^3\,\mathrm{kg\,m^{-3}}$?

(e) A calibration graph is drawn of T^2 (y-axis) against density ρ.
(i) Express the gradient and intercept in terms of A and B.
(ii) Sketch the graph and explain the significance of the point where it crosses the y-axis.

(f) Explain how you could use the calibration graph to find ρ from a measurement of the resonant frequency of the tube.

Q 37 The passage below is taken from the marketing material supplied by a manufacturer of electrically heated showers.

Most electric showers draw cold water direct from the main supply and heat it as it is used – day or night. Not only are they particularly useful for those who do not have a stored water supply, but they are versatile because every home can have one.

(a) Write a word equation to describe the energy transfer that takes place in an electric shower.

(b) Rewrite the equation using the appropriate formulae.

(c) The technical data supplied by one manufacturer states that their most powerful shower system is fitted with a $10.8\,\mathrm{kW}$ heating element and can deliver up to 16 litres of water per minute.

Show that the showering temperature is about $25\,^\circ\mathrm{C}$ if the temperature of the mains water is $15\,^\circ\mathrm{C}$ and the shower is used at its maximum settings.

(Specific heat capacity of water $= 4200\,\mathrm{J\,kg^{-1}\,K^{-1}}$. Mass of 1 litre of water $= 1.0\,\mathrm{kg}$.)

(d) The marketing material includes the statement:

Please remember that during the colder months, flow rates may need to be reduced to allow for the cooler temperature of incoming cold water.

Calculate the flow rate required for an output of $25\,^\circ\mathrm{C}$ when the incoming water temperature is $5\,^\circ\mathrm{C}$.

Q 38 Next to the 3000-year-old Drombeg Circle in Ireland is a stone-lined pit known as a Fulacht Fiadh. It is believed that this was used as a cooking place for meat caught by hunters. The pit was filled with water. Large stones were heated in a fire and then placed in the water to bring it to the boil and cook the meat.

In experiments to test this idea it was found that the water in the pit started to boil after 22 heated stones had been added. The total mass of the added stones was 198 kg and the mass of water was 513 kg.

(a) Show that this gives a minimum temperature for the fire of about 900 °C.

Specific heat capacity of water = 4200 J kg^{-1} °C^{-1}. Average specific heat capacity of stone = 1100 J kg^{-1} °C^{-1}. Initial temperature of water = 18 °C. Temperature of boiling water = 100 °C.

(b) Explain why the temperature of the fire would be higher than the calculated value.

Q 39 Some students measured the specific latent heat of vaporisation of ethanol using the apparatus shown in Figure 4.61. Table 4.9 lists the results.

	1st experiment	2nd experiment
Voltage V/V	11.8	10.4
Current I/A	2.3	1.6
Time Δt/minutes	45.0	45.0
Mass of ethanol m/g	90.0	56.0
Room temperature θ/°C	23.0	23.0

Table 4.9 Data for Question 39

One student wrote:

To make the experiment more accurate, we did the experiment twice with a different heater voltage and current, but for the same time. We also checked the room temperature did not change. This meant we could subtract the effect of energy transferred to the surroundings, because it was the same in both experiments. For each experiment:

$$IV\Delta t = L\Delta m + \Delta E_s$$

where ΔE_s is the energy transferred to the surroundings.

(a) Explain why the energy transferred to the surroundings was the same in each experiment although the energy transferred from the heater to the ethanol was different.

(b) Use the data from Table 4.9 to calculate the specific latent heat of vaporisation of ethanol.

5.3 Achievements

Now you have studied this chapter you should be able to achieve the outcomes listed in Table 4.10.

TABLE 4.10 ACHIEVEMENTS FOR THE CHAPTER *BUILD OR BUST?*

	Statement from Examination Specification	Section(s) in this chapter
1	know and understand the distinction between base and derived quantities and their SI units	2.2, 2.3, 4.2
2	demonstrate their knowledge of practical skills and techniques for both familiar and unfamiliar experiments	2.1, 2.2, 2.4, 4.1, 4.2, 4.3
3	be able to estimate values for physical quantities and use their estimate to solve problems	1.2
4	understand the limitations of physical measurement and apply these limitations to practical situations	2.1, 2.2, 2.4, 4.1, 4.2, 4.3
5	be able to communicate information and ideas in appropriate ways using appropriate terminology	2.1, 4.2, 4.4, 5.1
6	understand applications and implications of science and evaluate their associated benefits and risks	5.1
8	understand the ways in which society uses science to inform decision making	1.2
144	be able to use the equations $\Delta E = mc\Delta\theta$ and $\Delta E = L\Delta m$	4.1, 4.3
145	CORE PRACTICAL 12: Calibrate a thermistor in a potential divider circuit as a thermostat	4.2
146	CORE PRACTICAL 13: Determine the specific latent heat of a phase change	4.3
181	understand that the condition for simple harmonic motion is $F = -kx$, and hence understand how to identify situations in which simple harmonic motion will occur	2.2, 2.3
182	be able to use the equations $a = -\omega^2 x$, $x = A\cos\omega t$, $v = -A\omega\sin\omega t$, $a = -A\omega^2\cos\omega t$, and $T = \dfrac{1}{f} = \dfrac{2\pi}{\omega}$ and $\omega = 2\pi f$ as applied to a simple harmonic oscillator	2.2
183	be able to use equations for a simple harmonic oscillator $T = 2\pi\sqrt{\dfrac{m}{k}}$, and a simple pendulum $T = 2\pi\sqrt{\dfrac{l}{g}}$	2.1, 2.2
184	be able to draw and interpret a displacement–time graph for an object oscillating and know that the gradient at a point gives the velocity at that point	2.2
185	be able to draw and interpret a velocity–time graph for an oscillating object and know that the gradient at a point gives the acceleration at that point	2.2
186	understand what is meant by resonance	2.1
187	CORE PRACTICAL 16: Determine the value of an unknown mass using the resonant frequencies of the oscillation of known masses	2.4
188	understand how to apply conservation of energy to damped and undamped oscillating systems	2.2, 3.1
189	understand the distinction between free and forced oscillations	2.1
190	understand how the amplitude of a forced oscillation changes at and around the natural frequency of a system and know, qualitatively, how damping affects resonance	3.1
191	understand how damping and the plastic deformation of ductile materials reduce the amplitude of oscillation	3.1, 3.2

Answers

Q 1   Distance = 0.5 × circumference = πr

$$\text{time} = \frac{\text{distance}}{\text{speed}} = \frac{\pi \times 6.378 \times 10^6 \, \text{m}}{3 \times 10^3 \, \text{m s}^{-1}} = 6680 \, \text{s}$$

= 111 minutes (to the nearest minute)

(In fact faster waves also travel through the Earth as well as around the surface, so they will arrive sooner than this. The waves in the crust will also gradually get smaller as energy is transferred to other forms, so in practice these surface waves will have little effect on the far side of the Earth.)

Q 2   **(a)** stress, $\sigma = \dfrac{\text{force}}{\text{area}} = \dfrac{F}{A}$

strain, $\varepsilon = \dfrac{\text{extension}}{\text{original length}} = \dfrac{x}{l}$

$E = \dfrac{\sigma}{\varepsilon}$

and so $\varepsilon = \dfrac{\sigma}{\varepsilon} = \dfrac{(F/A)}{E}$

$x\varepsilon = \dfrac{Fl}{AE}$

$$= \frac{2 \times 10^{-3}\,\text{kg} \times 9.81\,\text{N kg}^{-1} \times 10 \times 10^{-2}\,\text{m}}{\pi \times (3\pi \times 10^{-6}\,\text{m})^2 \times 73.1 \times 10^9\,\text{N m}^{-2}}$$

$= 0.95 \times 10^{-3}\,\text{m} = 0.95\,\text{mm}$

(b) Maximum mass,

$m = \text{tensile strength} \times \dfrac{A}{g}$

$$= \frac{1000 \times 10^6\,\text{N m}^{-2} \times \pi \times (3 \times 10^{-6}\,\text{m})^2}{9.81\,\text{N kg}^{-1}}$$

$= 2.9 \times 10^{-3}\,\text{kg} = 2.9\,\text{g}$

Q 3   **(a)** Increasing the elastic modulus appears to increase the wave speed. You can deduce this from various numbers in Table 4.1; for example, a shear modulus of zero corresponds to a wave speed of zero for S waves; shale and sandstone have similar densities, but the one with the larger bulk modulus has the larger P wave speed.

(b) Increasing density appears to reduce the speed. For example, sandstone and salt have the same bulk modulus; salt is less dense and has a higher wave speed.

Q 4   The phenomenon is called refraction. The rays bend away from the normal and the wavelength increases. See Figure 4.67

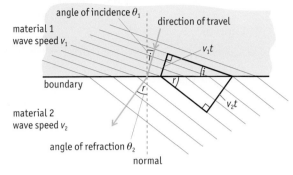

Figure 4.67   Diagram for the answer to Question 4

Q 5   **(a)** Along path (i) the wave has travelled directly; Along (ii) the wave met the interface at an angle greater than the critical angle and undergone total internal reflection; Along path (iii) the wave has been refracted into a deeper layer.

(b) The time of arrival of the waves will give information about the paths taken, and so the depth of the Moho (or other boundaries between layers) can be found.

Q 6   refractive index, $n = \dfrac{v_1}{v_2}$

From Figure 4.8, $n = \dfrac{13.5\,\text{km s}^{-1}}{7.5\,\text{km s}^{-1}} = 1.80$

Snell's law: $n = \dfrac{\sin i}{\sin r}$

$\sin r = \dfrac{\sin i}{n} = \dfrac{\sin 30°}{1.80} = 0.278$

$r = 16.1°$

Q 7   **(a)** The speed must increase with depth – the waves follow paths similar to path (iii) in Figure 4.12, only they curve gradually rather than changing direction abruptly.

(b) Density will increase with depth.

(c) On its own, an increase in density will produce a *reduction* in wave speed (see Equation 1). To bring about an increase in speed the elastic modulus must also increase with depth, and the increase must outstrip the increase in density.

Q 8

(a) 2 and π: no units (they are dimensionless numbers).

Units of m: kg

Units of k: $N m^{-1} = kg\,m\,s^{-2}\,m^{-1} = kg\,s^{-2}$

Units of $\dfrac{m}{k}$: $\dfrac{kg}{kg\,s^{-2}} = s^2$

Units of $\sqrt{\dfrac{m}{k}}$: s

(b) (i) From Equation 13, max. speed occurs when $\sin \omega t = 1$; i.e.:

$v_{max} = A\omega = 0.30\,m \times 5.0\,s^{-1} = 1.5\,m\,s^{-1}$

(ii) $f = \dfrac{1}{T} = \dfrac{\omega}{2\pi} = \dfrac{5.0\,s^{-1}}{2\pi} = 0.80\,Hz$
(2 sig. fig.)

(c) $\omega^2 = \dfrac{k}{m}$ so k

$= m\omega^2 = 0.40\,kg \times (5.0\,s^{-1})^2$

$= 10\,kg\,s^{-2} = 10\,N\,m^{-1}$

(d) When $t = 0.20\,s$, $\omega t = 1.0\,rad$
(i) $x = A\cos(1.0\,rad) = 0.162\,m$
(ii) $a = -\omega^2 x = -(5.0\,s^{-1})^2 \times 0.12\,m = 4.1\,m\,s^{-2}$

Q 9

(a) (i) $v_{max} = A\omega$ (from Equation 14) and

$\omega = 2\pi f$ (from Equation 10) so:

$v_{max} = 2\pi f A = 2\pi \times 50 \times 10^3\,s^{-1} \times 2 \times 10^{-6}\,m$

$= 0.63 \approx 0.6\,m^{-1}$

(ii) From Equation 19,
$a_{max} = \omega^2 A$
$= (2\pi \times 50 \times 10^3\,s^{-1})^2 \times 2 \times 10^{-6}\,m$
$= 2 \times 10^5\,m\,s^{-2}$

(b) $a \approx 2 \times 10^4\,g$. The peak force will be 20 000 times the weight of an oscillating part – so the material needs to withstand a high stress.

(c) Since $a_{max} = \omega^2 A = (2\pi f)^2 A$, the acceleration (and hence the internal stresses) are proportional to frequency squared.

Q 10

(a) (i) From Equation 29, $\omega = \sqrt{\dfrac{9.81\,m\,s^{-2}}{0.60\,m}} = 4.0\,s^{-1}$

(ii) From Equation 9, $T = \dfrac{2\pi}{\omega} = 1.55\,s \approx 1.6\,s$

(b) Using Equation 13, speed is maximum when $\sin \omega t = 1$, so:

$v = A\omega = 5.0\,cm \times 4.0\,s^{-1}$

$= 20\,cm\,s^{-1} = 0.20\,m\,s^{-1}$

(c) An angle of 60° is far too large for the small-angle approximation to be valid. The restoring force will not be directly proportional to displacement (it will depend on the tangent of the angle). The pendulum will not oscillate with SHM. Its period cannot be predicted using the SHM recipe.

Q 11

(a) From the Worked example on page 196:

$k = 50\,N\,m^{-1}$, $A = 3.0 \times 10^{-2}\,m$

$E_{tot} = \dfrac{1}{2}kA^2$

$= \dfrac{1}{2} \times 50\,N\,m^{-1} \times (3.0 \times 10^{-2}\,m)^2$

$= 2.25 \times 10^{-2}\,J$

$m = 0.1\,kg$, $\omega = 22.36\,s^{-1}$

$E_{tot} = \dfrac{1}{2}m\omega^2 A^2$

$= \dfrac{1}{2} \times 0.1\,kg \times (22.36\,s)^2 \times (3.0 \times 10^{-2}\,m)^2$

$= 2.25 \times 10^{-2}\,J$

(b) From Question 8, $k = 10\,N\,m^{-1}$, $A = 0.30\,m$, $m = 0.40\,kg$, $\omega = 5.0\,s^{-1}$.

$E_{tot} = \dfrac{1}{2}k\,A^2 = \dfrac{1}{2} \times 10\,N\,m^{-1} \times (0.30\,m)^2$

$= 0.45\,J$

$E_{tot} = \dfrac{1}{2}m\omega^2 A^2$

$= \dfrac{1}{2} \times 0.40\,kg \times (5.0\,s^{-1})^2 \times (0.30\,m)^2$

$= 0.45\,J$

Q 12

(a) $E_p = \dfrac{1}{2}kx^2 = \dfrac{1}{2} \times 1.0 \times 10^2\,N\,m^{-1} \times (0.15\,m)^2$

$= 1.1\,J$

(b) Assuming that the energy transfer to the missile is 100% efficient; i.e. that E_k gained equals E_p lost by catapult:

$\dfrac{1}{2}mv^2 = 1.1\,J$

$v = \sqrt{\dfrac{2 \times 1.1\,J}{2.0 \times 10^{-3}\,kg}} = 33\,m\,s^{-1}$

(c) $\dfrac{1}{2}mv^2 = \dfrac{1}{2}kx^2$ so $v = x\sqrt{\dfrac{k}{m}}$

(d) (i) Since $v \propto x$, the speed is one-third that in (b); i.e. $11\,\mathrm{m\,s^{-1}}$

   (ii) $v \propto 1/\sqrt{m}$ so multiplying m by a factor of two will lead to v being divided by $\sqrt{2}$; i.e.

$$v = \frac{33\,\mathrm{m\,s^{-1}}}{\sqrt{2}} = 23\,\mathrm{m\,s^{-1}}$$

Q 13 (a) Speed $= \dfrac{2 \times \text{length of bar}}{\text{time taken}}$

$$= \frac{2 \times 0.7\,\mathrm{m}}{512 \times 10^{-6}\,\mathrm{s}} = 2734\,\mathrm{m\,s^{-1}}$$
$$\approx 2700\,\mathrm{m\,s^{-1}}$$

(b) It suggests that the circuit contains elements of resistance and capacitance – the decay resembles that of an RC discharge.

Q 14 (a) Young modulus E has SI units $\mathrm{N\,m^{-2}}$. $1\,\mathrm{N} = 1\,\mathrm{kg\,m\,s^{-2}}$. Density ρ has SI units $\mathrm{kg\,m^{-3}}$.

Base units of $\dfrac{E}{\rho}$: $\dfrac{\mathrm{kg\,m\,s^{-2}\,m^{-2}}}{\mathrm{kg\,m^{-3}}} = \mathrm{m^2\,s^{-2}}$

Base units of $\sqrt{\dfrac{E}{\rho}}$: $\mathrm{m\,s^{-1}}$

(b) The Young modulus of oak must be significantly greater than that of pine; despite oak having a higher density, it also has a higher wave speed. Similarly, metals in general have higher Young moduli than woods – their waves speeds are higher despite their higher densities.

(c) From Equation 1, $E = \rho v^2$.
   (i) Pine: $E = 500\,\mathrm{kg\,m^{-3}} \times (3313\,\mathrm{m\,s^{-1}})^2$
     $= 5.49 \times 10^9\,\mathrm{N\,m^{-2}}$
   (ii) Copper: $E = 8933\,\mathrm{kg\,m^{-3}} \times (3650\,\mathrm{m\,s^{-1}})^2$
      $= 1.19 \times 10^{11}\,\mathrm{N\,m^{-2}}$

Q 15 (a) $f = \dfrac{1}{2\pi}\sqrt{\dfrac{k}{m}}$ so:

$$k_{\text{horiz}} = m \times (2\pi f)^2$$
$$= 40 \times 10^3\,\mathrm{kg} \times (2\pi \times 0.60\,\mathrm{Hz})^2$$
$$= 5.7 \times 10^5\,\mathrm{N\,m^{-1}}$$

(b) (i) $k \propto f^2$, so $k_{\text{vert}} = k_{\text{horiz}} \times \left(\dfrac{10\,\mathrm{Hz}}{0.60\,\mathrm{Hz}}\right)^2$
$$= 1.6 \times 10^8\,\mathrm{N\,m^{-1}}$$

   (ii) Compression $\Delta h = \dfrac{F}{k_{\text{vert}}}$

$F = mg$, so:
$$\Delta h = \frac{40 \times 10^3\,\mathrm{kg} \times 9.81\,\mathrm{N\,kg^{-1}}}{1.6 \times 10^8\,\mathrm{N\,m^{-1}}}$$
$$= 2.5 \times 10^{-3}\,\mathrm{m}\ (2.5\,\mathrm{mm})$$

(c) (i) Use Equation 19: $a = -\omega^2 x$, where $\omega = 2\pi f$.

$x = \dfrac{a}{(2\pi f)^2}$ (dropping the minus as we are only interested in size not direction)

$$x = \frac{12\,\mathrm{m\,s^{-2}}}{(2\pi f \times 3.0\,\mathrm{s^{-1}})^2} = 3.38 \times 10^{-2}\,\mathrm{m} \approx 3.4 \times 10^{-2}\,\mathrm{m}$$

   (ii) If the amplitude scales up by a factor of 3, then it will be $0.1\,\mathrm{m}$ – which would be alarming.

Q 16 (a) Linear dimensions are all multiplied by 3, so volume (and hence mass) will be multiplied by 27. Actual mass $= 27 \times 40$ tonnes $= 1080\,\mathrm{t}$

(b) As f depends on k/m, stiffness would also have to be multiplied by 27.

(c) Horizontal oscillations will increase in amplitude as you go up the building, but the vertical amplitude will remain more or less constant with height. Furthermore, any building is under vertical compression because of its weight (largest stress at the bottom). With vertical oscillations the acceleration will increase and decrease this compressive stress, but will be unlikely to put materials under tension (when they are more likely to fail).

Q 17 (a) We need to find $f = \dfrac{1}{2\pi}\sqrt{\dfrac{k}{m}}$

The suspended mass is unknown but it is not needed. Let the mass be m and its weight mg.

Force extending spring, $F = mg$

Extension, $x = 0.20\,\mathrm{m}$

So $k = \dfrac{F}{x} = \dfrac{mg}{x}$

Hence $\dfrac{k}{m} = \dfrac{g}{x}$

$$f_{\text{spring}} = \frac{1}{2\pi} \times \sqrt{\frac{g}{x}} = \frac{1}{2\pi} \times \sqrt{\frac{9.81\,\mathrm{m\,s^{-2}}}{0.20\,\mathrm{m}}}$$
$$= 1.1\,\mathrm{Hz}$$

(b) The overall length $l = 0.80\,\mathrm{m}$. From Equation 7,

$$f_{\text{spring}} = \frac{1}{2\pi} \times \sqrt{\frac{g}{l}} = \frac{1}{2\pi} \times \sqrt{\frac{9.81\,\mathrm{m\,s^{-2}}}{0.80\,\mathrm{m}}}$$
$$= 0.56\,\mathrm{Hz}$$

(c) $\dfrac{f_{\text{spring}}}{f_{\text{pend}}} = 2:1$

Both frequencies are proportional to $\sqrt{\dfrac{1}{\text{length}}}$

The lengths are in a ratio 4:1, so the frequencies must be in the ratio $\sqrt{4} : \sqrt{1}$, i.e. 2:1.

(d) There are two spring oscillations to every pendulum oscillation. Starting off a spring oscillation can gradually set off a pendulum oscillation so that one complete spring cycle takes place in half a pendulum cycle. The system gradually switches from one kind of oscillation to the other and then back again. It can be thought of as a kind of resonance, but where the frequencies are in the ratio 2:1 and it is not clear which is the driving system and which the driven.

Q 18 **(a)** From Equation 8:

$$k = \frac{F}{x} = \frac{700\,\text{N}}{3 \times 10^{-2}\,\text{m}} = 2.3 \times 10^4\,\text{N}\,\text{m}^{-1}$$

(b) From Equation 5,

$$f_0 = \frac{1}{2\pi}\sqrt{\frac{k}{m}} = \frac{1}{2\pi} \times \sqrt{\frac{2.3 \times 10^4\,\text{N}\,\text{m}^{-1}}{400\,\text{kg}}} = 1.2\,\text{Hz}$$

(c) At resonance, time t between bumps is the same as the natural period of oscillations. $T = 1/f_0$

Distance x between bumps is $x = vt$ where

$$v = 70\,\text{km}\,\text{h}^{-1} = \frac{70 \times 10^3\,\text{m}}{3600\,\text{s}}$$

$$x = \frac{v}{f_0} = \frac{70 \times 10^3\,\text{m}}{3600\,\text{s} \times 1.2\,\text{Hz}} = 16\,\text{m}$$

Q 19 **(a)** A heavily damped suspension system takes a long time to deform so the car's body would rise and fall as the wheels went over a bump. People in the car would feel the bump as a single up-and-down movement. The suspension would feel 'hard'.

(b) The suspension would only be lightly damped. Going over a bump would make the car body and its occupants oscillate and it would take some time for the oscillations to die away. The suspension would feel 'soft'.

Q 20 **(a)** Use $E_p = \frac{1}{2}Fx$

$$= \frac{1}{2} \times 1.0 \times 10^5\,\text{N} \times 50 \times 10^{-3}\,\text{m} = 2.5 \times 10^3\,\text{J}$$

(b) Assuming that $E_p = E_k = \frac{1}{2}mv^2$:

$$v = \sqrt{\frac{2E_p}{m}} = \sqrt{\frac{2 \times 2.5 \times 10^3\,\text{J}}{5.5 \times 10^3\,\text{kg}}} = 0.95\,\text{m}\,\text{s}^{-1}$$

(c) The decay *is* exponential. You could use any of the following tests: constant half-life; equal fractions (of A) in equal times; a graph of $\ln(A)$ against n is linear.

(d) At maximum amplitude A in each cycle, $E_p = kA^2$

Using the values for cycle 0:

$$2500\,\text{J} = k \times (50 \times 10^{-3}\,\text{m})^2$$

$$k = \frac{2500\,\text{J}}{0.0025\,\text{m}^2} = 1.0 \times 10^6\,\text{J}\,\text{m}^{-2}$$

Use this value of k to find E_p for each value of A in Table 4.6, to 2 sig. fig.

Values shown in Table 4.10.

Cycle number, n	Amplitude $A/10^{-3}$ m	Stored energy at end of cycle/J
0	50	2500
1	36	1300
2	26	680
3	19	360
4	13	170
5	10	100

Table 4.10 Answer to Question 19(d)

(e) Fractions lost are: energy, 0.48; amplitude, 0.28. The energy decreases by a larger fraction because energy is proportional to the square of the amplitude (e.g. if amplitude halves, energy falls to one-quarter of its initial value).

(f) (i) 1200 J

(ii) 300 J over first quarter cycle

(iii) The kinetic energy as the tower passes through its midpoint is thus 2.2×10^3 J. Using the same method as in **(b)**, this gives $v = 0.89\,\text{m}\,\text{s}^{-1}$.

(g) (i) $v_{av} = \dfrac{0.89\,\text{m}\,\text{s}^{-1}}{2} = 0.45\,\text{m}\,\text{s}^{-1}$

(ii) Displacement, $x = 50 \times 10^{-3}$ m, so:

$$\frac{T}{4} = \frac{x}{v_{av}} = \frac{50 \times 10^{-3}\,\text{m}}{0.45\,\text{m}\,\text{s}^{-1}} = 0.11\,\text{s}$$

(iii) $T = 0.44$ s

(iv) $f = \dfrac{1}{T} = \dfrac{1}{0.44\,\text{s}} = 2.3\,\text{Hz}$

You could probably imagine a tower oscillating at around this frequency; it is not implausible and it is within the range of earthquake frequencies.

Q 21 Rearranging Equation 7:

$$l = \frac{g}{4\pi^2 f^2}$$

$$= \frac{9.81\,\text{m s}^{-2}}{4\pi^2 \times (0.15\,\text{s}^{-1})^2}$$

$$= 11.0\,\text{m}$$

Q 22 $\Delta E = 1\,\text{kg} \times 4200\,\text{J kg}^{-1}\,^\circ\text{C}^{-1} \times 20\,^\circ\text{C} = 84\,000\,\text{J}$

$$\text{Power} = \frac{84\,000\,\text{J}}{120\,\text{s}} = 700\,\text{W}$$

Q 23 Energy from radiant heater: $\Delta E = P\Delta t$

$$\Delta E = 2 \times 10^3\,\text{W} \times 12 \times 60 \times 60\,\text{s}$$

$$= 8.64 \times 10^7\,\text{J}$$

Energy from storage heater: $\Delta E = mc\Delta\theta$

$$\Delta\theta = 70\,^\circ\text{C} - 25\,^\circ\text{C} = 45\,^\circ\text{C}$$

$m = \rho V$ where V is the volume of the block and ρ its density. Therefore:

$$\Delta E = \rho Vc\Delta\theta$$

$$V = \frac{\Delta E}{\rho c\Delta\theta}$$

$$= \frac{8.64 \times 10^7\,\text{J}}{2700\,\text{kg m}^{-3} \times 0.85 \times 10^3\,\text{J kg}^{-1}\,^\circ\text{C}^{-1} \times 45\,^\circ\text{C}}$$

$$= 0.84\,\text{m}^3$$

Q 24 $\Delta E = mc\Delta\theta$

$$P = \frac{\Delta E}{\Delta t} = \frac{\Delta m}{\Delta t} c\Delta\theta$$

where $\dfrac{\Delta m}{\Delta t}$ is the flow rate.

$$\frac{\Delta m}{\Delta t} = \frac{P}{c\Delta\theta}$$

$$= \frac{1.6 \times 10^3\,\text{W}}{4.2 \times 10^3\,\text{J kg}^{-1}\,^\circ\text{C}^{-1} \times 12\,^\circ\text{C}}$$

$$= 0.032\,\text{kg s}^{-1}$$

Q 25 **(a) (i)** The larger the radiator's area, the cooler the radiator.

(ii) The higher the sink temperature, the higher the radiator temperature.

(b) Approximately $1\,\text{m}^2$

Q 26 **(a)** Use Equation 46 ($\Delta E = mc\Delta\theta$) and deal with changes that take place in one minute.

For each wide face, $\Delta\theta = 7\,^\circ\text{C}$ and $m = 2400\,\text{kg}$:

$$\Delta E = 2400\,\text{kg} \times 4179\,\text{J kg}^{-1}\,^\circ\text{C}^{-1} \times 7\,^\circ\text{C}$$

$$= 70\,207\,200\,\text{J}\ (7.02 \times 10^7\,\text{J})$$

For each narrow face $\Delta\theta = 8\,^\circ\text{C}$ and $m = 240\,\text{kg}$.

$$\Delta E = 240\,\text{kg} \times 4179\,\text{J kg}^{-1}\,^\circ\text{C}^{-1} \times 8\,^\circ\text{C}$$

$$= 8\,023\,680\,\text{J}\ (8.02 \times 10^6\,\text{J})$$

So the total energy transferred in one minute is:

$$2 \times (70.2 + 8.02) \times 10^6\,\text{J} = 1.56 \times 10^8\,\text{J}$$

(b) Power $P = \dfrac{\Delta E}{\Delta t}$, $\Delta t = 1\,\text{min} = 60\,\text{s}$

$$P = \frac{1.56 \times 10^8\,\text{J}}{60\,\text{s}}$$

$$= 2.6 \times 10^6\,\text{W} = 2.6\,\text{MW}$$

Q 27 **(a)** SI units of U: $\text{W m}^{-2}\,\text{K}^{-1}$ (note that K not $^\circ$C is the base unit for temperature).

$$1\,\text{W} = 1\,\text{J s}^{-1} = 1\,\text{N m s}^{-1} = 1\,\text{kg m s}^{-2}\,\text{m s}^{-1} = 1\,\text{kg m}^2\,\text{s}^{-3}$$

Base units of U: $\text{kg m}^2\,\text{s}^{-3}\,\text{m}^{-2}\,\text{K}^{-1} = \text{kg s}^{-3}\,\text{K}^{-1}$

(b) From Equation 47 with $\Delta\theta = 10\,^\circ\text{C}$ and $A = 6\,\text{m}^2$:

$$P = 2.2\,\text{W m}^{-2}\,^\circ\text{C} \times 6\,\text{m}^2 \times 10\,^\circ\text{C} = 132\,\text{W}$$

Q 28 Use Equation 48 with $\Delta m = \rho V$

Volume of snow, $V = 1.6\,\text{m} \times 55\,\text{m}^2$

$$\Delta E = L\Delta m = L\rho V$$

$$= 3.34 \times 10^5\,\text{J kg}^{-1} \times 200\,\text{kg m}^{-3} \times 1.6\,\text{m} \times 55\,\text{m}^2$$

$$= 5.9 \times 10^9\,\text{J}$$

Q 29 $\Delta E = L\Delta m$ and $\Delta E = IV\Delta t$ so

$$L = \frac{IV\Delta t}{\Delta m}$$

$$= \frac{2.9\,\text{A} \times 11.9\,\text{V} \times 20 \times 60\,\text{s}}{18 \times 10^{-3}\,\text{kg}}$$

$$= 2.3 \times 10^6\,\text{J kg}^{-1}$$

Q 30 **(a)** $\Delta E_{\text{in}} = P\Delta t$

$$= 3.4 \times 10^3\,\text{W} \times 3 \times 3600\,\text{s}$$

$$= 3.67 \times 10^7\,\text{J}$$

From Equation 46:

$$\Delta\theta = \frac{\Delta E}{mc} = \frac{3.67 \times 10^7\,\text{J}}{152\,\text{kg} \times 950\,\text{J kg}\,^\circ\text{C}^{-1}} = 254\,^\circ\text{C}$$

(b) Brick near the heating element will be much hotter. Near the surface, in contact with the surrounding air, the brick will be much cooler.

(c) Using Equation 47:

$$P_{out} = UA\Delta\theta$$

$$= 11.2\,\text{W m}^{-2}\,°\text{C} \times 0.23\,\text{m}^2 \times 254\,°\text{C} = 654\,\text{W}$$

Q 31 **(a)** The large reflector focuses the solar radiation onto the water pipes. The smaller reflector 'beams' radiation from the heated water back into the room. Without the smaller reflector the insulator would absorb most of the radiation not emitted directly into the room.

(b) Insulation helps ensure that the absorbed energy is stored in the water and is not transferred directly into the room. Glass fibre or rock wool (such as that used in loft insulation) could be used. These materials trap air within and prevent it from circulating. The materials themselves are also poor conductors, as is the air. They therefore reduce both convection and conduction.

(c) **(i)** Warm water is less dense than cold water, so during the day the water circulates within the system by convection: heated water moves upwards to the tank while cooler water descends to the blind.

(ii) At night, when warm water is required to move downwards from tank to blind, a pump is needed.

(d) Energy transfer will be by mainly by convection (warmed air near the blind rises, setting up convection currents in the room) and also by radiation. Transfer by conduction will be minimal as air is a poor conductor.

(e) Ideally the tubes should be good absorbers of solar radiation (mostly visible and infrared), and good emitters of infrared radiation – preferably their surfaces should be dull black. They should also be good thermal conductors in order to transfer energy to and from the water.

Copper is a better thermal conductor than glass so energy would transfer more easily through the tube walls. However, depending on its density and specific heat capacity, more energy might be wasted in heating copper tubes at the expense of the water.

(f) **(i)** The surface area $A = 4\,\text{m}^3$ so the total amount of solar energy intercepted must be $15\,\text{MJ m}^{-2} \times 4\,\text{m}^2 = 60\,\text{MJ}$. Energy transferred to water is 70% of 60 MJ, i.e. 42 MJ ($= 42 \times 10^6$ J).

(ii) From Equation 46:

$$\Delta E = mc\Delta\theta \quad \text{so} \quad \Delta\theta = \frac{\Delta E}{mc}$$

$$\Delta\theta = \frac{42 \times 10^6\,\text{J}}{6000\,\text{kg} \times 4200\,\text{J kg}^{-1}\,°\text{C}^{-1}} = 1.7\,°\text{C}$$

Q 32 **(a)** The resistance of the thermistors decreases with increasing temperature.

(b) **(i)** Power transferred from heater = $240\,\text{V} \times 15\text{A}$
$$= 3600\,\text{W}$$

(ii) In 1 s, 0.1 kg of gas is heated through 13 °C. Energy transferred in 1 s is found from $\Delta E = mc\Delta\theta$:
$$\Delta E = 0.1\,\text{kg} \times 2200\,\text{J kg}^{-1}\,°\text{C}^{-1} \times 13\,°\text{C}$$
$$= 2860\,\text{J}$$
So power transferred to gas = $2860\,\text{J s}^{-1}$
$$= 2860\,\text{W}$$

(iii) The 'missing' power is transferred to the thermistors and the rest of the electric circuit, and to the pipeline walls and through them to the surroundings. Using the law of conservation of energy, this 'missing' power must be 3600 W – 2860 W = 740 W.

(c) **(i)** Since the 'missing' power is again 740 W, power transferred to gas = 2500 W – 740 W = 1760 W.

(ii) In 1 s, 1760 J is transferred to the gas, producing a temperature rise of 13 °C. $\Delta E = mc\Delta\theta$ so the mass, m, flowing past the heater in 1 s is given by:
$$m = \frac{\Delta E}{c\Delta\theta}$$
$$= \frac{1760\,\text{J}}{2200\,\text{J kg}^{-1}\,°\text{C}^{-1} \times 13\,°\text{C}} = 0.06\,\text{kg}$$
so the flow rate is 0.06 kg $^{-1}$

Q 33 **(a)** C

(b) **(i)** 3.5 °C (without PCM, temperature is 15 °C, with PCM it is 18.5 °C)

(ii) 2.5 °C (without PCM, temperature is 28 °C with PCM it is 25.5 °C)

(c) **(i)** 13:00
(ii) 16:00
(iii) between 21:00 and 22:00

(d) Use Equation 46 with $\Delta\theta = 24\,°\text{C} - 16\,°\text{C} = 6\,°\text{C}$

$$\Delta E = 8400\,\text{kg} \times 1080\,\text{J kg}^{-1}\,°\text{C}^{-1} \times 6\,°\text{C}$$

$$= 5.4 \times 10^7\,\text{J}$$

(e) (i) To reach melting point,
$$\Delta\theta = 18\,°C - 16\,°C = 2\,°C$$

Using Equation 46:

$$\Delta E = 8400\,kg \times 1160\,J\,kg^{-1}\,°C^{-1} \times 2\,°C$$

$$= 1.9 \times 10^7\,J$$

(ii) To melt the PCM:
$$\Delta E = mL$$
$$= 300\,kg \times 189 \times 10^3\,J\,kg^{-1}$$
$$= 5.7 \times 10^7\,J$$

(f) From (e), energy required to reach 18 °C and melt all the PCM is $1.9 \times 10^7\,J + 5.7 \times 10^7\,J = 7.6 \times 10^7\,J$. This is more than the $5.4 \times 10^7\,J$ used to heat the control room. So when $5.4 \times 10^7\,J$ has been transferred to the test room, not all the PCM will have melted, therefore the temperature will be the melting point: 18 °C.

CHAPTER 5

REACH FOR THE STARS

Why a chapter called *Reach for the Stars?*

Ever since human beings with modern brains have walked the planet, they've been doing cosmology – asking themselves questions about their ultimate origins: how the world came into being and what will happen to it. Yet, science has only been able to tackle such questions since the 20th century; and the pace of discovery has been astounding. Halfway through the 20th century, Cambridge radio astronomer and astrophysicist Peter Scheuer said: 'there are only two-and-a-half facts in cosmology'. Now we have experimental and observational evidence (Figure 5.1) to support our own myths of creation. And we can even begin to speculate on almost unimaginable questions, such as the existence of other Universes.

This chapter will show how physics has helped modern-day cosmologists unlock some of the secrets of the Universe and its origins: What makes stars shine? Do black holes exist? Is the Universe infinite?

The images shown in Figures 5.1(a) to (c) were obtained with the Hubble Space Telescope.

Figure 5.1 (a) The Eagle Nebula (Pillars of Creation), a region where stars are forming

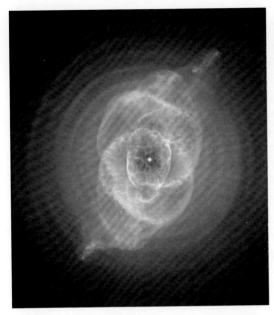

Figure 5.1 (b) The alluring Cat's Eye Nebula produced by the death of a star (photo courtesy of NASA, ESA, HEIC, and The Hubble Heritage Team (STScI/AURA))

Figure 5.1 (c) Some of the most distant galaxies yet observed

OVERVIEW OF PHYSICS PRINCIPLES AND TECHNIQUES

In this chapter you will study several areas of physics that relate to astronomy and cosmology. You will learn about the behaviour of light that enables astronomers to interpret observations of distant objects, the nuclear reactions that power stars, and the use of radioactivity to date rocks that make up our planet. You will also consider the ideas about molecules that help us to understand how stars form, and about the gravitational forces that keep stars and planets in their orbits and will determine the ultimate fate of the Universe. This chapter will also give you several opportunities to look back over your work from previous chapters and review your knowledge and understanding.

In this chapter you will extend your knowledge of:

- motion in a circle from *Probing the Heart of Matter*
- energy and work from *Higher, Faster, Stronger, Transport on Track, Probing the Heart of Matter* and *Build or Bust?*
- radiation from *Technology in Space* and *The Medium is the Message*
- radioactivity and nuclear reactions from *Probing the Heart of Matter*
- waves from *The Sound of Music, Technology in Space* and *Build or Bust?*
- inverse-square law fields from *Probing the Heart of Matter*.

1 In the beginning

1.1 Big questions

Figure 5.2 The supernova of 1054 AD is thought to be depicted on the walls of Chaco Canyon, New Mexico

Humanity has studied the stars since the dawn of civilisation. People have been making notes on cosmological observations for tens of thousands of years – as far back as 20,000 BC, when a bone fragment from sub-Saharan Africa was used to illustrate a lunar calendar. Throughout the course of human history there have been a few occasions that have truly caught our attention. Such an event occurred around the 4th July 1054 AD, where a star within the constellation Taurus dramatically exploded in a supernova; it was so bright that it could be seen during the daytime for around 3 weeks. The event was chronicled by a Christian physician called Ibn Butlan in Baghdad, by the Chinese – via the Suzhou – star map, and by the Japanese. The most beautiful record of the supernova, albeit slightly controversial in its interpretation, was in New Mexico in a star map in Chaco Canyon (Figure 5.2). It depicts the moon in a waning crescent position and the supernova to the left of it with the hand considered to represent part of the constellation of Orion. This supernova left behind a remnant that we now observe as the Crab Nebula.

The Crab Nebula has once again come to our attention in modern times due to the significant advancement of data gathering techniques and depiction. It has produced some of the most iconic astronomical images of our time (Figure 5.3). Our imaging of astronomical objects is now so sophisticated that we can produce images of wavelengths outside of the visible spectrum. The images in Figure 5.3 show how different the Crab Nebula appears at radio, infrared, visible, ultraviolet and X-ray wavelengths. The colours selected for wavelengths outside the visible range therefore do not represent true colours. The montage shows the filament-like structure of the nebula at visible wavelengths, and a more diffuse emission at the ultraviolet and radio wavelengths. At X-ray wavelengths a disc-like structure can be seen, with jets coming out from either side; this is a pulsar emitting synchrotron radiation as it spins. The Crab Nebula is now known to measure 9 light years from side to side, to be around 6500 light years away, and to contain a pulsar in the centre: a neutron star only 20 km across, 1.4 times the mass of the Sun, rotating at a staggering 30 times per second.

Figure 5.3 The Crab Nebula imaged at different wavelengths

Firstly, how do we know the size and distance of the Crab Nebula and pulsar? What was powering the star that originally exploded in the 1054 AD supernova? How does the expansion of the nebula link to gravitational forces and movement of gases in the nebula? What happened to make the star explode? Why and what is the pulsar in the centre of the nebula? Finally, where does the Crab Nebula fit into the history of the Universe? This chapter aims to probe these questions and start you on a journey asking some of the deepest questions about the origins and future of the Universe.

2 Cosmic measuring tapes

2.1 Astronomical units

In the 3rd century BC Aristarchus proposed the idea that the Earth revolves around the Sun. However, it was the **geocentric** model, placing the Earth at the centre of the Solar System, which persisted right up until the 16th century AD when Polish astronomer, mathematician and cleric Nicolaus Copernicus revived the **heliocentric** (Sun-centred) model.

Over the next century, Johannes Kepler elaborated on this model and Galileo added further observations, made using the recently invented telescope.

During the 18th and 19th centuries, astronomers grew to realise that the Sun was not at the centre of the Universe. By the early 20th century it was understood that our Sun was just one of many stars in just one of many billions of galaxies in an expanding Universe.

Measuring the distances to stars and galaxies may not seem the most interesting of activities for an astronomer, but its importance cannot be overestimated. The main reason for this is that it allows astronomers to determine the scale of the Universe, which in turn has a bearing on its past and future. Another important reason for knowing the distances to stars is that the fundamental physical properties of the stars can be found. This is essential for the study of stellar evolution.

How can we measure distances across the vast expanse of space? Astronomers have a toolkit of over twenty techniques, each of which is used for a particular range of distances. These are explored in the next sections, starting with the measurement of relatively small distances within the Solar System.

Question: Which is the star nearest the Earth?

Answer: It's not Proxima Centauri, a faint star in the constellation Centaurus and second closest to Earth. The answer is the Sun.

This is an old joke, but it is still easy to trick people with this question. Because the Sun looks so different from anything else in the sky, we sometimes forget that it is really a star. In fact the Sun is a very ordinary, average-sized typical star. Compared with the billions of other stars in the Universe there is nothing special about it – but it is very much closer to Earth.

Proxima Centauri is a distance of 4.24 light years from Earth. A **light year** is the distance light travels in a vacuum in one year. The average distance between the Earth and the Sun is 1.50×10^{11} m. This is known as **1 astronomical unit, 1 AU**.

ACTIVITY 1 INTO THE SUN

Use the internet to research some information about the Sun. In particular, look for information about space missions and telescopes that are studying physics processes and conditions in the Sun. Go to **www.pearsonhotlinks.co.uk** for some useful links.

QUESTION

Q 1 **(a)** How long does it take light to travel from the Sun to Earth, given that light travels at 3.00×10^8 m s^{-1}?

(b) Calculate the distance to the Sun in light years, given that 1 year = 3.16×10^7 s.

(c) Calculate the ratio $\dfrac{d_{PC}}{d_{Sun}}$ where d_{PC} is the distance from Earth to Proxima Centauri and d_{Sun} is the distance from Earth to the Sun.

Then express your answer as an order of magnitude – that is, state the nearest power of ten $(10, \ldots 10^4, 10^5, 10^6, \ldots)$.

(d) Calculate the following distances in metres.
  (i) Sun to Neptune, 250 light minutes
  (ii) Diameter of the Milky Way galaxy, 100 000 light years.

STUDY NOTE

This activity should remind you about aspects of the chapter *Technology in Space* and help you to look ahead to later sections of this chapter.

PRACTICAL SKILLS REFERENCE

Scientific questions and information research

See Practical skills notes 1.1–1.3

The Sun and planets, plus other objects such as moons and comets, make up the Solar System, our little neighbourhood within the Milky Way galaxy. Objects within the Solar System and other nearby stars look bright/dimmer, larger/smaller depending how far away they are.

2.2 Size and distance within the Solar System

The first technique in our measurement toolkit is simple geometry. It can be used to measure distances to celestial objects which are close enough to appear as a disc rather than just a point of light.

It is difficult to imagine the vastness of the Sun. We know that it is much further away than the Moon, although the Sun appears to be just about the same size as the Moon when we observe it from Earth – in other words, the Sun and the Moon have a very similar **angular diameter**. The Moon is at a distance of about 3.85×10^8 m from Earth. The Sun is approximately 400 times further away than the Moon – but it is much bigger.

Figure 5.4 shows how angular diameter is defined; the ratio of actual diameter D to distance d defines the angle α, measured in radians, that the object subtends at the observer:

$$\alpha = \frac{D}{d} \tag{1}$$

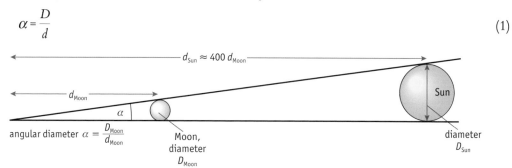

Figure 5.4 The Sun and Moon have similar angular diameters. Image not to scale.

Angular diameters and angular separations are used quite a lot by astronomers, since they can be measured directly, unlike true sizes and separations.

MATHS REFERENCE

Degree and radians

See Maths note 6.1

The small angle approximations

See Maths note 6.6

Unlike the Earth, the Sun does not have a solid surface. When we talk about the 'surface' of the Sun we usually mean the edge of the ball of light that we see. This is the photosphere – a layer of extremely hot gases that emits most of the Sun's electromagnetic radiation. The photosphere has a diameter of 1.4×10^9 m. The Sun's outer atmosphere, the corona, extends out into space sometimes by more than twice the radius of the photosphere, and the flares that erupt from the surface of the Sun arch in loops bigger than the Earth (Figure 5.5). We can observe the solar corona during a total eclipse, when the Moon blocks light from the photosphere.

Figure 5.5 (a) The solar corona, showing the inner streamer belt along the Sun's equator; the field of view in this coronagraph encompasses 8.4 million kilometres; (b) a solar flare arches above the photosphere

It is relatively easy to measure the angular diameter of the Sun's photosphere. In Activity 2, you do this using a pinhole camera. It works best with a big camera, so that the image of the Sun can be measured precisely.

⚙ ACTIVITY 2 MEASURING THE SUN AND MOON

Use a large pinhole camera to measure the angular size of the Sun, as shown in Figure 5.6, and use an 'eclipse' method to measure the angular diameter of the Moon (Figure 5.7). Compare your two results.

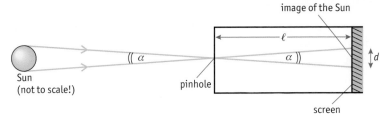

Figure 5.6 Measuring the angular diameter of the Sun

Figure 5.7 Measuring the angular diameter of the Moon

QUESTIONS

For these questions, you will need to use information from the previous paragraphs.

Q 2 Calculate the approximate angular diameters, in radians, of the Sun and Moon, as observed from Earth.

Q 3 The Earth has a radius of about 6.4×10^3 km. Calculate the volumes of the Earth and the Sun. To the nearest order of magnitude, how many Earth volumes would fit into the Sun?

2.3 Stellar parallax

The only star that appears as a disc from Earth is our Sun. All other stars, even Proxima Centauri, are too far away and look like points of light. Measuring their distance using angular diameter is therefore impossible. However, another geometric measuring technique is available to astronomers, at least for nearby stars.

As the Earth rotates on its axis we see the stars move across the sky, as is demonstrated in Activity 3. Figure 5.8 also reveals the stars' apparent motion.

PRACTICAL SKILLS REFERENCE

Carrying out practical work
See Practical skills notes 3.1–3.2

Analysis and interpretation of data
See Practical skills notes 4.1–4.2

SAFETY NOTE

Never look directly at the Sun. Sunglasses are not sufficient protection.

Figure 5.8 Long-exposure photograph of the night sky

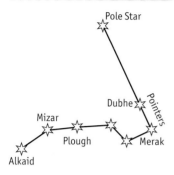

Figure 5.9 Finding Polaris

ACTIVITY 3 STAR WATCH

On a clear night, find the major constellation known as the Plough. Figure 5.9 shows you how to find Polaris, the Pole Star, using two of the stars in the Plough.

Alternatively, use Stellarium, which is is a free open-source planetarium for your computer (see **www.pearsonhotlinks.co.uk**). It shows a realistic sky, just like you see with the naked eye, binoculars or a telescope. If you set your coordinates, you can see the night sky from your own location at any time now or in the future.

Star Walk is an interactive star chart for use on a smartphone or tablet.

Draw a circle on a sheet of A4 paper and put the Pole Star in the centre. Observe the position of the Plough and draw it onto your diagram. Do the same thing every hour for, say, three hours.

Note how the Plough changes its position relative to the Pole Star and the horizon over the three-hour period. Find the approximate angle through which the Plough rotates in 1 hour and relate this to the length of the day.

An important thing to note from Activity 3 is that although the stars appear to move across the sky as the Earth rotates, the relative motion of the stars – that is, the motion of one star with respect to another – seems to be zero. Another way of saying this is that the star patterns of the constellations are fixed. One might expect that as the Earth moves around its orbit, the nearer stars would move relative to the more distant background stars in much the same way that nearby objects move more rapidly across the field of view compared to more distant objects; this effect is seen when you drive along a motorway, for example. The effect is called **parallax**.

Since there was no observable stellar parallax, some of Newton's contemporaries in the 17th century assumed that stars were infinitely distant and the observation led some astronomers to doubt the heliocentric (Sun-centred) model of the Universe that had been proposed by Copernicus in the 16th century.

Astronomical instruments became accurate enough to detect the parallax of nearby stars in 1838, when Frederich Bessel detected the parallax motion of 61 Cygni. Figure 5.10 shows the idea.

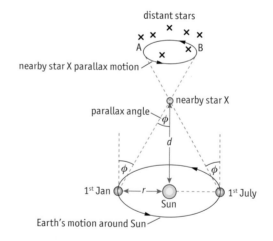

Figure 5.10 Stellar parallax

On 1st January, star X appears in position A against the background stars, but by 1st July, six months later, it appears in position B. The parallax angle, ϕ, is defined as half the total angular motion, and so the star appears to have moved through an angle of 2ϕ in six months (due only to the orbital motion of the Earth). Knowing the distance from Earth to Sun, r, we can calculate the distance, d, to the star:

$$d = \frac{r}{\phi} \tag{2}$$

where ϕ, the **parallax angle**, is in radians. This equation relies on ϕ being small. In practice, for stars, it always is; the nearest star to the Sun is Proxima Centauri, which has a parallax less than 1".

Parallax measurements give rise to the unit of distance commonly used by professional astronomers, the **parsec** (pc), which is short for *parallax second*. One parsec is defined as the distance of an object that would have a parallax of one arcsecond when observed from Earth.

$$1pc = 3.09 \times 10^{16}\,m = 3.26 \text{ light years}$$

Obtaining accurate stellar parallaxes is considered so important that the European Space Agency (ESA) designed and launched a special satellite called Gaia (Global Astrometric Interferometer for Astrophysics) in 2013, to chart a 3D map of the Milky Way galaxy. Over its five-year mission it will measure the positions of 1 billion stars with unprecedented precision, building on the work of the earlier Hipparcos mission. After some initial challenges, Gaia, which is in orbit around the Sun, started measurements, and in September 2014 discovered a supernova in a galaxy 500 million light years away.

MATHS REFERENCE

Angular units

See Maths note 6.7

Figure 5.11 The Gaia satellite

QUESTIONS

Data:
Earth–Sun distance, $r = 1.49 \times 10^{11}\,m$,
speed of light, $c = 3.00 \times 10^8\,m\,s^{-1}$,
1 year = $3.1536 \times 10^7\,s$,
1 rad = 2.06×10^5 arcsec,
1 parsec = $3.1 \times 10^{16}\,m$

 (a) The star Alpha Centauri is 4.351 light years from the Solar System. Calculate its parallax in radians and in arcsec.

(b) Use the definition of the parsec to confirm that it has the value quoted above.

 The smallest parallax that Gaia's predecessor, Hipparcos, could measure was 0.002". Determine the distance of the furthest star that Hipparcos could measure accurately. Express your answer in metres, in light years and in parsecs.

2.4 Brightness and distance

The reason that, when viewed from Earth, the Sun appears so much brighter than the other stars is because it is much closer. As you move further away from the source of any radiation, the radiation becomes less intense. This is because the radiation is spreading out in all directions, and so the amount that will land on you will become less as you move away. Figure 5.12 shows that if you double the distance from a star, then the area over which the radiations spread is multiplied by four; if you multiply the distance by three, then the area is multiplied by nine.

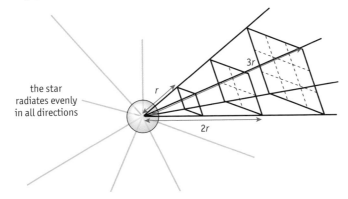

the star radiates evenly in all directions

Figure 5.12 Radiation spreads out as it travels

Figure 5.12 shows that the **intensity**, I, of the radiation must obey an **inverse-square law**. An inverse-square law is one in which two quantities (say x and y) are related by an equation of the type:

$$y = \frac{\text{constant}}{x^2}$$

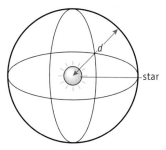

Figure 5.13 Radiation from a star

Intensity (sometimes called energy flux) is the rate at which energy is transferred across a unit area perpendicular to the light beam. Intensity is not quite the same as observed brightness, since the brightness we observe depends only on the visible component of the radiation, whereas intensity refers to the whole electromagnetic spectrum.

The **luminosity**, L, of a radiation source (such as a star) is defined as the rate at which the source radiates energy – the total energy it loses every second; that is, its radiated power. The SI unit of luminosity is the watt.

The further away the star, the lower the intensity seen by an observer. Figure 5.13 shows a star at the centre of an imaginary sphere of radius d.

The radiation emitted from the star in one second will spread out so that it is shared evenly over the surface of the sphere. Travelling at a steady speed, if it is not absorbed or scattered en route, the radiation will pass through the imaginary sphere in such a way that:

$$\text{Intensity} = \frac{\text{total energy emitted from source per second}}{\text{surface area of sphere}}$$

$$I = \frac{L}{4\pi d^2} \tag{3}$$

Intensity is measured in watts per square metre (W m^{-2}).

Equation 3 is important. It expresses the inverse-square law for intensity, and it enables us to calculate the distance to a star if we can measure its intensity at Earth and know its luminosity (or find its luminosity if we know the distance).

ACTIVITY 4 EXPLORING THE INVERSE-SQUARE LAW

Measure the conductance of a light-dependent resistor (LDR) when illuminated by a small light bulb at different distances. Comment on the extent to which the results are consistent with an inverse-square law for radiation intensity.

QUESTIONS

Q 6 The Sun is the main source of energy for all the planets and other bodies in the Solar System. By calculating intensity at a planet we can begin to have some idea of the environmental conditions on that planet. Table 5.1 shows the distance to the Sun from various bodies in the Solar System.

Body	Distance from Sun/10^9 m
Mercury	57.9
Venus	108.2
Earth	149.6
Mars	227.9
Pluto	5900.0

Table 5.1 Distances to bodies from the Sun

(a) For each body in Table 5.1, calculate the ratio $\dfrac{I_{\text{body}}}{I_{\text{Earth}}}$, where I_{Earth} is the intensity at Earth and I_{body} in the energy flux at the planet in question. (Hint: use Equation 3. You don't need to know the Sun's luminosity, L, because when you calculate the ratio it cancels.)

(b) How will the intensity of radiation from the Sun affect the temperature on each planet?

(c) What other factors do you think might affect the temperature on the surface of a planet?

Q 7 In the 17th century Christiaan Huygens (Figure 5.14) worked out the approximate distance to the star Sirius by the 'faintness means farness' principle. He made a small hole in the blackened window of a dark room and adjusted the opening until the beam of sunlight appeared to match the brightness of the star as he remembered it. He then measured the diameter of his hole, D_a, and the apparent diameter of the Sun, D_s, as it appeared at the window. He found that the hole had a diameter $\dfrac{D_a}{26\,644}$. Since the brightness of the beam depends on the area of the light beam, he then reasoned that:

$$\frac{I_{Sir}}{I_{Sun}} = \frac{\text{area of aperture}}{\text{apparent area of Sun}} = \left(\frac{1}{26\,644}\right)^2$$

(a) Huygens assumed that Sirius and the Sun were equally luminous. Use his assumption and Equation 1 to calculate, as he did, the ratio of distances from the Earth, $\dfrac{d_{Sir}}{d_{Sun}}$.

(b) In Question 1 you calculated the distance to the Sun as 1.58×10^{-5} light years. Given that Sirius's true distance is about 8 light years, calculate the real ratio $\dfrac{d_{Sir}}{d_{Sun}}$.

(c) Compare your answers from parts (a) and (b). At a time when no one had any idea how far it was to the stars, Huygens's work was brilliant, but not above criticism. Suggest an explanation for his result being so far out, and criticise his experimental technique.

Figure 5.14 Christiaan Huygens

Photometry

Equation 3 describes how the intensity of radiation from a source depends on the distance to the source and its luminosity. If we know (or assume) the luminosity of a star and measure its intensity, then we can calculate its distance – or, knowing the distance, we can calculate its luminosity. As you will see in Section 2.5, it is fairly easy to deduce a star's luminosity from its visual appearance, and so Equation 3 underlies another method of determining stellar distances, known as **photometry**.

The starlight may be absorbed or scattered en route by clouds of gas or dust, or in our own atmosphere, which limits the accuracy of this technique.

A very simple way to measure intensity using an oil spot on a piece of paper was devised in the middle of the 19th century by Robert Bunsen (1811–1899). Ordinary paper becomes extremely translucent when it is made greasy, so it is easily lit from behind. If a piece of paper is held in front of a lamp, any oil spots show up brightly. However, if the paper is also being lit from the observer's side, the ordinary white paper is bright. Bunsen made the assumption that if the oil spot and the paper in such a situation appeared to be of equal brightness, then the intensities of the radiation falling on the front and the back of the paper were equal.

In Activity 6 you will use Bunsen's method to find the Sun's luminosity. The essence of the experiment is very simple. You need a clear sunny day to carry it out. A piece of paper, dotted with oil, is held between a lamp and an observer. The lamp illuminates the oil from below while the Sun illuminates the paper from above (see Figure 5.15). The distance between the paper and the lamp is adjusted until the paper and the oil spot appear equally bright.

We know that:

$$I_{Sun} = I_{lamp}$$

where I refers to the intensity of radiation on the paper. From Equation 3:

$$\frac{I_{Sun}}{4\pi d^2_{Sun}} = \frac{I_{lamp}}{4\pi d^2_{lamp}}$$

which can be rearranged to find L_{Sun} if the other quantities are known.

Figure 5.15 Bunsen's method of measuring luminosity

PRACTICAL SKILLS REFERENCE

Scientific questions and information research

See Practical skills notes 1.1–1.3

Carrying out practical work

See Practical skills notes 3.1–3.2

Analysis and interpretation of data

See Practical skills notes 4.1–4.2

Conclusion and evaluation

See Practical skills notes 5.1–5.2

SAFETY NOTE

Never look directly at the Sun. Sunglasses are not sufficient protection.

ACTIVITY 5 THE SUN'S LUMINOSITY

Using Bunsen's method, determine the Sun's luminosity. You will need to know the distance to the Sun: $d_{Sun} = 1.50 \times 10^{11}$ m.

Your measurement will be subject to experimental uncertainty, and your calculation of luminosity is based on several assumptions. Try to think what these assumptions are, and suggest how you might correct for them.

2.5 Luminosity and temperature

At first you might think that measuring the temperature of a star is an impossible task – but it turns out that temperature is one of the easiest quantities to determine using remote observations. Photographs such as Figure 5.16 reveal that stars are coloured and, as you will see in Activity 6, temperature affects both the luminosity and the colour of a radiation source. Once a star's luminosity has been determined, we can use Equation 3 (Section 2.4) to estimate its distance: we have another cosmic measuring technique.

Figure 5.16 'True colour' photograph of part of the southern sky

ACTIVITY 6 COLOUR, LUMINOSITY AND TEMPERATURE

Switch on an electric heater or a laboratory radiant heater in a dimly lit room and observe the change in its appearance as it gets hotter. Notice in particular the changes in colour and in brightness.

What is happening? Discuss your observations with other students. There is a transfer of energy to the heating element, so it gets hotter. But why do we see what we do? And why does the element eventually stop getting hotter, even though energy is still being transferred to it at roughly the same rate?

Compare the colour and brightness of the electric heater with some hotter objects, such as a candle flame, a tungsten filament in a light bulb and the Sun. (Remember not to look directly at the Sun.)

In Activity 6 you saw that as objects get hotter they go from emitting invisible (infrared) radiation to emitting red, orange, yellow and white light; the hotter the whiter. Also, as the object gets hotter the intensity of the radiation increases: the hotter the brighter. You might argue that red objects are not necessarily red hot – an object, such as a pillar box, will appear red if it is illuminated because it reflects red light. However, a red-hot object glows red in the dark because of the light that it *emits*. This radiation given off from a body because of its temperature (which is determined by the thermal motion of its constituent atoms and molecules) is called **black-body radiation**. A body does not have to be black to emit black-body radiation – the term distinguishes bodies that are emitting radiation because they are hot from those that are coloured by dyes and paints. A perfectly black body will glow orange, red or white if you get it hot enough.

To the naked eye, stars mostly appear as faint pinpoints of white light. It is much easier to see their colours using binoculars or using a telescope to gather more light. However, most people can see the differences in colour when they look carefully at some of the brighter stars such as those in Orion (Figure 5.17). Even with the naked eye it is possible to see that Betelgeuse, one of the brightest stars in the constellation, has a red colour, whereas Rigel is blue-white. You can see more colour photographs of stars by searching the NASA gallery (see **www.pearsonhotlinks.co.uk**).

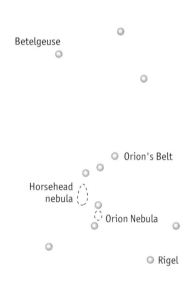

Betelgeuse

Orion's Belt

Horsehead
nebula

Orion Nebula

Rigel

Figure 5.17 Orion: (a) colour photograph, (b) the brightest stars in the constellation

⚙ ACTIVITY 7 COLOURS OF THE STARS

On a clear winter's night, find the constellation Orion and identify the stars Betelgeuse and Rigel, as shown in Figure 5.17. (You could use Stellarium or StarWalk to help you find the right part of the sky; see Activity 3.) Give your eyes about 10 minutes to adapt to the darkness and then you should be able to see a distinct difference in the colours of these stars. Which of these two stars is the hotter?

When we view stars in the night sky, their light is weak and our eyes do not respond to colour well in these conditions. Some people will inevitably find it difficult to discern the colours of different stars. You will see the colours more clearly if you use binoculars. You could try to take a series of colour photographs of Orion with different exposure times, which should bring out the colour difference.

QUESTION

Q 8 Based on their names alone, put the following types of star in order of increasing temperature: red giant, blue supergiant, white dwarf, brown dwarf.

Black-body radiation

A better name for a **black body** would be an **ideal thermal radiator**. It is an object whose radiation spectrum depends only on its temperature and not on its composition. Such a body is also a 'perfect absorber' – it absorbs all radiation that falls on it, and no radiation is reflected. This second feature makes the term 'black body' seem more sensible. The radiation spectrum of a true black body is a smooth curve, like those shown in Figure 5.18.

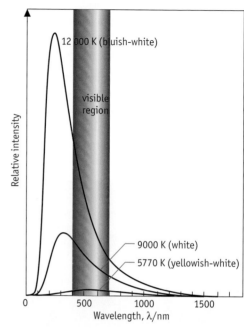

Figure 5.18 Black-body radiation curves (note the different vertical scales)

STUDY NOTE

Notice that Figure 5.18 uses a linear scale on the wavelength axis, whereas the scale in Figure 5.19 is logarithmic.

Figure 5.18 shows how the distribution of radiation (in other words, the spectrum) emitted from a black body varies as the body's temperature rises. The curves all have the same sort of shape but, as you observed in Activity 6, the intensity of the radiation increases dramatically as the temperature rises. Notice, too, that the position of the maximum shifts to shorter wavelengths.

To be a black body, an object must be at the same temperature throughout. In practice, very few objects behave exactly like true black bodies, but some are fairly close. Figure 5.19 shows that the Sun, for example, is not a true black body, particularly at wavelengths less than about 10^{-6} m, but the overall hump-shaped spectrum is quite similar to that of a black body.

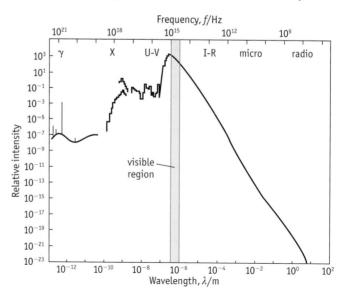

Figure 5.19 Radiation spectrum of the Sun

The radiation from a true black body is described by two laws, which can loosely be summarised as 'the hotter the whiter' and 'the hotter the brighter'.

As the temperature of a black body is raised, the peak of its radiation spectrum shifts to shorter wavelengths. This is described by **Wien's law** (sometimes called 'Wien's displacement law'):

$$\lambda_{max} T = 2.89 \times 10^{-3} \, \text{m K} \tag{4}$$

STUDY NOTE

The unit of constant in Equation 4 is metre kelvin (not millikelvin).

where T is the temperature on the absolute (Kelvin) scale and λ_{max} is the wavelength at which the brightness is maximum.

Raising the temperature also increases the amount of radiation emitted from the object's surface. The total energy radiated each second from each square metre of surface is proportional to the fourth power of its absolute temperature, T, and is described by by **Stefan's law**:

$$I = \sigma T^4 \tag{5}$$

STUDY NOTE

For more on the Kelvin temperature scale, see Section 4.2.

where the constant, σ, is known as the **Stefan constant** or the **Stefan–Boltzmann constant**. I is measured in W m^{-2}.

To find the luminosity of the object, multiply the emitted intensity by the object's surface area. For a sphere of radius r, the surface area, A, is:

$$A = 4\pi r^2$$

and so:

$$L = 4\pi r^2 \sigma T^4 \tag{6}$$

QUESTIONS

Q 9   Deduce the SI base units of the Stefan constant.

Q 10   **(a)** The star Spica A has a luminosity approximately 2000 times that of the Sun (i.e. 8×10^{29} W) and a surface temperature 3×10^4 K. What, approximately, is its radius? Express your answer as a multiple of the Sun's radius ($r_{Sun} \approx 7 \times 10^8$ m).

(b) Repeat your calculations for Barnard's Star, which has $L \approx 4 \times 10^{23}$ W and $T \approx 2.7 \times 10^3$ K.

(Use $\sigma = 5.67 \times 10^{-8} \, \text{W m}^{-2} \text{K}^{-4}$.)

Q 11   Predict the colour of:

(a) a very hot star that has a surface temperature of $10\,000$ K

(b) a giant star with a surface temperature of about 2700 K.

'The hotter the whiter'

In Activity 6: Colour, luminosity and temperature, you saw that the radiation emitted by a black body depends on its temperature – the hotter the body, the brighter and whiter it appears, as summarised in Figure 5.18. The fact that the distribution of radiation over different wavelengths depends on the temperature enables astronomers to measure the surface temperatures of stars just by analysing their radiation.

Astronomers classify stars according to their surface temperature, as shown in Table 5.2. (One interesting point to note about the colour of stars is that they are unlikely to appear to be green. Any star with a green peak wavelength is likely to have sufficient red and blue to appear white.) The spectral type indicates the temperature range, and was originally based on the appearance of absorption lines in stellar spectra.

The spectroscopic observations taken at Harvard University from 1870 onwards were analysed by a team of ladies known as the 'Harvard Computers' (Figure 5.20a) among whom were Williamina Fleming, Henrietta Swan Leavitt, Antonia Maury and Annie Jump Cannon. Their work, in particular that of Annie Cannon (Figure 5.20b), who classified the spectra of about 200 000 stars (a monumental task requiring great skill and dedication), led to a great step forward in the understanding of stars and how they evolve.

Surface temperature/K	Colour	Spectral type	Example
>30 000	blue	O	Alnitak
11 000–30 000	bluish-white	B	Rigel
7500–11 000	bluish-white	A	Vega
6000–7500	white	F	Canopus
5000–6000	yellow-white	G	Sun
3500–6000	yellow-orange	K	Arcturus
<3500	red	M	Betelgeuse

Table 5.2 Spectral classifications of stars

When astronomers talk about an O-type star they mean a star with a temperature in excess of 30 000 K. As you will see in Part 4, such stars are much more massive and brighter than our Sun, but live short lives. On the other hand, the M-type stars glow a dull red because their surface temperature is low. The Sun is a G-type star with a yellowish-white colour. Astronomers sometimes use a rather ridiculous (but memorable) mnemonic for the sequence of letters (O, B, A, F, G, K, M): Oh Be A Fine Guy/Girl Kiss Me.

(a)

(b)

Figure 5.20 **(a)** The 'Harvard Computers' **(b)** Annie Jump Cannon

'The hotter the brighter'

In the early years of the 20th century, American astronomer Henry Norris Russell (1877–1957) (Figure 5.21a) studied stars that were close enough for their parallaxes to be measured, and he looked for a relationship between stellar luminosity and colour (that is, their surface temperature). Around the same time, Danish astronomer Ejnar Hertzsprung (1873–1967) (Figure 5.21b) looked at stars that were grouped closely together in clusters (such as those in Figure 5.22). He reasoned that stars found clustered closely together must be all at the same distance, therefore their intensities would be a good indication of the relative luminosities, and so he looked for a relationship between colour and intensity.

Figure 5.22 Pleiades star cluster

Both Hertzsprung and Russell constructed graphs that were essentially plots of luminosity against temperature for the stars they observed. This type of diagram is called a **Hertzsprung–Russell diagram** (or HR diagram) and it was of tremendous importance in helping astronomers to understand stars; it is still very widely used as a way of displaying and analysing stellar data.

Figure 5.23 shows the axes for a Hertzsprung–Russell diagram. Notice that the luminosities are expressed in units of the Sun's luminosity, L_{Sun}, rather than watts, as this is more convenient than dealing with the huge numbers involved in stellar luminosity. The range of luminosities is large, from about $10^{-4} L_{Sun}$ to about $10^6 L_{Sun}$. To cope with this range of values, a logarithmic scale is used. Although the range of surface temperatures is not as large, it is still convenient to use a logarithmic scale. Note that the temperature axis (horizontal axis) goes backwards. The high temperatures are plotted on the left of the axis and the low temperatures are plotted on the right. However, if we think in terms of wavelength, the short wavelengths (blue) are on the left and the longer wavelengths (red) are further right, which seems reasonable.

(a)

(b)

Figure 5.21 (a) Henry Norris Russell, **(b)** Ejnar Hertzsprung

> **MATHS REFERENCE**
>
> Using log scales
> See Maths note 8.6

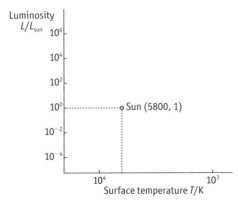

Figure 5.23 Axes for a Hertzsprung–Russell diagram

In other texts, you might come across HR diagrams plotted with other quantities. On the *y*-axis you might find 'magnitude', which is related to the logarithm of intensity. On the *x*-axis you might find 'spectral type', which is, as you know, related to temperature, or 'colour index', which is also related to temperature. However, we will stick with luminosity and temperature.

PRACTICAL SKILLS
REFERENCE

Analysis and interpretation
of data

See Practical skills notes
4.1–4.2

Conclusion and evaluation

See Practical skills notes
5.1–5.2

⚙ ACTIVITY 8 A HERTZSPRUNG–RUSSELL DIAGRAM

Use logarithmic graph paper and a table of stellar luminosities and temperatures to construct a Hertzsprung–Russell diagram for a large number of stars. Include the Sun on your diagram. Comment on any grouping of stars that are apparent from your graph.

From the HR diagram you plotted in Activity 8, three distinct groups of stars can be identified. The main group, containing the greatest number of stars and running roughly diagonally across the diagram from top left to bottom right, is called the **main sequence**. Most of the stars we see in the night sky are in this band. A second group of stars inhabits the top right of the HR diagram. These stars are cool but also very luminous; they must be very large in order to emit such large amounts of radiation from their surface. These are the **red giant** stars. A third group is found at the bottom left of the HR diagram. These are the **white dwarf** stars. Despite being very hot they have low luminosity, so must be small. Because they are so faint, white dwarf stars are hard to find – indeed, there are no such stars visible with the naked eye from Earth – and it is likely that the data you used in Activity 8 included few, if any, white dwarfs.

Once the patterns apparent in the HR diagram became known, astronomers found that they had a powerful tool. For example, it enabled them to estimate stellar distances from measurements of intensity and temperature, as you will see later and in Question 14.

In Sections 4.2 and 4.3 we shall see how the HR diagram can help us to understand the evolution of stars, and we will also see how these three regions of the HR diagram are linked together to give a coherent account of the birth, life and death of a star. However, it is not easy to piece together the life history of a star from a 'snapshot' picture of a large number of stars at a particular point in time, as Question 12 indicates. To help you answer Question 12, you might like to look at some more detailed HR diagrams that can be found online.

QUESTION

Q 12 **(a) (i)** Describe in words how luminosity varies with temperature for stars on the main sequence.

   (ii) Making the very rash assumption that all main-sequence stars are the same size, suggest a possible approximate mathematical relationship between *L* and *T* for stars on the main sequence.

(b) Comment on your answer to **(a)(ii)**. Is it possible to say anything about the size of stars on the main sequence? For example, how might a hot, highly luminous star at one end of the main sequence compare in size with a star at the other end?

(c) What, if anything, can be said about the masses of the stars on the main sequence?

(d) Does the HR diagram give any clues about the way stars form, or the way they change with age? If so, what?

Estimating distances

Astronomers can use the HR diagram to find the luminosity of a star from its temperature (or colour). This is most obvious if a star lies on the main sequence; subtle differences in a star's spectral lines (see Part 3) enable astronomers to distinguish between dwarf, giant and main-sequence stars of the same temperature. Then, by measuring the intensity of the star's radiation received on Earth, the star's distance can be estimated.

$$I = \frac{L}{4\pi d^2}$$

(Equation 3)

There are reasons why this is only an estimate: the radiation from some stars might have been absorbed or scattered more than others en route to us, and the proportion of radiation in the visible part of the electromagnetic spectrum also varies between stars.

QUESTION

Q 13 Table 5.3 lists the luminosities of some stars and the intensity of radiation we receive from them. Assuming that the intensity refers to the complete wavelength range, and that no radiation is absorbed or scattered en route, complete the table.

Star	Luminosity, L/W	Intensity, I/W m^{-2}	Distance, d/m
Sirius	9.9×10^{27}	1.17×10^{-7}	
Rigel	3.8×10^{31}	5.37×10^{-7}	
Betelgeuse	5.3×10^{30}	1.11×10^{-8}	
Proxima Centauri	2.8×10^{23}	1.09×10^{-11}	
Sun	3.83×10^{26}	1.37×10^{3}	

Table 5.3 Star data for Question 13

Of the stars listed in Table 5.3, the one that appears brightest to us is the one that produces the greatest intensity at the Earth – the Sun, by a long way – and the one that seems faintest is Proxima Centauri. From the answer to Question 13 we can see that the Sun is actually much less luminous than some of the other stars listed, and that although Sirius appears brighter than Betelgeuse in the night sky, it actually is much less luminous.

2.6 Standard candles

The technique descibed in Section 2.5 for estimating distances is difficult to use for distant star clusters and galaxies because the individual stars cannot be distinguished as separate objects. To work out distances to such faraway objects, astronomers adopt another method using bright objects whose luminosity is already known or can be reliably estimated. Such objects are called **standard candles** (Figure 5.24).

In the early 20th century, one of the 'Harvard computers', American astronomer Henrietta Swan Leavitt (Figure 5.25) found that certain variable stars, called Cepheid variables, which grow bright and then dim in a regular cycle (Figure 5.26), appear to be standard candles. Cepheid variables are bright stars whose luminosity varies with periods ranging from about 2 to 100 days. Polaris, the Pole Star, is a famous example. There is a strong correlation between the periods of Cepheids and their luminosities. All Cepheids of a given period have the same luminosity, so the luminosity of a Cepheid can be deduced from its period. Before Cepheids could be used as standard candles, their distance and therefore their luminosity had to be determined using an independent method. Based on such independent distance measurements, astronomers now have a formula for determining the Cepheids' luminosity from their periods and can in this way find their distance.

Figure 5.24 Standard candles: an object of given luminosity looks dimmer as its distance increases

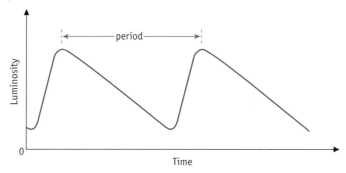

Figure 5.26 Light curve of a Cepheid variable star

Figure 5.25 Henrietta Swan Leavitt

Cepheid variables are no use beyond about 30 million light years away because their intensity becomes too low and they cannot be seen. For such vast distances, astronomers use supernova explosions. One particular type of supernova, known as Type 1a, occurs when a white dwarf acquires material from a nearby star, which triggers a huge explosion. Crucially, they always happen in the same way and produce the same luminosity which, at its peak, is about 50 million times that of the Sun. Type Ia supernovae have become very important for studying the expansion of the Universe, as you will see in Part 5.

STUDY NOTE

Supernovae are discussed in Part 4 of this chapter.

Other standard candle methods arc based on assuming that all galaxies of the same type are equally luminous. The greater the distance, the more difficult it is to find standard candles.

2.7 Summing up Part 2

This part of the chapter has described how distances to faraway objects can be estimated, and has shown how the methods for estimating distance draw on various principles and laws of physics.

The geometric parallax method involves measuring angles, and methods using photometry rely on the inverse-square law for radiation.

Photometric methods involve knowing or deducing an object's luminosity. Luminosity depends on an object's size and on its temperature, as decribed by the laws of black-body radiation. The Hertzsprung–Russell (HR) diagram shows the relationship between stellar temperature and luminosity.

Bright objects whose luminosity is known, or reliably estimated, may be used as standard candles for determining distances. Cepheid variable stars are used as standard candles as they are bright and their luminosity can be deduced from their period. Type Ia supernovae are important standard candles as they can be observed at extremely large distances.

STUDY NOTE

In Part 4 of this chapter you will extend your knowledge of techniques for estimating cosmological distances using the measurement of redshift.

ACTIVITY 9 SUMMING UP PART 2

Look back through your work on this part of the chapter and make sure you understand all the terms printed in **bold**.

Then make a chart summarising the various techniques for estimating distances. For each method:

- make brief notes on the underlying physical principles
- list the relevant equations
- state the approximate distance range over which it can be used.

You could link each section together to make a 'space line' (like a timeline); such a line is sometimes referred to as a 'distance ladder'.

QUESTION

Q 14 When the Orion constellation is observed with binoculars, two stars with distinct colours can be seen: Betelgeuse is red and Rigel is blue.

(a) Taking the frequency of red light to be 4.3×10^{14} Hz, and that of blue light to be 10×10^{14} Hz, estimate the surface temperature of each star.

(b) Explain why your calculations are only estimates.

3 The Sun and the Solar System

3.1 Dating the Solar System

Without the Sun there could be no life as we know it. The Sun has been worshipped by many civilisations (for example, see Figure 5.27), and astronomers were held in high esteem because of their power to predict cosmic events such as solar eclipses. However, this sometimes put them in conflict with religious leaders. When Galileo Galilei (1564–1642) first observed spots on the 'immaculate' Sun, the Church leaders were so shocked that they questioned Galileo's sanity rather than entertain the possibility that the Sun could be 'blemished'.

Humans have always been interested in where we came from and where we are going. All civilisations and cultures have stories that explain our place in the history of the world and the Universe. Because the Sun is so obviously crucial to our life, there has always been a great deal of speculation and thought concerning the age of the Sun (for example, see Figure 5.28). We cannot imagine that life began before the Sun shone on the Earth, and we know that if the Sun ceases to shine, life as we know it will come to an end.

ACTIVITY 10 THE SUN IN HISTORY

Use the internet to find out what your ancestors may have been told about the Sun. The Stanford Solar Center website (see **www.pearsonhotlinks.co.uk**) contains a wealth of information about the Sun's folklore of many cultures.

If the Bible is taken literally, the Sun would be around 10 000 years old. However, we have known since the 19th century that life on Earth developed from extremely simple organisms by the process of evolution. If we assume that in the past evolution changed species at the same slow pace as now, we can see that life on Earth must be much more than 10 000 years old: it must have taken millions of years to develop the complex organisms we observe on Earth today, if life began as simple replicating molecules.

Geologists nowadays can use fossils to determine the relative ages of rocks back to about 600 million years ago. If the development of life depends on the Sun, then the Sun must be more than 600 million years old.

In 1796 the French scientist Pierre Simon Laplace suggested that the whole Solar System formed from one large rotating cloud of gas and dust, now known as the solar nebula. Most of the material contracted to form the Sun, but other fragments condensed ultimately to form the planets (Figure 5.29). Since Laplace proposed his idea, powerful telescopes have allowed us to observe discs of matter surrounding young stars, making the solar nebula theory widely accepted. This gives us another possibility for dating the Sun – if we can find some of the rocks that condensed while the Sun was forming, and if we can date them, then we can date the Sun.

Figure 5.27 Mayan Sun temple

Figure 5.28 Medieval picture of the Sun

PRACTICAL SKILLS REFERENCE

Scientific questions and information research

See Practical skills notes 1.1–1.3

Figure 5.29 Artist's impression of the formation of the Sun and planets

Isotopes

If we want to find out where a rock has come from, we might study the minerals, the fossils or the basic elements present in the rock. Yet we can do better than that: it is possible to date rocks by studying the **isotopes** they contain.

Isotopes of an element are atoms that are chemically identical but differ in the number of neutrons. They contain exactly the same number of protons in the nucleus of the atom; this is what identifies them as a particular element. Because of the different numbers of neutrons, isotopes of the same element have very slightly different masses. Only by using a mass spectrometer, which measures the mass of individual atoms, can we tell isotopes apart. By analysing the atoms in a rock according to mass, we can determine the relative proportions of each isotope in the rock. This gives us a clue to the origin of the rock.

Let's take a small atom, lithium, as an example. Lithium has three protons in the nucleus (3 is its **proton number**, symbol Z, also called **atomic number**) and must therefore be surrounded by three electrons to be electrically neutral overall. In the nucleus of a lithium atom, along with the three protons, are three or four neutrons. Protons and neutrons are collectively known as **nucleons**. Lithium's **nucleon number** (symbol A) is therefore 6 or 7. These two isotopes can be represented like this:

$$^{6}_{3}\text{Li or } ^{7}_{3}\text{Li}$$

The upper number next to each symbol is the nucleon number and the lower number is the proton number, which is also its charge in multiples of the proton charge. These numbers together specify a **nuclide**: they define which isotope of which chemical element we are referring to.

Both lithium isotopes behave the same chemically (since chemical reactions depend on the electron arrangement), but $^{7}_{3}\text{Li}$ (lithium-7) has a slightly larger mass than $^{6}_{3}\text{Li}$ (lithium-6).

Figure 5.30 shows three isotopes of oxygen that exist on Earth. Table 5.4 lists the symbols for chemical elements and their proton number. Note that the hydrogen nucleus $^{1}_{1}\text{H}$ is simply a proton, and so it is sometimes represented by the symbol p.

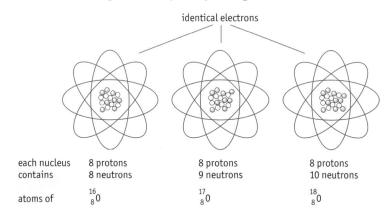

each nucleus contains	8 protons 8 neutrons	8 protons 9 neutrons	8 protons 10 neutrons
atoms of	$^{16}_{8}\text{O}$	$^{17}_{8}\text{O}$	$^{18}_{8}\text{O}$

Figure 5.30 Schematic representations of oxygen isotopes

STUDY NOTE

You met the atomic mass unit in the chapter *Probing the Heart of Matter*.

Since masses of atoms and nuclei are sometimes expressed as multiples of the proton mass, and since protons and neutrons have virtually the same mass, the nucleon number is also called the **mass number**. The mass of each nucleon is close to 1 atomic mass unit (1 u) and so, rounded to the nearest whole number of atomic mass units, the mass of a nucleus is equal to its mass number.

By measuring the oxygen isotope compositions in different rocks, we find that rocks, the atmosphere and water on the Earth have oxygen-isotope compositions that indicate they formed from the same kind of original material.

Proton number	Name	Chemical symbol	Proton number	Name	Chemical symbol	Proton number	Name	Chemical symbol
1	hydrogen	H	36	krypton	Kr	71	lutetium	Lu
2	helium	He	37	rubidium	Rb	72	hafnium	Hf
3	lithium	Li	38	strontium	Sr	73	tantalum	Ta
4	beryllium	Be	39	yttrium	Y	74	tungsten	W
5	boron	B	40	zirconium	Zr	75	rhenium	Re
6	carbon	C	41	niobium	Nb	76	osmium	Os
7	nitrogen	N	42	molybdenum	Mo	77	iridium	Ir
8	oxygen	O	43	technetium	Tc	78	platinum	Pt
9	fluorine	F	44	ruthenium	Ru	79	gold	Au
10	neon	Ne	45	rhodium	Rh	80	mercury	Hg
11	sodium	Na	46	palladium	Pd	81	thallium	Tl
12	magnesium	Mg	47	silver	Ag	82	lead	Pb
13	aluminium	Al	48	cadmium	Cd	83	bismuth	Bi
14	silicon	Si	49	indium	In	84	polonium	Po
15	phosphorus	P	50	tin	Sn	85	astatine	At
16	sulfur	S	51	antimony	Sb	86	radon	Rn
17	chlorine	Cl	52	tellurium	Te	87	francium	Fr
18	argon	Ar	53	iodine	I	88	radium	Ra
19	potassium	K	54	xenon	Xe	89	actinium	Ac
20	calcium	Ca	55	caesium	Cs	90	thorium	Th
21	scandium	Sc	56	barium	Ba	91	protactinium	Pa
22	titanium	Ti	57	lanthanum	La	92	uranium	U
23	vanadium	V	58	cerium	Ce	93	neptunium	Np
24	chromium	Cr	59	praseodymium	Pr	94	plutonium	Pu
25	manganese	Mn	60	neodymium	Nd	95	americium	Am
26	iron	Fe	61	promethium	Pm	96	curium	Cm
27	cobalt	Co	62	samarium	Sm	97	berkelium	Bk
28	nickel	Ni	63	europium	Eu	98	californium	Cf
29	copper	Cu	64	gadolinium	Gd	99	einsteinium	Es
30	zinc	Zn	65	terbium	Tb	100	fermium	Fm
31	gallium	Ga	66	dysprosium	Dy	101	mendelevium	Md
32	germanium	Ge	67	holmium	Ho	102	nobelium	No
33	arsenic	As	68	erbium	Er	103	lawrencium	Lr
34	selenium	Se	69	thulium	Tm	104	rutherfordium	Rf
35	bromine	Br	70	ytterbium	Yb	105	dubnium	Bb

Table 5.4 Chemical elements, their symbols and their proton numbers

Figure 5.31 Layer from a larger carbonaceous chondrite meteorite seen under a microscope

What about rocks that were not formed on the Earth? Meteorites are simply extraterrestrial lumps of rock that land on Earth. They may be made of rock, metal, or a mixture of both. For dating the Solar System the most important meteorites to study are those known as carbonaceous chondrites (Figure 5.31). These meteorites have chemical compositions very similar to that of the Sun, leading us to suspect that they have the composition of the solar nebula. Carbonaceous chondrites are relatively unaltered by processes that have affected other planetary matter. Their oxygen isotope compositions are very different from those measured on materials from Earth. We can therefore be certain that they formed in a different part of the solar nebula.

Radioactive decay

Not all isotopes of an element are stable. Some undergo **radioactive decay**, which is a general name for processes in which nuclei rearrange themselves to become more stable by emitting radiation that has high enough energy to cause ionisation. There are three types of radioactive emission: **alpha** α, **beta** β and **gamma** γ.

> ### ⚙ ACTIVITY 11 α, β AND γ RADIATION
>
> Using your knowledge from earlier work, or using information from textbooks or websites, draw up a table to summarise the properties of α, β and γ emission. Your table should include:
>
> - the nature of the emission
> - the sign of any electric charge
> - the mass
> - the typical speed.

What happens when an isotope undergoes radioactive decay? Take carbon, for example. Much of the material that makes up your body is based on chains of carbon atoms, the vast majority of which are stable carbon-12 atoms. However, an unstable version of carbon – carbon-14, which is chemically identical to carbon-12 – also forms a tiny percentage of the carbon in your body.

Both carbon isotopes contain six protons, but carbon-14 contains two extra neutrons compared with carbon-12. This makes its nucleus unstable: there are too many neutrons for the nucleus to stay together permanently, and so it will arrange itself to become more stable. A neutron converts into a proton plus an electron, flinging out the electron from the nucleus at high speed, accompanied by an electron anti-neutrino. As the lower number represents the charge in multiples of the proton charge, we can extend the convention used for representing nuclei and in this way represent the reaction using symbols:

$$\ _{0}^{1}\text{n} \rightarrow \ _{1}^{1}\text{p} + \ _{-1}^{0}\text{e}^{-} + \overline{\nu}_{e} \tag{R1}$$

This changes the nucleus into that of another element – one that has one more proton and one fewer neutron than the original carbon-14 nucleus. This is the element with a proton number $6 + 1$, i.e. it is nitrogen, $_{7}^{14}\text{N}$ (Figure 5.32). We can write this reaction using symbols:

$$\ _{6}^{14}\text{C} \rightarrow \ _{7}^{14}\text{N} + \ _{-1}^{0}\text{e}^{-} + \overline{\nu}_{e} \tag{R2}$$

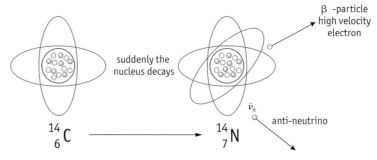

Figure 5.32 Radioactive β^{-} decay of carbon-14

The nucleus changes from one element to another, a process called **transmutation**. For example, when carbon-14 decays it becomes nitrogen, with all the usual chemical properties of nitrogen and none of the properties of carbon.

Note that the reaction equations R1 and R2 both balance:

● the sum of the lower numbers is unchanged, showing conservation of charge

● the sum of the upper numbers is unchanged, showing conservation of nucleons (or, more generally, conservation of baryons).

The electrons that orbit the nucleus of the decaying atoms are disturbed during the decay then settle back into orbit around the new **daughter nucleus**.

The high-speed electron ejected from the nucleus is known as a beta-minus particle, β^-, and the whole reaction is one example of the process called **beta-minus decay** (or often simply beta decay). The high-speed electron travels at almost the speed of light, and has enough energy to ionise atoms it encounters, knocking out some of their electrons. The release of energy that accompanies radioactive decay is vast compared with energies involved in chemical reactions.

The decay of carbon-14 is of enormous scientific interest because it allows us to date plant- or animal-derived materials. Living plants absorb carbon-14 from the air as part of their carbon dioxide intake. Once they die, the carbon-14 concentration falls as the carbon-14 decays. By measuring the carbon-14 concentrations we can estimate the age of organic matter.

Beta-minus decay is one type of **radioactive decay**, a general name for processes in which nuclei rearrange themselves to become more stable by emitting a particle that has high enough energy to cause ionisation.

There is also a process known as **beta-plus decay**, which involves the emission of a positron (anti-electron) and a neutrino while a proton in the nucleus turns into a neutron:

$$^1_1p \rightarrow\, ^1_0n +\, ^0_1e^+ + \nu_e \tag{R3}$$

(The same result is produced if one of the orbiting electrons is captured by the nucleus and combines with a proton to produce a neutron.) An example of a positron emitter is magnesium-23, which decays into sodium-23:

$$^{23}_{12}Mg \rightarrow\, ^{23}_{11}Na +\, ^0_1e^+ + \nu_e \tag{R4}$$

Some massive nuclei are unstable because they contain too many protons or neutrons. Such nuclei become more stable by **alpha decay**, in which they emit an alpha particle (α), which consists of two neutrons and two protons (and is identical to a helium nucleus, ^4_2He). Uranium-238 is one example of an alpha emitter. The equation for the alpha decay of uranium-238 is:

$$^{238}_{92}U \rightarrow\, ^{342}_{90}Th +\, ^4_2He \tag{R5}$$

Figure 5.33 shows the radioactive alpha decay of uranium-238; again, notice that the total electric charge and nucleon number are each conserved.

STUDY NOTE

You used charge conservation and baryon conservation in the chapter *Probing the Heart of Matter*.

STUDY NOTE

You might have found out about carbon-14 dating in the chapter *Digging Up the Past*.

STUDY NOTE

You met the positron, neutrinos and anti-neutrinos, and their symbols, in the chapter *Probing the Heart of Matter*.

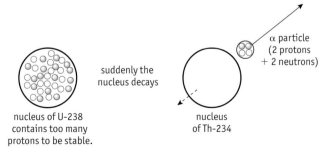

suddenly the nucleus decays

α particle (2 protons + 2 neutrons)

nucleus of U-238 contains too many protons to be stable.

nucleus of Th-234

Figure 5.33 Radioactive alpha decay of uranium-238

The third main type of radioactive decay is **gamma decay**, in which a nucleus becomes more stable by emitting a photon of gamma radiation. Gamma radiation is very-high-energy electromagnetic radiation. In common with all electromagnetic radiation it travels at a speed of $3.00 \times 10^8 \, \text{m s}^{-1}$ in a vacuum. Gamma radiation has a very short wavelength (about 10^{-14} m or less); each photon carries a large amount of energy and is very penetrating. In gamma decay there is no change to the proton number or neutron number, so no transmutation into another element, just a loss of energy from the nucleus. Gamma emissions usually accompany alpha or beta decay because these decays often produce a new nucleus that is in an excited state.

Alpha, beta and gamma emissions all cause ionisation of air and the other materials through which they pass. This ionisation can be very dangerous to life because of the disruption to chemical bonds and the production of ions that readily take part in chemical reactions. Large doses of radiation cause death very quickly, but even tiny doses can damage the DNA in replicating cells so that the cells cannot replicate correctly and become cancerous.

We are always surrounded by very-low-level **background radiation** from naturally occurring rocks and cosmic rays, but we should avoid any unnecessary extra exposure to ionising radiations. When carrying out experiments involving radioactivity, you should:

- pay particular attention to the safety instructions in order to minimise your own and other people's exposure to radiation;

- remember that background radiation is present all the time and will register on any detector that you are using as well as any additional radiation that you are trying to measure, so you should record the **background count rate** and subtract it from your measurements.

Ionising ability and penetration

As alpha, beta and gamma radiations pass through material, they transfer energy to the atoms and molecules they encounter, causing ionisation. The range of the radiations depends on their initial energy and on their ionising ability.

Alpha particles are very effective ionisers, due to their large mass and charge. They can readily interact with atoms and molecules, causing ionisation. They therefore lose a lot of energy over a short distance and have only a short range, even in air.

Beta particles travel close to the speed of light. Their low mass and high speed mean that they produce less ionisation than alpha particles over a given distance. Although beta particles are typically emited with lower energies than alpha particles, they can travel much greater distances through matter.

Gamma radiation is uncharged and moves at the speed of light. It produces relatively little ionisation so it loses little energy as it passes through matter. It is therefore very penetrating.

SAFETY NOTE

Before attempting any of Activities 12, 13, or 14, make sure that you know the safety precautions you must follow when using radioactive materials.

PRACTICAL SKILLS REFERENCE

Carrying out practical work
See Practical skills notes 3.1–3.2

Analysis and interpretation of data
See Practical skills notes 4.1–4.2

⚙ ACTIVITY 12 INVESTIGATING α, β AND γ RADIATION

Set up a Geiger–Müller (GM) tube with a data logger or scaler-ratemeter and measure the background count rate (the number of ionising events detected per second) in your laboratory.

Investigate the properties of α, β and γ radiation using sealed laboratory sources. They can all be detected with a suitably adjusted GM tube. Investigate their range in air and find out how easy it is to stop each type of radiation.

Send a stream of α, β and γ radiation through a strong magnetic field and observe any deflection.

⚙ ACTIVITY 13 ABSORPTION OF GAMMA RADIATION

By investigating the absorption of gamma rays by various thicknesses of lead, determine the thickness of lead that will halve the intensity of gamma radiation. This is known as the **half-thickness** of gamma radiation in lead.

Cloud chambers are designed to show the tracks of ionising radiation (Figure 5.34). These tracks are rather like miniature versions of the vapour trails we see in the sky when jets fly high overhead, and they form as droplets condense in a supersaturated vapour. Radioactive emissions ionise air and this causes the vapour to condense, and so the vapour trails show the particles' tracks.

Figure 5.34 Tracks in a cloud chamber

⚙ ACTIVITY 14 USING A CLOUD CHAMBER

Set up the cloud chamber according to the manufacturer's instructions. Using information from Activity 12, decide which type of radiation will produce the thickest tracks.

Look for tracks from the cloud chamber's own source and tracks due to background radiation.

Watch the tracks appear and answer the following questions.

● Can you predict where or when a track will form?
● Are all the tracks the same length?

STUDY NOTE

Look back at the work that you did on absorption of radiation in *The Medium is the Message.*

PRACTICAL SKILLS REFERENCE

Scientific questions and information research

See Practical skills notes 1.1–1.4

Planning and experimental design

See Practical skills notes 2.1–2.4

Carrying out practical work

See Practical skills notes 3.1–3.2

Analysis and interpretation of data

See Practical skills notes 4.1–4.2

Conclusion and evaluation

See Practical skills notes 5.1–5.2

STUDY NOTE

You will have studied cloud chamber or bubble chamber tracks in the chapter *Probing the Heart of Matter.*

PRACTICAL SKILLS REFERENCE

Carrying out practical work

See Practical skills notes 3.1–3.2

QUESTIONS

Q 15 Carbon dating, by measuring carbon-14 concentrations, is a valuable tool in archaeology. Why couldn't it be used to date the Earth or the Sun?

Q 16 The core of the Earth is heated by the energy released by the radioactive decay of naturally occurring isotopes. The most significant of these are uranium-235, uranium-238, thorium-232 and potassium-40. For each of these isotopes write its abbreviated symbol, including the nucleon and proton numbers, and state the number of neutrons in the nucleus.

Q 17 (a) One of the radioactive decay processes commonly used for dating rocks is the decay $^{87}_{37}$Rb (rubidium) to form $^{87}_{38}$Sr (strontium). What type of decay occurs? Write a full, balanced equation to show the decay process.

(b) Some radioactive isotopes have daughter products that are themselves unstable, leading to a **radioactivity series**. Part of the actinium series is shown in Figure 5.35. The symbols by each arrow indicate the radiation(s) emitted in each decay.
   (i) Write a full balanced equation for each decay of bismuth (Bi) shown in Figure 5.35, deducing the nucleon and proton numbers for Tl (tellurium), Po (polonium) and Pb (lead).
   (ii) Based on your answer to (i) write an equation for a possible mode of decay from Tl to Pb.

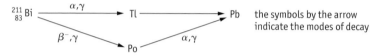

Figure 5.35 Part of the actinium radioactive decay series

Activity, exponential decay and half-life

The minerals we find in rocks originally crystallised when the rock was molten. If some of the atoms in the mineral are radioactive isotopes they will, in time, decay. Chemical analysis of old rocks may reveal elements that could not have been present in the original crystal; therefore they must have formed as daughter products by radioactive decay. By measuring the relative proportions of parent and daughter nuclei, we can find out how much time has passed since the mineral first crystallised. This method of dating rocks depends on our understanding of how rapidly a sample of radioactive material will decay.

A radioactive nucleus is an unstable nucleus that could decay at any moment. The decay, or disintegration, is **spontaneous**; it is independent of any external stimulus such as physical or chemical conditions. It is also **random**, in that it is impossible to predict when a particular nucleus will decay. Laws of probability govern the chance that it will decay at a given moment. If you toss a coin you know that each time there is a 50% chance of it coming up heads, and that within a few tosses it almost certainly will come up heads. If you roll a die there is a one in six chance of a given number coming up. Atoms of radioactive isotopes behave in rather the same way: in any one second there is a certain probability that the isotopes will decay. Some isotopes are extremely unstable, so there is a high probability that in one second a nucleus will decay. Some isotopes are relatively stable, so any given nucleus is very unlikely to decay in any one second.

Activity 15 uses a model to demonstrate how the number of radioactive nuclei varies with time.

⚙ ACTIVITY 15 MODELLING RADIOACTIVE DECAY

Use dice, coins or a computer model to simulate the random decay of radioactive nuclei. Plot a graph showing how the number of remaining 'nuclei' changes with time.

In Activity 15 you will have seen that the number of 'nuclei', ΔN, decaying in a given time interval, Δt, is proportional to the number of nuclei, N, remaining in the sample. Expressing this mathematically for a very small time interval (effectively 'at an instant'):

$$\frac{dN}{dt} \propto N \qquad\qquad (7)$$

and so we can write:

$$\frac{dN}{dt} = -\lambda N \qquad (7a)$$

where λ is called the **decay constant**. The minus sign is there because N decreases as t increases.

The rate of decay, $\frac{dN}{dt}$, is called the activity, A. It is the number of disintegrations per unit time (e.g. per second or per minute) and is therefore also the number of alphas, betas or gammas emitted per unit time, which can quite easily be measured using a GM tube or similar instrument. The SI unit of activity is the Becquerel, Bq. 1 Bq is one disintegration per second.

$$A = \frac{dN}{dt} (= -\lambda N) \qquad (8)$$

Equation 7a can be rearranged:

$$\frac{dN}{N} = -\lambda dt \qquad (7b)$$

Integrating both sides:

$$\int \frac{dN}{N} = -\lambda \int dt \qquad (7c)$$

we now have:

$$\ln N = -\lambda t + c \ (c \text{ is a constant})$$

If the initial number of nuclei is N_0, i.e. if $N = N_0$ when $t = 0$, then by Equation 9:

$$\ln N_0 = c$$

Substituting for c in Equation 9 gives:

$$\ln N = -\lambda t + \ln N_0 \qquad (9a)$$

$$\ln N - \ln N_0 = -\lambda t \qquad (9b)$$

$$\ln \left(\frac{N}{N_0} \right) = -\lambda t \qquad (9c)$$

$$\frac{N}{N_0} = e^{-\lambda t} \qquad (10)$$

$$N = N_0 e^{-\lambda t} \qquad (10a)$$

Equation 10 expresses the exponential nature of radioactive decay: the number of parent nuclei decreases exponentially with time.

Since the number of unstable nuclei decays exponentially, so does the activity – with the same decay constant. Differentiating Equation 10 gives:

$$A = \frac{dN}{N} = -\lambda N_0 \, e^{-\lambda t} \qquad (11)$$

We can then use Equation 7a to substitute an expression for A_0; that is, the activity when $t = 0$:

$$A = A_0 \, e^{-\lambda t} \qquad (12)$$

Another way to see this is by analogy with the decay of charge and current when a capacitor discharges through a resistor. As you saw in the chapter *The Medium is the Message*, the stored charge, Q, decays exponentially, and so does the current I, where:

$$I = \frac{dQ}{dt}$$

> **MATHS REFERENCE**
>
> Exponential changes
> See Maths note 9.1
>
> Exponential functions
> See Maths note 9.2
>
> Exponentials and logs
> See Maths note 9.3

The 'life' of a radioactive isotope (the time for all nuclei to disintegrate) is infinite, so the decay of a radioactive isotope is usually described in terms of its **half-life**, $t_{\frac{1}{2}}$ – that is, the time for the number of nuclei of that isotope to halve. Some isotopes are very unstable and have short half-lives (a few seconds or even less), whereas more stable isotopes have much longer half-lives – even millions of years.

The half-life is related to the decay constant. From the definition of half-life, when $N = \dfrac{N_0}{2}$, $t = t_{\frac{1}{2}}$. Putting these values into Equation 10 we have:

$$\frac{N_0}{2} = N_0\, e^{-\lambda t_{1/2}}$$

Dividing by N_0 and taking the reciprocal of both sides:

$$2 = e^{\lambda t_{1/2}}$$

Taking natural logs of both sides:

$$\ln(2) = \lambda t_{\frac{1}{2}}$$

$$t_{\frac{1}{2}} = \frac{\ln(2)}{\lambda} \tag{13}$$

Note that $t_{\frac{1}{2}}$ can be expressed in any unit of time, for example seconds, minutes or years, and the corresponding units for λ are then s^{-1}, min^{-1} or yr^{-1}.

QUESTION

Q 18 **(a)** The activity of a particular sample of a radioactive nuclide falls from $1.0 \times 10^{11}\,Bq$ to $2.0 \times 10^{10}\,Bq$ in 10 hours.

  (i) Calculate the decay constant for the nuclide in s^{-1}.

  (ii) Now use your answer to **(i)** to find the nuclide's half-life. Give your answer in hours.

(b) Find the half-life and decay constant for your radioactive decay simulation from Activity 15. Suggest suitable units.

⚙ ACTIVITY 16 MEASURING HALF-LIFE

Measure the half-life of a short-lived radioactive isotope.

Measure the background count and subtract it from your measurements.

Plot a suitable graph showing how the activity of the sample changes with time, and in this way show that the decay is indeed exponential.

The laboratory sources you may have used in Activities 12, 13 and 14 are likely to have had long half-lives. This is so their activity is at a safe level (λ is small) and they do not need replacing often. Examples include:

$$^{241}_{95}\text{Am } (\alpha \text{ emitter}) \; t_{1/2} = 432 \text{ years}$$

$$^{90}_{38}\text{Sr } (\beta^{-} \text{ emitter}) \; t_{1/2} = 28.8 \text{ years}$$

Equation 10a, $N = N_0\, e^{-\lambda t}$, is known as the **exponential law of radioactive decay**.

The fact that radioactive decay is exponential indicates that the decay of individual nuclei really is governed by probability. This is quite a disturbing (and interesting!) idea. After all, in most situations that you meet in physics, something either definitely happens or definitely does not; on an atomic scale, things look rather different. It is also interesting that, even though the decay of each nucleus happens by chance, we can still make very definite predictions and measurements of the behaviour of a large number of nuclei – the half-life of a sample, for example.

MATHS REFERENCE

Using log graphs

See Maths note 8.7

Exponential changes

See Maths note 9.1

Exponential functions

See Maths note 9.2

Exponentials and logs

See Maths note 9.3

SAFETY NOTE

Before attempting this activity, make sure that you know the safety precautions you must follow when using radioactive materials.

PRACTICAL SKILLS REFERENCE

Carrying out practical work

See Practical skills notes 3.1–3.2

Analysis and interpretation of data

See Practical skills notes 4.1–4.2

Conclusion and evaluation

See Practical skills notes 5.1–5.2

Dating

Carbon dating of once-living matter was mentioned earlier in this section. Radiometric dating of rocks (or of archaeological artefacts) essentially involves knowing, or assuming, the amount of an isotope of known half-life initially present in the sample, measuring the amount present now, and in this way calculating the time elapsed since the sample was formed. The following example illustrates the principle, and shows how the units of $t_{1/2}$ and λ are interrelated.

WORKED EXAMPLE

Q Uranium-238 ($^{238}_{92}U$) decays by alpha emission to form thorium-234 ($^{234}_{90}Th$). The decay constant for this process is $1.54 \times 10^{-10}\,yr^{-1}$.

If you had 1.00 tonne of pure uranium-238:

(a) How long would it take before the activity of the uranium-238 has halved?

(b) How much would you expect to remain 1.00 billion years later (i.e. after $1.00 \times 10^9\,yr$)?

(c) How long would it take for the mass of uranium-238 to decrease to 0.80 tonne?

A **(a)** The time to halve (i.e. the half-life) does not depend on the initial activity. Using Equation 7:

$$t_{1/2} = \frac{\ln(2)}{\lambda}$$

$$= \frac{\ln(2)}{1.54 \times 10^{-10}\,yr^{-1}} = 4.5 \times 10^9\,yr$$

(b) Here we need to use Equation 10 with $t = 1.00 \times 10^9\,yr$.

$$\lambda t = 1.54 \times 10^{-10}\,yr^{-1} \times 1.00 \times 10^9\,yr = 0.154$$

(Note that the quantity λt has no units.)

$e^{-\lambda t} = 0.857$, and so $N = 0.857\,N_0$, and there will be 0.857 tonne of uranium-238 remaining.

(c) Now we need Equation 9 with $N = 0.80N_0$:

$$-\lambda t = \ln(N/N_0)$$

$$\lambda t = \ln(N_0/N)$$

$$t = \frac{\ln(N_0/N)}{\lambda} = \frac{\ln(1/0.80)}{1.54 \times 10^{-10}\,yr^{-1}}$$

$$= 1.45 \times 10^9\,yr$$

When we are dating rocks we need to study isotopes with relatively long half-lives if they are still to be detectable today. Table 5.5 shows some of the isotopes commonly used by geologists for dating rocks. Note: the daughter products listed for ^{238}U and ^{235}U are not the immediate products, but are the stable products produced by a series of decay reactions. The stated half-lives and decay constants refer to the complete series of decays.

Parent isotope	Daughter isotope	Half-life, $t_{1/2}/10^6\,yr$	Decay constant, $\lambda/10^{-10}\,yr^{-1}$
$^{238}_{92}U$	$^{206}_{82}Pb$		1.552
$^{235}_{92}U$	$^{207}_{82}Pb$	704	
$^{40}_{19}K$	(See Question 22)		5.810
$^{87}_{37}Rb$	$^{87}_{38}Sr$	48 800	

Table 5.5 Some isotopes commonly used for dating rocks

If a sample initially contained none of the daughter isotope, then dating is straightforward. Figure 5.36 shows how the amounts of ^{235}U and ^{207}Pb change with time in a rock that initially contained no ^{207}Pb. If we measure the current proportions of ^{235}U and ^{207}Pb, we can calculate the time since the sample crystallised. The proportions of ^{235}U and ^{207}Pb are equal after one half-life, and as time goes on the proportion of lead increases.

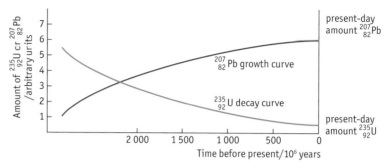

Figure 5.36 Proportions of ^{235}U and ^{207}Pb changing with time

Figure 5.37 Zircon crystals

Radiometric dating techniques typically assign ages of over 4 billion years to many Earth rocks. By comparing the ^{87}Rb and ^{87}Sr ratios in zircon crystals (Figure 5.37) found in ancient rocks in Australia, the rocks have been dated as 4200 million years old. Some deep-sea sediments give results that suggest they are even older. Using similar techniques, the oldest Moon rocks and meteorites have been shown to be about 4600 million years old. If we accept the solar-nebula theory of the formation of the Solar System, it is fair to give the Sun a similar age.

Table 5.6 shows a slightly different way of relating the age of a sample to its half-life. The 'amount' could be either the mass or the number of nuclei.

Number of half-lives elapsed	Amount of parent isotope remaining		
0	100%		
1	50%	$\frac{1}{2}$	$\left(\frac{1}{2}\right)^1$
2	25%	$\frac{1}{4}$	$\left(\frac{1}{2}\right)^2$
3	12.5%	$\frac{1}{8}$	$\left(\frac{1}{2}\right)^3$

Table 5.6 Half-life and age

QUESTIONS

In answering these questions, you will need to refer to Table 5.6 and to information in the text.

Q 19 Study Table 5.6 and write a mathematical expression relating the number of half-lives that have elapsed to the amount of parent isotope remaining.

Q 20 Calculate the missing decay constants and half-lives needed to complete Table 5.5.

Q 21 Using your answer to Question 19, calculate the proportion of ^{87}Rb you would expect to find remaining in a sample that is analysed 24 400 million years after it crystallised.

Q 22 A grain of mineral, which contained uranium but was assumed to have contained no lead when it crystallised, was found to have 15 times as much $^{207}_{82}$Pb as $^{235}_{92}$U. How old is it?

Q 23 Some recently discovered meteorites are believed to have come from Mars – the pockets of gas trapped in the rock closely match that of the Martian atmosphere. The crystallisation ages of these rocks (the time since they crystallised) are found to be 0.2–1.3 billion years.

(a) If we assume that Mars formed from the solar nebula at the same time as Earth, what crystallisation age would we expect from these rocks?

(b) Mars has many extinct volcanoes, some very high. Suggest a way in which the meteorites got to Earth.

(c) If Mars formed at the same time as Earth, how might we explain the younger crystallisation ages of the Martian meteorites?

3.2 What fuels the Sun?

The active Sun

Until the invention of telescopes, people thought that the Sun was a globe of pure fire and light. Galileo shook the world when he discovered imperfections – dark spots – on the Sun (Figure 5.38). In the 18th and 19th centuries most scientists believed that the Sun had a cool, dark interior surrounded by a burning shell: sunspots were explained as either mountains of cool material poking out of the burning clouds, or holes in the burning clouds.

We now know that sunspots are relatively cool regions of the photosphere. They are related to large-scale upheavals in the photosphere that involve the Sun's magnetic field and prevent hot material reaching the photosphere. The amount of this so-called solar activity, and the numbers of sunspots, vary regularly with time and follow an approximately 11-year cycle.

Figure 5.38 Sunspots

> **⚙ ACTIVITY 17 SUNSPOTS**
>
> By projecting an image of the Sun on to a screen, study and record the positions of sunspots and in this way demonstrate the Sun's rotation.
>
> Use the Internet to find images of the Sun produced at solar observatories.
>
> Compare your images of the Sun with images obtained in quiet and active periods of the sunspot cycle. Where are we now in the cycle?

In the 19th century the technique of spectroscopy was developed. This allowed a chemical analysis of distant objects by studying their light. It was soon discovered that the Sun emitted the continuous spectrum of an intensely hot object, but that there were dark lines in the spectrum where light had been absorbed by elements in a cooler outer atmosphere beyond the photosphere (Figure 5.39). These are named Fraunhofer lines, after the German physicist Joseph von Fraunhofer (1787–1826) who first recorded them.

Figure 5.39 Absorption lines in the solar spectrum

PRACTICAL SKILLS REFERENCE

Scientific questions and information research

See Practical skills notes 1.1–1.3

Carrying out practical work

See Practical skills notes 3.1–3.2

SAFETY NOTE

Never look directly at the Sun. Sunglasses are not sufficient protection.

Figure 5.40 Kirchhoff, Bunsen and Roscoe

By shining continuous spectra through vaporised materials in the laboratory and comparing them with the spectra of the Sun, the German scientist Gustav Kirchhoff (1824–1887) was able to detect many elements in the outer atmosphere of the Sun: sodium, calcium, magnesium and iron; he even discovered new elements that were only later detected on Earth. Kirchhoff worked with Robert Bunsen and Henry Roscoe (see Figure 5.40). Figure 5.41 shows a spectroscope dating from 1861, similar to those made and used earlier by Bunsen.

Figure 5.41 A spectroscope from 1861

PRACTICAL SKILLS REFERENCE

Carrying out practical work

See Practical skills notes 3.1–3.2

ACTIVITY 18 THE SUN'S SPECTRUM

Use a spectroscope to observe the solar spectrum in sunlight reflected from a matt white surface or from clouds. Do not point a spectroscope directly at the Sun. Alternatively, use a spectrometer fitted with a diffraction grating. You should be able to see the full 'rainbow' of colours plus the Fraunhofer lines.

SAFETY NOTE

Be careful not to look directly at the Sun.

Solar power

In December 1838 an English astronomer, Sir John Herschel (1792–1871), was visiting South Africa. At noon, while the Sun was almost directly overhead (12° north), he used the Sun's rays to heat a carefully measured amount of water for 10 minutes. By measuring the rise in temperature of the water he calculated the energy received by the Earth from the Sun, and in this way he was able to calculate the total power output of the Sun.

QUESTION

 Q 24　We currently believe the luminosity of the Sun to be about 3.84×10^{26} W. In 1838 the unit of power was the horse: 1 horsepower (hp) = 746 W. A steam engine with a 1000 hp engine would have seemed incredibly powerful. What is the approximate luminosity of the Sun in horsepower?

It was soon shown that such a vast energy output from the Sun was difficult to explain. Scientists began to wonder where such energy came from, and in 1842 Julius Robert Mayer (1818–1878) proposed the law of conservation of energy. Scientists realised that the Sun must have a fuel source. But what kind of fuel could sustain such a power output?

As far as the 19th-century scientists knew, the richest source of energy was coal, so the idea was proposed that the Sun was a huge piece of coal. However, if the Sun consisted of anthracite (a type of coal), a layer of 20 feet in thickness would have to be burned every hour to account for the energy released by the Sun. This would have consumed the whole of the Sun in less than 5000 years, and no one then believed the Solar System to be as young as that.

Mayer's idea involved meteorites falling into the Sun. German physicist Hermann von Helmholtz (1821–1894) correctly proposed that gravitational collapse of a vast cloud heated the early Sun, but could not account for its continuing radiation billions of years later.

Nuclear fusion

With the discovery of radioactivity and the nuclear atom at the beginning of the 20th century, physicists and astronomers began to realise that the source of the Sun's energy could more realistically be explained by some kind of nuclear process. In 1919 Rutherford showed that it was possible to change one element into another by bombarding it with very rapidly moving particles, and in a classic book of 1926 Sir Arthur Eddington proposed that the Sun's energy output might be accounted for by the combination of hydrogen nuclei to form helium. We now believe that he was correct – the **fusion** of nuclei releases huge amounts of energy, and can fully explain why the Sun is still shining billions of years after it formed.

To understand why nuclear fusion can release energy, it can be helpful to carry out a thought experiment. Imagine assembling a nucleus from its constituent protons and neutrons (which in practice would be extremely difficult to achieve). As these nucleons become very close together, once they are close enough for the strong nuclear force to exceed electostatic repulsion, they are attracted to one another by the strong force, and as they are drawn together and become more tightly bound they lose 'nuclear potential energy', rather like an object falling towards Earth loses gravitational potential energy. Figure 5.42 shows the energy that would be released per nucleon if each nucleus were to be created in this way. This energy is called the nuclear **binding energy**; it is usually measured in MeV per nucleon. $1\,\text{MeV} = 1.6 \times 10^{-13}\,\text{J}$.

Figure 5.42 Nuclear binding energy: the energy released in the formation of nuclei from separate nucleons

Remember that binding nucleons together involves a release of energy. Another way to think of the binding energy is that it is the energy that you would need to supply in order to separate a nucleus into its nucleons – perhaps 'unbinding energy' would be a better term.

The binding energy per nucleon is a measure of the stability of a nucleus (the greater the binding energy, the more difficult it is to break the nucleus apart). Iron-56 is the most tightly bound nucleus of all. Three nuclei, $^{4}_{2}\text{He}$, $^{12}_{6}\text{C}$ and $^{16}_{8}\text{O}$, lie significantly above the main curve, showing that they are more stable than other nuclei of similar mass; they are all 'multiples' of an alpha particle.

When two light nuclei fuse to produce a more massive nucleus, additional encrgy is released. In the thought experiment, you can picture this as pulling apart each reacting nucleus into its separate nucleons and then reassembling them to make a single, larger nucleus – this involves a net release of energy. Nuclear **fission** (the splitting of a very massive nucleus into lighter fragments) also releases energy. You can also picture the massive nucleus being pulled apart and reassembled as two lighter nuclei.

STUDY NOTE
For more on nuclear stability, see STA Additional Sheet 1.

STUDY NOTE
Nuclear fission is discussed later.

The Sun is composed mostly of hydrogen with some helium. Figure 5.42 shows that if hydrogen nuclei fuse to form helium, there is a release of several MeV per nucleon. The release of energy manifests itself mainly in the emission of high-energy (gamma-ray) photons and neutrinos and in the kinetic energy of the helium nuclei. While a few MeV might not be a very large amount of energy on its own, remember that this is the energy released per nucleon. In just 1 g of hydrogen there would be about 6×10^{23} nuclei, and so the energy released by the fusion of 1 g of hydrogen would be many billions of joules – far more than the energy involved in any chemical reaction using a similar amount of material.

It would be brilliant if we could control a nuclear fusion reaction on the Earth to generate electricity, because hydrogen (in water, H_2O) is plentiful, and there would be no polluting waste products. Unfortunately, to get the hydrogen nuclei to fuse requires them to be moving close together at high speeds. This is because of their positive charge: unless they are moving very fast, the electrostatic repulsion ensures that they never become close enough for the strong force to attract them. Also, to give the hydrogen nuclei a good chance of colliding with one another, the density, too, needs to be high.

It is very difficult to produce the high temperatures and densities required for fusion on Earth, but the core (centre) of a star (such as the Sun), with the enormous pressure due to the weight of overlying material, is able to reach a temperature of over 10^7 K, high enough to sustain hydrogen fusion. Such high temperatures can be produced in laboratories on Earth (Figure 5.43), but it is difficult to maintain them for very long and there is the additional problem of controlling very hot hydrogen plasma (ionised gas), since it would vaporise the walls of any solid container. Usually complex magnetic fields are used to contain the plasma within the reactor.

Figure 5.43 An experimental nuclear fusion power plant under construction in France as part of the ITER international collaboration

The precise details of the fusion reactions that occur in a star depend on its temperature and pressure. It was only as recently as 1938 that the details of the nuclear reactions were worked out. Hans Bethe and Charles Critchfield showed that in the Sun a proton–proton chain reaction (known as the ppI chain) releases energy. The chain has several stages, and the overall effect is:

$$4^1_1H \rightarrow {}^4_2He + 2e^+ + 2\nu_e + 2\gamma \qquad (R6)$$

The two positrons annihilate with two electrons to produce more gamma photons:

$$2e^+ + 2e^- \rightarrow 4\gamma \qquad (R7)$$

In stars like the Sun, the core temperature and pressure are just high enough to sustain the reaction, and so it actually proceeds quite slowly. Since its formation about 5 billion years ago, the Sun has converted only about half the hydrogen in its core, and so it will be able to go on shining for about another 5 billion years.

Nuclear fission

Very massive nuclei can also release energy by the process of nuclear **fission**. Fission plays no part in supplying the energy output of stars, but it is worth mentioning for completeness. Some unstable massive nuclei can decay by splitting into two less massive nuclei, usually accompanied by a few neutrons. For example, an isotope of fermium can decay into xenon and palladium, like this:

$$^{256}_{100}\text{Fm} \rightarrow {}^{140}_{54}\text{Xe} + {}^{112}_{46}\text{Pd} + 4{}^{1}_{0}\text{n} \qquad \text{(R8)}$$

As you can see from Figure 5.42, the two lighter nuclei are more tightly bound than the initial nucleus. Spontaneous fission reactions such as this are rare, and tend to occur only in nuclei (such as $^{256}_{100}\text{Fm}$) that are produced artificially rather than occurring in nature. However, as with fusion, the energy released per reacting particle is vast, far outstripping that in chemical reactions, and since the mid-20th century people have sought to exploit this energy in nuclear reactors for generating electricity and in nuclear bombs. In both of these, a commonly used reaction is the fission of uranium-235. If a nucleus of ^{235}U absorbs an extra neutron, it becomes unstable and undergoes fission in reactions such as:

$$^{235}_{92}\text{U} + {}^{1}_{0}\text{n} \rightarrow {}^{236}_{92}\text{U} \rightarrow {}^{141}_{56}\text{Ba} + {}^{92}_{36}\text{Kr} + 3{}^{1}_{0}\text{n} \qquad \text{(R9)}$$

Figure 5.44 Nuclear power station

If the neutrons produced are absorbed by three further nuclei of ^{235}U, then they too will undergo fission, releasing further neutrons: a chain reaction. In a nuclear reactor (Figure 5.44), other materials are used to absorb some of the neutrons and to ensure that there are just enough to sustain the reaction at a steady rate. However, in a nuclear bomb, once the reaction is triggered to start, most of the neutrons go on to produce fission of another ^{235}U nucleus, leading to a runaway reaction – an explosion that causes huge damage not only because of the vast energy released, but also because of the ejection of highly radioactive materials into the surroundings (Figure 5.45). Uncontrolled reactions have led to a small number of nuclear accidents at power stations such as Chernobyl (Ukraine, 1986) and Fukushima (Japan, 2011), but fortunately such incidents have been rare.

Figure 5.45 Explosion of a nuclear bomb

Mass and energy

So much energy is released in a nuclear reaction that we can measure the associated loss of mass. Mass and energy are interrelated by Einstein's equation:

$$\Delta E = c^2 \Delta m \tag{14}$$

where ΔE is the energy released, c is the speed of light in a vacuum ($3.00 \times 10^8 \, \text{m s}^{-1}$) and Δm is the loss of mass, often called the **mass deficit** or **mass defect** when associated with nuclear reactions. A mass deficit always arises when a nucleus is formed from nucleons (as in the thought experiment we used earlier); it is the mass equivalent of the nuclear binding energy. For a given nucleus:

mass deficit = sum of masses of individual nucleons − mass of nucleus

Also:

binding energy = mass deficit × c^2

In a nuclear reaction (fusion, fission or radioactive decay) the overall reduction in mass is equivalent to the difference in binding energy between reactants and products.

There is also a binding energy associated with the electrons in an atom. Energy is needed in order to remove electrons from an atom, and the same amount of energy is released as free electrons are attracted to a nucleus to form an atom (photons are emitted as electrons drop to lower energy levels). However, this energy is only a few eV per electron, compared to a few MeV per nucleon, so to a very good approximation the overall binding energy of an atom is the same as that of its nucleus.

Chemical reactions involve rearrangement of electrons, and the energies released are very much smaller than the energies typical of nuclear reactions. So the associated changes of mass are also very much smaller – too small to measure. For example, when 1 kg of uranium-235 undergoes fission, the energy released is approximately 8×10^{13} J. The decrease in mass (calculated using Equation 14) is about 9×10^{-4} kg, which is measureable. Burning 1 kg of petrol releases about 5×10^7 J, which is equivalent to a mass of only 5.5×10^{-10} kg.

In a spontaneous reaction, energy is always emitted and the total mass of the products is always less than that of the reactants. An example using nuclear fission illustrates this.

WORKED EXAMPLE

Q Using data from Table 5.7, calculate the mass deficit when a nucleus of ^{235}U absorbs a neutron and undergoes fission into ^{141}Ba and ^{92}Kr, and in this way calculate the energy released.

$$1 \, \text{u} = 1.661 \times 10^{-27} \, \text{kg}$$

A

$$^{235}_{92}\text{U} + {}^{1}_{0}\text{n} \rightarrow {}^{141}_{56}\text{Ba} + {}^{92}_{36}\text{Kr} + 3{}^{1}_{0}\text{n}$$

$$\underbrace{235.04393 \, \text{u} + 1.008665 \, \text{u}}_{\substack{\text{initial mass} \\ = 236.0526 \, \text{u}}} \qquad \underbrace{140.91434 \, \text{u} + 91.92625 \, \text{u} + 3(1.008665)}_{\substack{\text{final mass} \\ = 235.8666 \, \text{u}}}$$

$$\Delta m = (236.0526 - 235.8666) \, \text{u}$$

$$= 0.1860 \, \text{u}$$

$$= 0.1860 \times 1.661 \times 10^{-27} \, \text{kg} = 3.090 \times 10^{-28} \, \text{kg}$$

$$\Delta E = c^2 \Delta m = (3.00 \times 10^8 \, \text{m s}^{-1})^2 \times 3.090 \times 10^{-28} \, \text{kg} = 2.78 \times 10^{-11} \, \text{J}$$

Particle	Mass/u
${}^{1}_{0}\text{n}$	1.008665
${}^{235}_{92}\text{U}$	235.04393
${}^{92}_{36}\text{Kr}$	91.92625
${}^{141}_{56}\text{Ba}$	140.91434

Table 5.7 Particles involved in a fission reaction of uranium-235

QUESTIONS

 25 The Sun has a mass of 1.99×10^{30} kg and is emitting energy at a rate of 3.84×10^{26} W.

    (a) How much energy will the Sun emit in 1000 years? (1 yr = 3.16×10^{7} s)

    (b) (i) If this energy were converted back into mass, how much mass would that be?
        (ii) What percentage of the Sun's mass is this?

    (c) What percentage of its mass will the Sun lose over its lifetime (roughly 10^{10} years)?

 26 **(a)** Write an equation that shows the overall effect of the ppI chain (Reaction R6) including the electron–positron annihilations (Reaction R7). Include sub- and superscripts to show that both charge and nucleon number are conserved.

    (b) Use the data in Table 5.8 to calculate the mass deficit associated with the production of one nucleus of $^{4}_{2}$He in the ppI fusion chain.

    (c) How much energy is therefore released when one nucleus of $^{4}_{2}$He is produced?

    (d) How many hydrogen nuclei must be consumed per second in order to maintain the Sun's luminosity of 3.84×10^{26} W?

    (e) The Sun's mass is 1.99×10^{30} kg. Given that when it formed from the solar nebula about 75% of its mass was hydrogen, and that about 15% of this hydrogen will eventually be converted to helium, estimate the Sun's lifetime. Express your answer in years. (1 year $\approx 3 \times 10^{7}$ s)

Particle	Mass/10^{-27} kg
$^{4}_{2}$He	6.645
$^{1}_{1}$H	1.673
e^{+}	9.110×10^{-4}
e^{-}	9.110×10^{-4}
n	0.000

Table 5.8 Particles involved in the ppI chain

 27 Without doing any calculations, sketch a graph showing how the mass per nucleon will vary with atomic number (i.e. the actual mass of the nucleus, not its mass number, divided by nucleon number).

 ⚙ ACTIVITY 19 SUMMING UP PART 3

Spend some time looking through your work on this part of the chapter. Make sure you understand all the terms highlighted in **bold**, and refer to the websites on **www.pearsonhotlinks.co.uk** to enhance your knowledge and understanding.

3.3 Summing up Part 3

In this part of the chapter, you have used some important concepts in nuclear physics: radioactive decay, fusion, fission, binding energy and mass deficit.

QUESTION

 28 To answer this question you will need to refer to Tables 5.4, 5.6 and 5.8.

    (a) The isotope $^{40}_{19}$K is used in dating rocks. It decays to form a stable isotope of argon gas, which becomes trapped in the rock.

       Write a balanced equation for the decay of ^{40}K to argon, and in this way calculate the nucleon number of the daughter isotope and the type of decay that might be involved.

    (b) A geophysicist finds that a certain rock sample contains 0.5 µg of ^{40}K and 3.3 µg of ^{40}Ar. Assuming that the rock initially contained no argon, and that all the argon produced has remained trapped in the rock, calculate the age of the sample.

    (c) Calculate the approximate number of ^{40}K nuclei in the sample and so calculate the activity due to their decay. Express your answer in suitable units. (1 yr $\approx 3 \times 10^{7}$ s)

4 Stars

We begin this part of the chapter with an exploration of how stars and other astronomical objects such as satellites and comets move, how they influence one another's motion through gravitational forces and what their motion enables us to work out about their properties.

We then turn to the evolution of stars: how they are born, live and die. This is a story of real scientific endeavour and achievement. Until the early part of the 20th century, it was thought to be impossible to find out anything about the nature of stars and how they evolved. However, since then, astronomers, physicists and mathematicians have made great progress in this area, to the extent that we can now describe the life story of a star with some degree of confidence.

Stellar evolution is interplay between gravitation and the behaviour of gases. There are links here with Part 3 of this chapter, where you saw how nuclear fusion reactions provide a star's power source, and Part 2, where you used the Hertzsprung–Russell diagram. For stable stars on the main sequence, such as our Sun, the gravitational forces pulling the material towards the centre of the star are in a fine balance with the outward gas pressure. This balance and how the star evolves is the subject of the last section of this part.

4.1 Heavenly bodies in motion

Observing orbits

In the early 17th century, Danish astronomer Tycho Brahe (1546–1601) and his German pupil Johannes Kepler (1571–1630) made many detailed observations of the planets' positions in the sky (see Figure 5.46). Kepler did a lot of work on analysing the data and looking for mathematical patterns, and was able to formulate three laws of planetary motion that summarised what he found. His first law stated that planets move in elliptical orbits, and the second that the line between a planet and the Sun sweeps out equal areas in equal times. The third law relates orbital radius to orbital period.

Figure 5.46 Tycho's Uraniborg observatory on the island of Hven, now in Sweden

The values in Table 5.9 are modern values of orbital radius and period. (Kepler had to deduce these from measurements of the planets' positions.) It is clear that orbital time is not proportional to the radius. If you assume that there is a power–law relationship:

$$T = Kr^p$$

then it is possible to analyse the data using a log–log graph and find values for the constants K and p.

MATHS REFERENCE
Using log graphs
See Maths note 8.7

⚙ ACTIVITY 20 ANALYSING ORBITS

Use the data from Table 5.9 to plot a graph of log (*T*) against log (*r*) and in this way estimate the relationship known as Kepler's third law of planetary motion.

PRACTICAL SKILLS REFERENCE

Analysis and interpretation of data

See Practical skills notes 4.1–4.2

Planet	Mean radius of orbit, $r/10^{10}$ m.	Planetary orbital period, *T*/Earth days
Mercury	5.791	87.97
Venus	10.82	224.7
Earth	14.96	365.3
Mars	22.79	687.0
Jupiter	77.86	4333
Saturn	143.35	10760

Table 5.9 Orbital radius and period

Kepler's laws are empirical. That is, they describe *how* the planets move, based on observation, but they give no underlying explanation of *why* they move as they do. For an explanation, we need to call on Newton's work.

Newton and gravity

Before the discoveries of the English scientist Isaac Newton (1642–1727), it must have seemed impossible to say anything meaningful about the masses of distant stars or even the Sun. However, Newton made fundamental breakthroughs in the science of gravitational physics that now enable astronomers to use these calculations as a matter of routine.

Arguably, Newton's greatest scientific work was *Philosophiæ Naturalis Principia Mathematica* or, in English, *The Mathematical Principles of Natural Philosophy* (Figure 5.47). Often referred to simply as the *Principia*, this book is regarded by many as the greatest feat of intellectual endeavour ever achieved. It was published in 1687 and contains a complete description of the laws of mechanics and gravitation, explaining the orbital motion of the planets and how the tides of the sea are generated – among many other things. These laws, discovered by Newton, are still used every day by scientists some 300 years after their formulation. They enable spacecraft to be launched to distant planets, and provide the means to measure the masses of stars and galaxies.

You probably know the famous tale of the falling apple. A descendant of the apple tree from which it fell can be seen today in the garden of Woolsthorpe Manor in Lincolnshire, Newton's childhood home (Figure 5.48). The important thing about this story is that it inspired the concept of **universal gravitation**. The word 'universal' means what it says. Newton's great insight was to realise there is an attractive force of gravity between all masses, no matter where they are in the Universe. Previously, it had been thought that the motion of objects on Earth was governed by rules quite different from those that applied to 'heavenly bodies'.

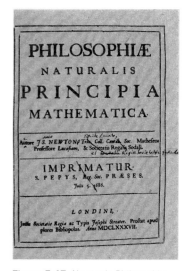

Figure 5.47 Newton's *Philosophiæ Naturalis Principia Mathematica*

Figure 5.48 Woolsthorpe Manor

Newton argued that the force of gravity exerted by a body on another must be proportional to its mass. He also argued that the force exerted by a body of mass m_1 on a body with mass m_2 must be equal in size to that exerted by the second on the first. From this it follows that the magnitude, F_{grav}, of the gravitational force must be proportional to the product m_1m_2:

$$F_{grav} \propto m_1m_2$$

By studying the motion of the Moon in its orbit about Earth, Newton was also able to deduce how F_{grav} depends on the separation of the objects. He did this by realising that the gravitational force between the Earth and the Moon must be responsible for the centripetal force needed to maintain the Moon's near circular motion. By an ingenious mathematical argument, he was able to show that the force between two bodies must diminish with distance according to an inverse-square law; that is:

$$F_{grav} \propto \frac{m_1m_2}{r^2}$$

where r is the distance between the *centres* of the two masses (see Figure 5.49). To make this relationship into an equation, we need a constant of proportionality. This is called the **gravitational constant** and given the symbol G. The full mathematical expression of Newton's law of universal gravitation is therefore:

$$F_{grav} = \frac{Gm_1m_2}{r^2} \tag{15}$$

Figure 5.49 Two masses, m_1 and m_2, separated by a distance r

QUESTION

Q 29 Use Equation 15 to deduce the SI units of G. Then express your answer using SI base units.

Measuring G experimentally is difficult, mainly because the gravitational force is very small unless at least one of the masses is very large. In principle it is possible to deduce G from measuring the force between two large bodies, such as the Earth and the Moon, but in practice this is difficult since determining the masses of such bodies relies on knowing G in the first place, as you will see shortly. However, it is possible to set up delicate laboratory experiments to measure the forces between quite small masses, and these yield a value of $G = 6.67 \times 10^{-11}\,\text{N}\,\text{m}^2\,\text{kg}^2$. In Activity 21 you will get a feel for the sizes of some gravitational forces.

⚙ ACTIVITY 21 UNIVERSAL GRAVITY

Table 5.10 lists some objects, their masses and separations. Use a spreadsheet to calculate the magnitude of the gravitational force between each pair, and so complete Table 5.10.

Objects	m_1/kg	m_2/kg	r/m	F_{grav}/N
Moon in orbit around Earth	7.35×10^{22}	5.97×10^{24}	3.85×10^8	
Earth in orbit around Sun	5.97×10^{24}	1.99×10^{30}	1.50×10^{11}	
Moon and Sun	7.35×10^{22}	1.99×10^{30}	1.50×10^{11}	
Person on surface of Earth	60.0	5.97×10^{24}	6.37×10^6	
Person on surface of Moon	60.0	7.35×10^{22}	1.74×10^6	
Person at Sun's photosphere (!)	60.0	1.99×10^{30}	6.96×10^8	
Two people standing together	60.0	60.0	1.00	
Electron in orbit around proton	9.11×10^{-31}	1.67×10^{-27}	5.3×10^{-11}	

Table 5.10 Data for Activity 21; the mass m_1 is that of the first object listed in each pair

Force and field

Gravitational field strength, g, is defined as the gravitational force per unit mass:

$$g = \frac{F_{grav}}{m} \tag{16}$$

Close to the Earth's surface, $g = 9.81\,\text{N kg}^{-1}$. The gravitational force on an object, that is, its weight, W, is related to its mass, m, via the gravitational field strength:

$$W = F_{grav} = mg \tag{17}$$

Gravitational field is defined as a region in which a mass experiences a gravitational force. This definition is very similar to the definition of **electric field**: a region in which a charged object experiences an electrostatic force. The direction of an electric field is defined to be that of the force acting on a positive charge, and the strength, E, of an electric field is defined as the force per unit charge. The electrostatic force, F_{elec}, on a charged particle is related to its charge, q, via the electric field strength:

$$F_{elec} = qE \tag{18}$$

There is a marked similarity between Equations 17 and 18. You have probably also noticed the similarity between the mathematical expression of Newton's law of gravitational field strength and Coulomb's law of electrostatic force:

$$F_{elec} = \frac{Q_1 Q_2}{4\pi\varepsilon_0 r^2} \tag{19}$$

From this relationship we can derive an expression for the field strength at a distance r from a point charge Q:

$$E = \frac{Q}{4\pi\varepsilon_0 r^2} \tag{20}$$

Similarly, we can use the definition of gravitational field strength to derive an expression for the gravitational field due to a point mass. Imagine placing a small mass m at a distance r from the centre of a mass M. The gravitational force acting on m is given by:

$$F_{grav} = \frac{GmM}{r^2} \tag{Equation 15}$$

The gravitational field strength is then:

$$g = \frac{Gm}{r^2} \tag{21}$$

Like force, gravitational field is a vector, and its direction is the same as that of the gravitational force. Just as we represented electric fields by drawing field lines, we can draw lines of gravitational field. Figure 5.50 shows the gravitational field around a planet (or any spherical mass). The field is represented by arrows, each arrow representing the direction and magnitude of the force on a 1 kg mass placed at the blunt end of the arrow. Notice that the field is spherically symmetric, which means that the field strength is dependent only on the radial distance from the mass and not on direction. Also notice that the length of the arrows decreases with increasing distance from the mass. This indicates that the field strength decreases with radial distance, according to the inverse-square law.

STUDY NOTE

You met gravitational field strength in the chapter *Higher, Faster, Stronger*.

STUDY NOTE

You met electric field strength in the chapters *The Medium is the Message* and *Probing the Heart of Matter*.

STUDY NOTE

Coulomb's law was introduced in the chapter *Probing the Heart of Matter*.

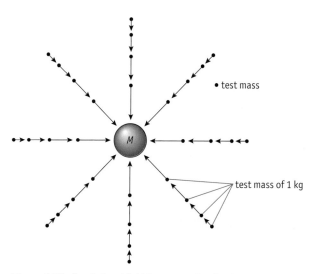

Figure 5.50 Gravitational field lines around a planet

	Gravitational	Electrical
Force acts on ...	all matter	
Direction of force	always attractive	
Force law between two objects formulated by ...		Charles Coulomb (1736–1806)
Equation for force law between two objects		$F = \dfrac{Q_1 Q_2}{4\pi\varepsilon_0 r^2}$
SI units of constant in above equation		
Field strength	$g =$	$E = \dfrac{F}{q}$
Field strength in a radial field	$g = \dfrac{GM}{r^2}$	$E =$
SI units of field strength		
Direction of field around isolated object	towards object	
Change in potential energy in uniform field	$\Delta E_{grav} = mg\Delta h$	

Table 5.11 For use in Activity 22: compare and contrast gravitational and electrical forces

QUESTIONS

Q 30 **(a)** Calculate the gravitational field strength at the surface of Mars, which has a mass of $6.42 \times 10^{23}\,\mathrm{kg}$ and a radius of $3.39 \times 10^6\,\mathrm{m}$.

(b) Calculate your own weight on the surface of Mars.

Q 31 Assuming that a planet with radius r has uniform density, ρ, show that the gravitational field strength at its surface is given by $g = \dfrac{4\pi r \rho G}{3}$.

$\left(\text{Hint: the volume, } V, \text{ of a sphere radius } r \text{ is } \dfrac{4\pi r^3}{3}\right).$

Landing on a comet

On 12 November 2014, at 16:03 GMT, a small lander only 1 m across, travelling at $20\,\mathrm{km\,s^{-1}}$, bounced onto the nucleus of a comet a distance of 673 million km from the Sun. The lander's name was Philae and it was sent by the Rosetta probe, which was in orbit around comet 67P/Churyumov–Gerasimenko (Figure 5.51), some 24 km above its surface. Launched in March 2004, Rosetta took over 10 years to get to the comet, travelling 6.4 billion kilometres as it underwent gravity-assisted slingshots around Mars and the Earth, finally orbiting the 4 km wide comet on 10 September 2014.

After the lander made contact with the comet, its harpoons failed to attach it to the surface and it bounced twice. At the first bounce, it reached a height of about 1 km above the comet's surface. What was its speed as it left the comet's surface? To answer this question we cannot simply use equations of uniformly accelerated motion, or $\Delta E_{grav} = mg\Delta h$, as we did in the earlier chapter *Higher, Faster, Stronger* when analysing motion under gravity near Earth's surface. The comet's small size means that the gravitational field strength varies noticeably with height, so we have to take account of that. When analysing situations such as this, the idea of **gravitational potential energy** is very useful.

Figure 5.51 Comet 67P/Churyumov–Gerasimenko taken from the Rosetta probe on 10 December 2014 (photo courtesy of ESA/Rosetta/NAVCAM)

In the chapters *The Medium is the Message* and *Probing the Heart of Matter* you met the idea of electrostatic potential as the electrostatic potential energy per unit charge relative to a chosen zero level. In *Probing the Heart of Matter* you saw that, in the inverse-square electric field due to an isolated charge, we conventionally define the zero level to be at an infinite distance from the charge and the potential is then proportional to $1/r$. Similarly, in an inverse-square gravitational field we choose to take the zero of potential energy to be at infinity. This convention gives a common reference point for all objects and it is a very special reference point: at an infinite separation, masses exert no influence on each other.

The work, ΔW, done in bringing an object, m, from infinity to a distance r from the centre of an object of mass M is given by:

$$\Delta W = -\frac{GmM}{r} \qquad (22)$$

Equation 23 defines the gravitational potential energy of the system:

$$E_{\text{grav}} = -\frac{GmM}{r} \qquad (23)$$

This leads us to the idea of **gravitational potential**, which is analogous to the idea of electrostatic potential. The gravitational potential, V_{grav}, at a point is defined as the work done per unit mass against the gravitational field in bringing the mass from infinity to that point:

$$V_{\text{grav}} = \frac{E_{\text{grav}}}{m} \qquad (24)$$

And so:

$$V_{\text{grav}} = -\frac{GM}{r} \qquad (25)$$

Gravitational potential has SI units J kg^{-1}. Note the minus sign in Equations 23 and 25. This is because work is done *by* the gravitational field, and there is a loss of gravitational potential energy as the masses come closer together. You can think of the small mass m 'falling' towards the larger mass M.

Figure 5.52 shows how gravitational potential varies with distance from an isolated mass.

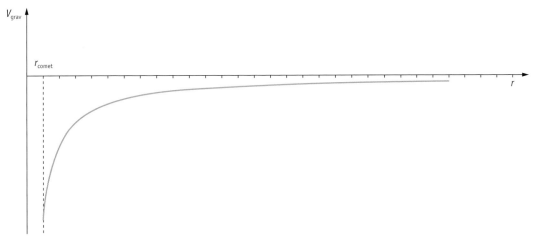

Figure 5.52 Variation of gravitational potential with distance

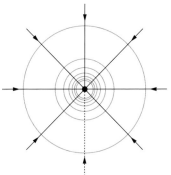

Figure 5.53 Gravitational equipotentials and field lines around an isolated mass

Just as we were able to map equipotential lines and surfaces in an electric field, we can think of gravitational equipotential lines and surfaces. Figure 5.53 shows the **gravitational equipotentials** and field lines around a point mass. Notice that the field lines are always at right angles to the equipotentials, and that when the equipotentials are plotted at equal energy intervals, the lines become further apart as the field becomes weaker.

We can now use gravitational potential energy to relate the lander's bounce height to its initial vertical velocity. As the lander, mass m, moves from a radius r_1 to r_2, the increase in gravitational potential energy is:

$$\Delta E_{\text{grav}} = \frac{GmM}{r_1} - \frac{GmM}{r_2}$$

$$= GmM\left(\frac{1}{r_1} - \frac{1}{r_2}\right) \tag{23a}$$

Energy conservation tells us that the gravitational potential energy gained must be equal to the kinetic energy lost as the lander moves upwards from the comet.

WORKED EXAMPLE

Q The lander bounced to a height of about 1 km. What was its approximate vertical velocity as it left the comet's surface?

A When lander, mass m, is leaving the surface of comet, mass M and radius r_1 (= 2 km), at speed v:

$$E = E_k + E_{\text{grav}} = \frac{1}{2}mv^2 - \frac{GmM}{r_1}$$

At the top of its bounce the lander was at a distance r_2 (= 3 km) from the comet's centre and $E_k = 0$. Energy is conserved and so, using Equation 23a:

$$\frac{1}{2}mv^2 = GmM\left(\frac{1}{r_1} - \frac{1}{r_2}\right)$$

Cancelling m, multiplying by 2 and taking the square root:

$$v = \sqrt{\left[2GM\left(\frac{1}{r_1} - \frac{1}{r_2}\right)\right]}$$

$$= \sqrt{2 \times 6.67 \times 10^{-11}\,\text{N}\,\text{m}^2\,\text{kg}^{-2} \times 3.1 \times 10^{12}\,\text{kg}\left(\frac{1}{2 \times 10^3\,\text{m}} - \frac{1}{3 \times 10^3\,\text{m}}\right)}$$

$$= \sqrt{2 \times 6.67 \times 10^{-11}\,\text{N}\,\text{m}^2\,\text{kg}^{-2} \times 3.1 \times 10^{12}\,\text{kg}\left(\frac{1}{6 \times 10^3\,\text{m}}\right)}$$

$$= 0.26\,\text{m}\,\text{s}^{-1} \approx 0.3\,\text{m}\,\text{s}^{-1}$$

MATHS REFERENCE

Reciprocals

See Maths note 3.3

QUESTIONS

Assume that the comet is a sphere with mass 3.1×10^{12} kg and radius 2 km, and the lander had a mass 100 kg.

Q 32 What gravitational force acted on the lander Philae as it was released from the Rosetta probe 24 km above the surface of the comet?

Q 33 What is the gravitational field strength of 67P:

(a) at the surface of the comet?

(b) at 24 km from its surface?

Express your answers as (very small) multiples of the Earth's surface gravitational field ($g_{\text{Earth}} = 9.81\,\text{N}\,\text{kg}^{-1}$).

Q 34 Explain why the comet's small mass made it difficult to land a probe on its surface.

Q 35 What is the gravitational potential 20 km above the surface of Comet 67P?

Q 36 Suppose the lander left the comet's surface with a vertical velocity of $10\,\text{cm}\,\text{s}^{-1}$ on its second bounce. Approximately how high would it rise?

As you have seen in Activity 22, there are many similarities between electrical and gravitational forces and fields, but there are also some important differences. Add to your table from Activity 22 using the ideas from this section on Comet 67P. Include equations for potential and potential energy, and the definition of potential in your table.

Explaining orbits

Newton's work on gravitation provided an explanation for Kepler's laws in terms of more general ideas. While Kepler had successfully described what happened in the Solar System, Newton's law of gravitation is, as we have already stated, universal. This means that it applies anywhere in the Universe and gives the gravitational force between any two masses, be they atoms or stars. The law of gravitation is particularly useful for determining the masses of stars and planets. This method is of great importance in astronomy, since it is the only means by which such masses can be directly measured.

Figure 5.54 shows two objects attracting one another gravitationally and moving in orbit. If the mass of one object is very much greater than the other, it remains essentially at rest while the less massive object orbits around it.

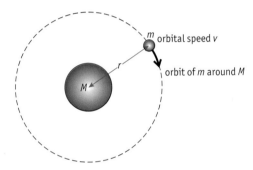

Figure 5.54 Two objects in orbit: $M \gg m$

To see how mass can be measured, consider the situation in Figure 5.54, where the small mass, m, is moving in a circle, radius r, at speed v. There must be a centripetal force, F, acting to keep m moving in a circle, where:

$$F = \frac{mv^2}{r} \tag{26}$$

Here, the force must be provided by the gravitational attraction, F_{grav}, between the two masses:

$$F_{grav} = \frac{GMm}{r^2} \tag{Equation 15}$$

Identifying the two forces as one and the same:

$$\frac{GMm}{r^2} = \frac{mv^2}{r} \tag{27}$$

Cancelling m and r and rearranging gives:

$$M = \frac{v^2 r}{G} \tag{28}$$

It is often useful to rewrite Equation 28 in terms of the period T (the time for one orbit) where:

$$v = \frac{2\pi r}{T}$$

and so:

$$M = \left(\frac{2\pi r}{T}\right)^2 \times \frac{r}{G} = \frac{4\pi^2 r^3}{T^2 G} \tag{29}$$

Rearranging Equation 29 to get a relationship between r and T yields:

$$T^2 = \frac{4\pi^2 r^3}{MG}$$

(29a)

Comparing this with your results from Activity 20 shows how Kepler's third law is related to Newton's law of gravitation. Kepler's law is in fact an inevitable consequence of Newton's, but that's not the way in which it was discovered.

Equation 29 tells us that to measure the mass, M, of the central object we need to know the radius of the orbit of its companion and the period of the orbit. It can be used to determine the masses of planets by observing the orbits of their moons, or the Sun's mass from measurements of the orbits of planets. Note that the mass, m, of the orbiting object does not feature in Equation 29 – we do not need to know m, nor can m be found using this method.

✪ ACTIVITY 24 IN ORBIT

Use an applet (see **www.pearsonhotlinks.co.uk**) to explore the motion of an object under an inverse-square law force.

By trial and error, find out how to change the shape of the orbit so that it is either a circle or a highly eccentric (elongated) ellipse. Try to produce an orbit that is not closed – that is, the object disappears off the screen and does not return.

Experiment with changing the force law. For example, try $F = -kr^{-1}$ and $F = -kr^{-3}$.

Satellites and geostationary orbits

The Earth has one natural satellite (the Moon) and many thousands of man-made satellites including the International Space Station (and a lot of debris). All of these satellites follow either a circular orbit around the centre of mass of the Earth or an elliptical orbit with the Earth at one focus. Newton's law of gravitation allows us to find what speed a satellite must travel at to maintain a particular circular orbit, using the orbital equations set out above. Rearranging Equation 28 shows how speed depends on orbital radius:

$$v^2 = \frac{GM}{r}$$

(28a)

In order to maintain their circular orbits, satellites must be released into a lower orbit at a higher speed and into a higher orbit at a lower speed. An interesting aspect of Equation 28a is that the speed is independent of the mass of the satellite, so all objects will travel at the same speed in the same orbit. This is very good news for astronauts on space walks as it means they will move at the same speed as their spacecraft; their difference in mass is not relevant.

Man-made satellites are used for a vast variety of purposes and many different orbits (Figure 5.55). However, communications and weather satellites usually travel in a geostationary orbit so that the satellite appears in a fixed position to an observer on Earth.

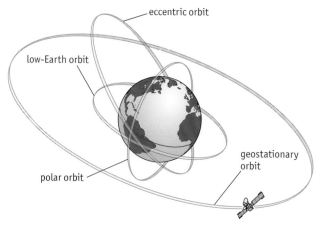

Figure 5.55 Different types of satellite orbit

This has the advantage that antennae used to communicate with the satellites can be pointed to a fixed position in the sky at all times. The geostationary satellite will have an orbital period, T, to match that of the Earth's rotation: 24 hours. Equation 29 shows that since T is 24 hours, the radius must also be a fixed value. This radius is approximately 42 300 km from the centre of the Earth. A geostationary orbit can therefore only be around the equator at a fixed height above it, as shown in Figure 5.54. As a geostationary orbit is so commercially advantageous, there is an organisation that allocates orbital slots for satellites from different countries to stop conflict occurring.

STUDY NOTE

For more on orbits, including a discussion of kinetic and potential energy, see STA Additional Sheet 2.

QUESTIONS

Q 37 All of the orbits in Figure 5.55 above are centred around one common location. State this location. Could a geostationary orbit ever be in a non-equatorial plane? Explain your answer.

Q 38 For a geostationary satellite, show that the orbit height above the Earth's surface is around 36 000 km. The radius of the Earth is 6400 km, $G = 6.67 \times 10^{-11}\,\text{Nm}^{-2}\text{kg}^{-2}$ and the mass of the Earth $M = 6.0 \times 10^{24}\,\text{kg}$.

Q 39 Using the data from Question 38, calculate:

(a) the speed at which the geostationary satellite must be travelling

(b) the gravitational field strength at that location.

Binary stars

Until now we have considered astronomical systems where one object is orbiting a much larger mass. However, if the two objects are comparable in mass, they both move as shown in Figure 5.56, where point X (their centre of mass) remains at rest and the two masses orbit so that they are always on opposite sides of X. The two objects might be a planet and one of its moons, a star and a planet, or two stars – in which case they are called a binary pair of stars or, more often, the two together are called a **binary star**. Many stars are either binary pairs or members of multiple-star systems, whereas our Sun – as a single star – is in a minority.

STUDY NOTE

You met the idea of centre of mass in the chapter *Higher, Faster, Stronger*.

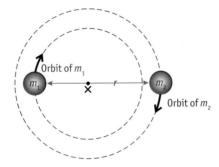

Figure 5.56 Two objects in orbit: $M \sim m$

Analysis of the situation in Figure 5.56 is more complicated than when $m \ll M$, but it yields the same result:

$$T^2 = \frac{4\pi^2 r^3}{MG}$$

(Equation 29a)

where M is now the total mass of the pair of objects ($M = m_1 + m_2$). (This analysis also yields expressions that enable the two masses to be determined separately.) So to measure the mass of a star, it must have an observable gravitational effect on another object, usually another star. The masses of isolated stars must be deduced by other means.

Studying binary stars and measuring the periods and separations is a job for the observational astronomer. It is possible for us to observe some binaries directly and to see both stars in the eyepiece of the telescope, but mostly they are unresolved – that is, they appear as a single star, even with the most powerful telescopes available. So how do astronomers know that they are looking at a binary star?

STUDY NOTE

You met the idea of resolution of an optical instrument in the chapter *Digging Up the Past*.

Unresolved binaries give themselves away in two ways. First, if the orbital plane of the binary system is in our line of sight, then one star will at some stage eclipse the other. When this happens there will be a change in the brightness of the star, just as when the Moon eclipses the Sun. Figure 5.57 shows the effect. The graph (known as **light curve**) represents the variation in observed brightness as the small bright star passes behind, and later in front of, the larger star.

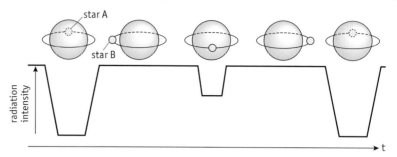

Figure 5.57 Light curve of an eclipsing binary star

The bright star Algol, in the constellation of Perseus, is an eclipsing binary. It is most easily observed in the autumn night sky. For most of the time it is about as bright as Polaris, the pole star, but every 2½ days it fades noticeably over about 4 hours. It stays dim for about 20 minutes, and then takes another four hours to return to its normal brightness. Algol's behaviour was first explained by a young English astronomer, John Goodricke of York (1764–1786) (Figure 5.58). A sculpture representing the two stars of Algol stands outside the college named after Goodricke at the University of York (Figure 5.59).

Figure 5.58 John Goodricke

Figure 5.59 Algol sculpture at Goodricke College

This method of identifying binary stars using the variation in brightness over time can also be used for identifying **exoplanets**: planets orbiting stars outside of our Solar System. As the planet passes in front of the star, the detected brightness periodically decreases. From this it is possible to determine the relative size of the planet and star. The Kepler space telescope, launched by NASA, uses this method to find exoplanets as well as eclipsing binary stars. In April 2014 the telescope found the first exoplanet similar to Earth in its size and distance from its star. You can find out more about detecting exoplanets and some relevant exercises at **www.pearsonhotlinks.co.uk.**

The second way that an astronomer knows that a star is a binary is to observe the spectrum of the star. Consider again the smaller star in Figure 5.56. As it moves towards or away from us, the lines in its observed spectrum will be Doppler shifted. As the star recedes the lines will be shifted to longer wavelengths, and as it approaches they will be shifted to shorter wavelengths. The expression used for the shift in frequency is:

$$\frac{\Delta f}{f_{em}} \approx \frac{v}{c}$$

(30)

STUDY NOTE

You met the Doppler effect in the chapter *Spare Part Surgery* in the context of Doppler ultrasound to measure blood flow rates.

where v is the relative speed between the source and detector, c is the wave speed, and:

$$\Delta f = f_{em} - f_{rec} \tag{31}$$

where 'em' and 'rec' refer to the emitted and received frequencies. The expression applies only when $v \ll c$. In the case of binary stars, c is the speed of light and v is the speed of the star along our line of sight. When studying spectra it is more convenient to measure wavelength, λ, than frequency, and the appropriate expression is then:

$$\frac{\Delta\lambda}{\lambda_{em}} \approx \frac{v}{c} \tag{32}$$

where:

$$\Delta\lambda = \lambda_{rec} - \lambda_{em} \tag{33}$$

So, using Equation 32, it is quite straightforward to measure the orbital speed. Orbital period, T, can be deduced simply from seeing how the amount of Doppler shift changes with time, and so all the information is available for determining the separation, r, and therefore the mass, M of the pair of stars.

Masses of stars

In the middle decades of the 20th century, measurement of large numbers of stellar masses provided astronomers with further clues to the nature and evolution of stars. One important finding was that the masses of main-sequence stars are strongly linked to their temperature and luminosity. Main-sequence stars with the same luminosity and temperature also have the same mass. The most luminous main-sequence stars are also the most substantial, with masses up to about 20 times that of the Sun. The faintest main-sequence stars, on the other hand, have masses less than half the Sun's mass.

The measurement of mass provides very good reasons for believing that main-sequence stars do *not* change their luminosity and temperature, since to do so they would need to change their mass by a considerable amount – it would be very difficult to see how they could do this. The main sequence, then, does not represent a star's evolution, even though its name tends to suggest otherwise.

In Sections 4.2 and 4.3 you will see how the 'story' of star formation and evolution has been developed to explain the features of the HR diagram that you saw in Section 2.5.

> **STUDY NOTE**
>
> For a derivation of the equations for the Doppler effect, see STA Additional Sheet 3.

QUESTIONS

Q 40   **(a)** From the graph in Figure 5.57, decide which of the two stars has the brighter surface and explain how you reached your decision.

        (b) Label a copy of Figure 5.57 to show the orbital period.

Q 41   **(a)** Sketch the light curve of a binary pair of stars, of equal sizes and luminosities, in a circular orbit viewed edge-on. Mark the orbital period on your sketch.

        (b) How would the light curve be affected if the orbit were elliptical, as in Figure 5.60?

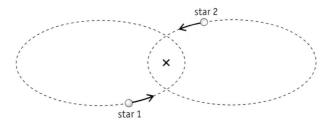

Figure 5.60 Binary pair of stars with elliptical orbits

Q 42   Use the wave equation $v = f\lambda$ to derive Equation 32 from Equation 30. (Hint: near the end you will need to use the fact that $\lambda_{rec} \approx \lambda_{em}$.)

Q 43   In February 1987 astronomers observed a huge stellar explosion (a supernova) in a nearby galaxy. About six months later, they measured the wavelength of a spectral line of helium as being 1.070 μm, whereas in the laboratory this line has a wavelength 1.083 μm.

Were the gases from the explosion approaching or receding, and how fast were they moving? Express your answer as a fraction of the speed of light, c, and in km s^{-1} ($c = 3.00 \times 10^8$ m s^{-1}).

Q 44   Analysis of the spectrum of an eclipsing spectroscopic binary of period $T = 8.6$ years shows that the maximum Doppler shift of a hydrogen absorption line with $\lambda_{em} = 656.28$ nm is 0.041 nm.

Assume the system consists of a small star in orbit about a much more massive companion. Calculate the radius of the small star's orbit (assuming it to be circular) and in this way determine the mass of the large star. ($c = 3.00 \times 10^8$ m s^{-1}; 1 year $= 3.16 \times 10^7$ s; $G = 6.67 \times 10^{-11}$ N m^2 kg^{-2})

4.2 Star formation

We have seen how some of the properties of stars are measured and calculated. In this section we will follow some aspects of the story of stellar evolution, from the gas clouds that are the stellar nurseries, and in Section 4.3 we will continue the story to the final stages of the lives of stars.

What lies between the stars? You might think 'empty space', but that would not be quite true. Within our galaxy, the space between the stars is filled with gas and dust (minute solid particles). By our own everyday standards, this **interstellar material** is very tenuous (that is, it has very low density), but we can still detect it. Figure 5.61 shows the Orion Nebula – visible to the naked eye as a fuzzy patch of light in Orion (see Figure 5.17). This and photographs of other **nebulae** (a general term for 'fuzzy objects' or clouds) show that between the stars there are huge clouds of hot gas, many light years across, that emit their own light, and cool, darker regions that show their presence by blocking out the light from more distant stars. And in between the visible clouds, there is gas that is too cold, or too tenuous, for us to detect with visible light.

Figure 5.61 Orion Nebula

The bright stars embedded in the Orion Nebula are believed to have formed quite recently, and so the nebula is a site of star formation. So how can a tenuous cloud of gas give rise to the conditions of very high density and temperature needed for the nuclear fusion reactions that power stars? Understanding what is involved requires some knowledge of gravitation (from Section 4.1) and of the behaviour of gases.

In the 17th century, scientists including Robert Boyle first investigated the physical properties of gases. They did experiments on the relationships between the pressure, volume and temperature of gases. We can repeat their pioneering experiments fairly easily in a modern physics laboratory.

Figure 5.62 Apparatus for demonstrating the change of volume with pressure of a gas

⚙ ACTIVITY 25 EXPERIMENTS ON GASES

Use a syringe and masses or the apparatus shown in Figure 5.62 to investigate how the volume of a fixed amount of gas changes as the pressure is increased. Use your results to suggest a simple mathematical relationship between the two variables.

If apparatus is available, also investigate the relationships between pressure and temperature at constant volume, and between volume and temperature at constant pressure.

The graphs in Figure 5.63 are based on experiments such as those in Activity 25 and show the relationships between pressure, p, volume, V, and temperature, T, for a fixed mass of gas as each variable in turn is held constant.

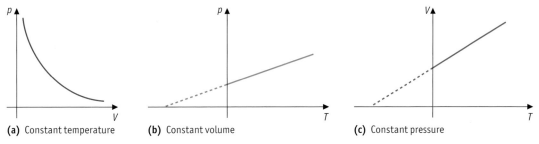

(a) Constant temperature   (b) Constant volume   (c) Constant pressure

Figure 5.63 Temperature, volume and pressure graphs for a fixed mass of gas

Temperature scales

Notice that the pressure–temperature graph and the volume–temperature graph are both straight lines. If extrapolated, both these lines should meet the horizontal axis at a temperature of −273.15 °C. You should realise that this could never be achieved in practice, since any real gas will liquefy and stop exerting a pressure well before zero temperature is reached, and it is impossible to imagine a gas with zero volume (what would happen to the molecules?). The extrapolation shows the behaviour of an **ideal gas** – one that does not liquefy, among other things. However, the theoretical idea can be put to good use in the definition of an **absolute temperature scale**. The **absolute zero** of temperature, zero kelvin (0 K – notice no degree symbol), is defined as the temperature at which the pressure of an ideal gas becomes zero. Absolute zero, 0 K, is equivalent to −273.15 °C.

The other fixed point on the absolute temperature scale is the 'triple point of water', which is the unique temperature at which pure water, pure ice and water vapour can coexist; this is 0.01 °C. Thus we have:

temperature in K = temperature in °C + 273.15   (34)

We have emphasised the connection between the Celsius and the Kelvin scales, as the gas laws use kelvin as a measure of temperature but everyday life uses degrees Celsius – so you will need to convert between the two. Note, though, that the kelvin is the same size as the degree Celsius, so a temperature *change* ΔT has the same numerical value expressed in either scale.

PRACTICAL SKILLS REFERENCE

Scientific questions and information research
See Practical skills notes 1.1–1.3

Planning and experimental design
See Practical skills notes 2.1–2.4

Carrying out practical work
See Practical skills notes 3.1–3.2

Analysis and interpretation of data
See Practical skills notes 4.1–4.2

Conclusion and evaluation
See Practical skills notes 5.1–5.2

Gas laws

Provided temperature is measured in kelvin, then the results shown in Figure 5.63 for a fixed mass of gas can be expressed as follows:

The **pressure law**: at constant volume, pressure is directly proportional to temperature:

$$p \propto T \tag{35}$$

Charles's law: at constant pressure, volume is directly proportional to temperature:

$$V \propto T \tag{36}$$

Boyle's law: at constant temperature (in any units!) pressure is inversely proportional to volume:

$$p \propto \frac{1}{V} \tag{37}$$

Collectively, these three laws are known as the **gas laws**. These laws are empirical; that is, they are purely experimental – they tell us how ideal gases behave under a variety of conditions, but they do not tell us *why* they behave in this way. The gas laws can be combined in the form of a single equation:

$$\frac{pV}{T} = \text{constant} \tag{38}$$

This is called the **ideal-gas equation** or the **equation of state for an ideal gas**. An **ideal gas** is defined as one that exactly obeys the gas laws. In reality, ideal gases do not exist – there is no gas that exactly obeys the gas laws, although if a gas is at a high enough temperature and low enough pressure to avoid liquefying, then it will approximate to an ideal gas.

Experiment shows that the constant is proportional to the quantity of gas. This might give the impression that the 'constant' is not constant, but remember that the ideal-gas equation applies to a fixed mass of gas, and the 'constant' is only constant if the quantity of gas is fixed. If the mass or the type of gas is changed, then the constant also changes. A complete statement of the ideal-gas equation is:

$$pV = NkT \tag{39}$$

where N is the number of particles in the gas and k is the **Boltzmann constant**: $k = 1.38 \times 10^{-23}\,\text{J}\,\text{K}^{-1}$.

Notice that the word 'particle' has been used rather than 'atom' or 'molecule'. In a monatomic material such as copper, sodium or argon, the particle would be an atom, whereas for a substance that exists in molecular form, such as hydrogen (H_2), the particle would be a molecule of hydrogen.

STUDY NOTE

The Boltzmann constant is sometimes given the symbol k_B.

STUDY NOTE

In chemistry the ideal-gas equation is usually written $pV = nRT$, where n is the number of moles and R the universal gas constant.

QUESTIONS

Q 45   Show that pV has SI units of J and that k has SI units of $\text{J}\,\text{K}^{-1}$.

Q 46   The Sun has a mass $M = 2 \times 10^{30}\,\text{kg}$ and a radius $r = 7 \times 10^8\,\text{m}$. It can be modelled as a spherical ball of hydrogen ions, H^+ (each of mass $m = 1.67 \times 10^{-27}\,\text{kg}$), and an equal number of electrons (whose mass is negligible compared to the ions). If the temperature inside the Sun is $T = 1 \times 10^7\,\text{K}$ (i.e. high enough for fusion), what is the pressure?

($k = 1.38 \times 10^{-23}\,\text{J}\,\text{K}^{-1}$)

Assume the temperature and pressure are uniform throughout the Sun.

Kinetic theory of gases

In the 19th century, around 200 years after the gas laws were established, scientists began to develop the **molecular kinetic theory of gases**. In this work, a theory was used to link the particle model of matter to the large-scale measurable properties of matter. Essentially, the ideal-gas equation was described in terms of the motion of invisible atoms. This was remarkable work at the time, as many scientists disputed the existence of atoms.

The kinetic theory of gases is one of the best theories produced in the 19th century. It was developed principally by Scottish physicist James Clerk Maxwell (1831–1879) (Figure 5.64a) and subsequently by Austrian physicist Ludwig Boltzmann (1844–1906) (Figure 5.64b). Maxwell was a remarkable scientist. As well as the kinetic theory, he produced the theory of electromagnetism, which led to his prediction of electromagnetic waves.

The motivation behind developing the theory was pure intellectual curiosity rather than any application (though it has come to be applied in many situations, including star formation, as you will see later). The theory was remarkable because it set out to link **macroscopic properties** (properties we can measure with instruments) to the invisible **microscopic properties** of matter (randomly moving atoms and molecules) (see Table 5.12). What is also remarkable about the molecular kinetic theory of gases is that it uses nothing more sophisticated than Newton's laws of motion to produce an elegant and testable theory about invisible particles of matter.

(a)

Macroscopic	Microscopic
Temperature	average kinetic energy of a particle
Pressure	forces due to collisions of particle: rate of change of momentum

Table 5.12 Macroscopic and microscopic properties

The elegance referred to is part of the beauty of physics – the final equation is neat. Many scientists have argued that mathematical beauty is necessary for an equation to be correct. Einstein said: 'The only physical theories that we are willing to accept are the beautiful ones'. Any scientific theory must also be testable. One test of the kinetic theory of gases is that it predicts the average speed of particles in a gas, which can be measured.

The theory is based on the idea that all matter (solids, liquids and gases) is made of discrete particles (molecular or atoms) in a state of random motion (see Figure 5.65). One way to understand the theory is to approach it 'backwards'; that is, start with an experimental estimate of particle speed and see how that leads to a calculation of gas pressure.

(b)

Figure 5.64 (a) James Clerk Maxwell; **(b)** Ludwig Boltzmann

Figure 5.65 Model of gas particles in random motion

ACTIVITY 26 PARTICLE SPEED AND GAS PRESSURE

Observe the release of bromine vapour into an evacuated tube and estimate the speed of the bromine particles.

Then calculate the force exerted by air particles moving at this speed as they bounce against the walls of your lab. Use this force calculation to predict the pressure exerted by a room full of air particles, and compare the result with known atmospheric pressure.

In the calculations for Activity 26 you had to make some assumptions about the air particles and some simplifications. You may not like assumptions and simplifications, but without them, scientists would not be able to make much progress. The final result from the kinetic theory makes predictions that can be tested. If the predictions did not agree with what actually happened, scientists would refine the theory to get a better agreement with experiments.

The assumptions that you made in Activity 26 are the same as those made by Maxwell and Boltzmann, namely that:

● particles in a gas collide frequently

● a very large number of particles are present

● the particles have negligible long-range forces between them

● the particles travel in straight lines between collisions and obey Newton's laws of motion

● the volume of the particles themselves is negligible compared with the volume of their container

● collisions between particles in a gas and walls of containers are perfectly elastic

● the time taken for collisions is negligible compared with the time between collisions.

These assumptions, along with ideas about force and momentum, lead to an equation that relates the pressure of a gas to the motion of its particles. For an algebraic version of the calculation from Activity 26, consider the situation shown in Figure 5.66, which contains a gas of N particles, each of mass m, moving randomly. We will first fix attention on just one particle, moving at velocity c, with components u, v and w in the x, y and z directions respectively – and imagine that all the other particles have been removed.

STUDY NOTE

You met elastic collisions in the chapter *Transport on Track*, only they involved much larger objects.

STUDY NOTE

Notice the use of vector notation to describe the particle's motion.

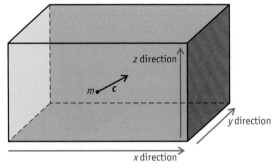

Figure 5.66 Container of gas with one particle singled out for attention

Suppose the particle collides elastically with the dark shaded face of the container. The x-component, u, of its velocity will be reversed, and the other components unchanged. It therefore undergoes a change of momentum in the x-direction, of magnitude ΔPx, where:

$$\Delta P_x = \text{final momentum} - \text{initial momentum}$$
$$= -mu - mu$$
$$= -2mu \qquad (40)$$

The particle continues to bounce back and forth, colliding repeatedly with the shaded face of the box. Between successive collisions it travels a distance $2x$ (there and back), so the time interval, Δt, between collisions is:

$$\Delta t = \frac{2x}{u} \qquad (41)$$

The repeated bombardment of the face of the box produces a force on it. Since force is equal to the rate of change of momentum, the force exerted by the particle on the shaded face of the container has magnitude:

$$F = \frac{\Delta P_x}{\Delta t} = \frac{2mu}{2x/u} = \frac{2mu^2}{x}$$

STUDY NOTE

Unfortunately lower-case letter p is conventionally used for both pressure and momentum. We will use upper case P for momentum here and lower case p for pressure.

Then by Newton's third law, the (outward) force on the wall is:

$$F_x = \frac{mu^2}{x}$$ (42)

We add a suffix 'x' because the force acts in the x-direction.

The pressure exerted on the shaded face is:

$$p_x = \frac{F_x}{yz}$$ (43)

because yz is the area of the face. Therefore we have:

$$p_x = \frac{mu^2}{xyz} = \frac{mu^2}{V}$$ (44)

where V is the volume of the container.

Now think of all N particles in the container, each with its own velocity that can be resolved into components in the x, y and z directions. Each particle contributes to the pressure exerted on the shaded face, so that the total pressure on that face is:

$$p_x = \frac{m}{V} \times (u^2_1 + u^2_2 + \ldots + u^2_N)$$ (45)

which we can write as:

$$p_x = \frac{Nm\langle u^2 \rangle}{V}$$ (46a)

where $\langle u^2 \rangle$ is the mean value of u^2 for all the particles — see notes below.

However, there is nothing special about the x-direction, so we could equally well have derived similar expressions by considering the other directions:

$$p_y = \frac{Nm\langle v^2 \rangle}{V}$$ (46b)

$$p_z = \frac{Nm\langle w^2 \rangle}{V}$$ (46c)

Yet the pressure, p, of a gas is the same in all directions, so we can drop the subscripts and write:

$$3p = p_x + p_y + p_z = \frac{Nm}{V} \times (\langle u^2 \rangle + \langle v^2 \rangle + \langle w^2 \rangle)$$ (47)

At this stage we can also note that the velocity components for any one particle are related by:

$$u^2 + v^2 + w^2 = c^2$$ (48)

and so:

$$(\langle u^2 \rangle + \langle v^2 \rangle + \langle w^2 \rangle) = \langle c^2 \rangle$$ (49)

where $\langle c^2 \rangle$ is the **mean square speed** of the particles. Equation 47 then becomes:

$$pV = \frac{Nm\langle c^2 \rangle}{3}$$ (50)

This equation relates properties that we can measure (p and V) to the motion of the particles, which we cannot see. It is also remarkably similar to the ideal-gas equation:

$$pV = NkT$$ (Equation 39)

which indicates that the kinetic theory can indeed explain the behaviour of gases. Using Equations 38 and 50 together, we can produce many useful relationships that describe the behaviour of gases, as you will see later.

Mean squares and rms speeds

In producing Equations 46 and 49, we used the idea of a mean square speed. This is the mean value of c^2, averaged over all the particles, which is *not* the same thing as finding the mean speed and squaring it. The square root of the mean square speed, called the **root-mean-square speed** (or **rms speed**), $\sqrt{\langle c \rangle^2}$, is a useful quantity when dealing with particles in a gas, since it is an 'average' speed that is directly related to the pressure exerted by the particles.

To find $\sqrt{\langle c \rangle^2}$, you need to:

- square all the individual speeds

- sum all the squares

- divide by the total number of particles

- take the square root of the result.

The rms speed is in general not the same as the 'ordinary' mean speed. For example, suppose we have three particles with speed $200 \, \text{m s}^{-1}$, five with speed $300 \, \text{m s}^{-1}$ and one with speed $700 \, \text{m s}^{-1}$:

$$\sqrt{\langle c \rangle^2} = \sqrt{\left(\frac{3 \times 2^2 + 5 \times 3^2 + 1 \times 7^2}{9} \right)} \times 100 \, \text{m s}^{-1}$$

$$\sqrt{\langle c \rangle^2} = 343 \, \text{m s}^{-1}$$

But the 'ordinary' mean speed is $311 \, \text{m s}^{-1}$.

Note that the average *velocity* of particles in a gas is zero (there is equal movement in all positive and negative directions), unless the gas as a whole is drifting in a particular direction.

QUESTION

Q 47 At a certain moment, five particles have the following speeds: $300 \, \text{m s}^{-1}$, $400 \, \text{m s}^{-1}$, $500 \, \text{m s}^{-1}$, $500 \, \text{m s}^{-1}$ and $600 \, \text{m s}^{-1}$. Calculate:

(a) their mean speed

(b) their rms speed.

⚙ ACTIVITY 27 ROOT MEAN SQUARES

Use a spreadsheet to explore root-mean-square speed.

Use the RAND() function to generate 100 different speeds. In one of the cells put the formula '= RAND()*1000'. This will generate 100 different speeds between 0 and 1000 m s^{-1}.

In the next column put 100 randomly chosen integers between 1 and 10. This represents the number of particles with that speed. To do this use the function '= INT(RAND()*10)'.

Use these values to find the rms speed and the average speed of the molecules in the spreadsheet. Compare the two values.

Results based on the kinetic theory

The kinetic theory and the gas laws together provide a powerful tool in furthering our understanding of gases. By rewriting and combining Equations 39 and 50 in various ways, we can make many deductions about the behaviour of gases and the particles of which they are made. Here you will derive just some of the many useful relationships. The fact that these all produce a self-consistent picture of gases, which ties in with experimental results, provides a good reason to believe that particles in real gases do behave more or less in the ways assumed by the kinetic theory.

Look first at Equation 50. On the right-hand side the product Nm – the number of particles times the mass of a single particle – gives the total mass, M, of the gas in the container. Noting that:

$$\frac{M}{V} = \rho$$

where ρ is the gas density, we can write Equation 50 as:

$$p = \frac{\rho\langle c^2\rangle}{3} \qquad (51)$$

So, we now have an expression that relates the rms speed to properties of the gas itself, regardless of the volume present. Note that Equation 51 does not require a gas of identical particles – it applies equally well to a mixture.

Look again at Equations 39 and 50. Both have pV on the left, so we can eliminate pV and write:

$$\frac{Nm\langle c^2\rangle}{3} = NkT$$

Cancelling N and rearranging gives another expression for the particles' mean-square speed, this time in terms of temperature:

$$\langle c^2\rangle = \frac{3kT}{m} \qquad (52)$$

It is only a short step from here to kinetic energy. The average kinetic energy per particle is:

$$\langle E_k\rangle = \frac{m\langle c^2\rangle}{2}$$

Using Equation 52 we can therefore write:

$$\langle E_k\rangle = \frac{3kT}{2}$$

$$\langle E_k\rangle = \frac{1}{2}m\langle c^2\rangle = \frac{3}{2}kT \qquad (53)$$

Notice that the average kinetic energy depends *only* on the temperature and not on the mass of the particles. This gives us a way to interpret absolute zero in terms of the behaviour of particles in an ideal gas: it is the temperature at which they cease to move.

It is worth thinking a bit more about the speed and kinetic energy of particles at any temperature above 0 K. The particles have a range of kinetic energies and collide frequently with one another. When they collide there is often a transfer of kinetic energy from one to another, but total momentum and total kinetic energy are conserved; that is, the collisions are elastic. The way the total energy is shared between the particles depends only on the temperature, and the speeds found in a particular gas depend only on the temperature and the mass of the particles in the gas. At a particular instant some particles will have high kinetic energy and others will only have low kinetic energy. The majority will have kinetic energy close to the average for that particular temperature. As the temperature increases the average kinetic energy will increase, as will the average speed of the particles.

Figure 5.67 shows the number of particles plotted on the vertical axis against the speed on the horizontal axis. A few particles have very small speeds and a few have very large speeds. The majority of the particles have intermediate speeds.

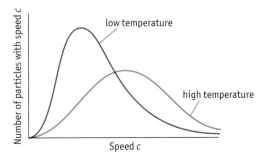

Figure 5.67 Distributions of particle speeds

Developing the kinetic theory

Science does not progress smoothly, and it often involves a lot of arguments and small fragments being pieced together. When condensed into a few paragraphs of a history book, it can seem that scientists push on remarkably quickly. However, the kinetic theory of gases was developed over 200 years after the ideal-gas equation. Many scientists contributed to deriving Equation 50. J. J. Waterson worked out that the pressure of a gas would depend on the square of its molecular speed: doubling the speed would double the momentum change in collisions, but halve the time between collisions, thus producing a factor of four increase in the rate of change of momentum (force). R. Clausius then derived Equation 50. However, Clausius did not have a mathematical grasp of what fraction of molecules would have a particular speed. It was James Clerk Maxwell in 1860 who worked out the distribution of speeds for a gas in thermal equilibrium, shown in Figure 5.67.

And don't forget Brownian motion and Einstein!

The first recorded observation of what we now call Brownian motion was made in 1785 by Jan Ingenhauz using charcoal dust. Then in 1827 the biologist Robert Brown published a scientific paper in which he described observing through a microscope the motion of small particles extracted from plant pollen jiggling about in water. He did not know what was causing it. It looked as if the particles had a life of their own, but he worked hard – observing many different particles immersed in water – to prove that it was not due to some biological motion. His hard work was rewarded, and from then on the random jiggling motion was referred to as 'Brownian motion'.

In a paper based on his PhD thesis and published on 30 April 1905, Albert Einstein mathematically described the collisions and motion of large particles buffeted by invisible molecules. This was the first of three brilliant papers he wrote in that year. This and the work of other scientists in the same field in the first few years of the 20th century convinced even hardened sceptics that atoms existed. Einstein derived an equation for diffusion, which describes how complex fluids such as milk spread into water, using the idea of atoms. Importantly, Einstein's paper also made predictions about the properties of atoms that could be tested. The French physicist Jean Perrin used Einstein's predictions to work out the diameter of atoms and remove any remaining doubts about their existence.

For showing that Einstein's mathematical prediction was right, Perrin later got a Nobel Prize. Einstein got his Nobel Prize in 1921, largely for his 1905 paper on the photoelectric effect, but he became most famous for his paper on relativity, most notable for $E = mc^2$; however, his most cited paper from 1905 is the one concerning Brownian motion. Research scientists have often referred to aspects of the paper in their own research. The paper has spawned a large number of related PhD theses over the last century. There is even an annual Brownian motion conference, where scientists share research work related to this jiggling motion. The speed of the buffeting motion is important for working out the rate of chemical reactions; it is used by scientists working on such varied topics as aerosol particles in pollution research and the properties of paints.

ACTIVITY 28 A GOOD THEORY

Discuss why some scientists might regard the molecular kinetic theory of matter as being an elegant and strong theory. What makes it so good?

And finally ...

Returning at last to our astronomical starting point, we can use the results of the kinetic theory to gain some insight into the low-density interstellar material such as the Horsehead Nebula (Figure 5.68), where we might look for star formation. This is the subject of Question 51.

Figure 5.68 Horsehead Nebula

QUESTIONS

Q 48 At atmospheric pressure (1×10^5 Pa) and room temperature, the density of air is about $1 \, \text{kg m}^{-3}$. What is the approximate rms speed of the air molecules?

Q 49 The mass of an oxygen molecule is 16 times that of a hydrogen molecule. At a given temperature, which molecules would have the greater rms speed, and by what factor?

Q 50 Use Equation 52 to find the rms speed of hydrogen ions ($m = 1.67 \times 10^{-27}$ kg) at a temperature of 1×10^7 K.

Q 51 Table 5.13 contains some data on various types of interstellar gas. (You can see pictures of the examples mentioned in many websites and astronomy books.)

(a) Find the rms speed of:
- **(i)** hydrogen molecules ($m = 2 \times 1.67 \times 10^{-27}$ kg) in a typical molecular cloud
- **(ii)** hydrogen ions in a supernova remnant that has $T = 10^6$ K.
  (Hint: there is a shortcut to **(ii)** using your answer to Question 50.)

(b) The number density of particles, n, is the number per unit volume, equal to $\dfrac{N}{V}$.
- **(i)** Use this information and the ideal gas law (Equation 39) to calculate the values missing from the 'pressure' column of Table 5.13.
- **(ii)** Comment on the range of values in this column.

Type of region	Typical diameter, d/light years	Typical number density of particles, n/m^{-3}	Typical temperature T/K	Composition	Typical pressure, p/10^{-13}Pa	Example(s)
Molecular clouds (also called dense clouds)	0.3–60	10^{10}	30	H_2, He, small and large molecules	40	Horsehead Nebula
Diffuse clouds	10–300	10^8	100	H_2, H, He, other atoms, diatomic molecules	1.4	(Not detected in visible light)
Supernova remnants	up to 3000	10^5	10^6	H^+ or H, He, He$^+$ other atoms and/or ions		Crab Nebula Vela Nebula
Regions of ionised hydrogen (also called HII regions)	3–60	10^8	8000	H^+, He$^+$, other singly-charged ions		Rosette Nebula Carina Nebula Lagoon Nebula

Table 5.13 Interstellar gas regions

Internal energy and specific heat capacity

In an ideal gas, the particles do not interact with one another (except when they collide). However, in a liquid or solid (and in a non-ideal gas) the particles exert forces on one another (in a solid rather as if they were joined by springs). So the energy of the particles cannot be described in terms of kinetic energy alone – their potential energy also varies – and so the **internal energy**, U, of a collection of particles is now the total of their kinetic and potential energies, which are distributed at random between the particles. However, it is still the case that the interactions between particles ensure that the overall distribution of kinetic energy depends only on the temperature, as described by Equation 53.

The molecular kinetic theory allows us to explain **specific heat capacity** in terms of microscopic properties of matter. Specific heat capacity, C, is related to the amount of energy, ΔE, that must be supplied to a mass M of a substance to produce a temperature rise of $\Delta\theta$:

$$\Delta E = MC\Delta\theta \text{ or } MC\Delta T \tag{54}$$

In an ideal gas, particles have only kinetic energy, and so the internal energy is related to temperature by:

$$U = N\langle E_k \rangle = \frac{3NkT}{2} \tag{55}$$

> **STUDY NOTE**
>
> Normally c is used for both specific heat capacity and particle speed. Here we have used C for specific heat capacity, to avoid confusion.
>
> We have used M here for the mass of substance, so as to avoid confusion with m that represents the mass of a particle.
>
> The symbols $\Delta\theta$ and ΔT are both used for temperature change.

Provided the gas is not allowed to expand, all the energy supplied, ΔE, is accounted for by an increase in the internal energy of the gas:

$$\Delta E = \Delta U = \frac{3Nk\Delta T}{2} \tag{56}$$

(If the gas is allowed to expand, it must do work pushing against the pressure of its surroundings. This requires energy. Some of the energy supplied is used to do work, so there is less energy available to increase the particles' kinetic energy and the temperature does not rise so much.)

Imagine two gases. Gas A has particles with twice the mass of those in gas B. What can you deduce about the temperature rise when equal masses of each gas are heated with the same energy? Sample B has twice as many particles as sample A. Therefore this same energy will be distributed amongst a smaller number of particles in A, giving gas A an average increase in particle kinetic energy that is twice that of B. The temperature rise in A will be twice as big: smaller N gives a larger ΔT for the same ΔE. The specific heat capacity of gas A is *half* that of gas B – it needs only half as much energy for a given rise in temperature.

More generally, we can express Equation 54 in terms of particles. If there are N particles each of mass m, then:

$$M = Nm$$

$$\Delta E = NmC\Delta T \tag{57}$$

From Equations 56 and 57 we then have:

$$NmC\Delta T = \frac{3Nk\Delta T}{2}$$

$$C = \frac{3k}{2m} \tag{58}$$

Molecules and real gases

All our discussion so far has been about ideal gases made from particles whose only motion is translational – that is, movement from place to place. However, molecules made from two or more atoms also have energy associated with their vibration and rotation. At any given temperature, a gas made of such molecules has more internal energy than a monatomic gas because, like translational kinetic energy, the energy of vibration and rotation must also increase with temperature. This means that the specific heat capacity of such a gas is greater than that of a monatomic gas.

The situation is complicated still further in real gases, because there are long-range forces between particles. Now, if the gas expands, the potential energy of its particles increases so there is less energy available for increasing the kinetic energy, and the temperature rise is less than would be expected for an ideal gas.

QUESTION

(a) 1000 J is used to heat 1.00 kg of helium at constant volume. Assuming helium behaves as an ideal monatomic gas, what is the resulting temperature rise?

(Mass of helium atom $m_{He} = 6.68 \times 10^{-27}$ kg; Boltzmann constant $k = 1.38 \times 10^{-23}$ J K^{-1})

(b) Argon is a monatomic gas. The mass of an argon atom (^{40}Ar) is 10 times that of a helium atom. Assuming argon behaves as an ideal gas, what would be the temperature rise if 1000 J were used to heat 1.00 kg of argon at constant volume?

(c) The mass of a water molecule (H_2O) is 4.5 times that of a helium atom. Discuss how you would expect the specific heat capacity (at constant volume) to compare with that of helium.

Conditions for nuclear fusion

Having established some important points about the behaviour of particles in gases, we now return to the question of star formation. Earlier, we stated that nuclear fusion requires temperatures over 10^7 K. We can now use the kinetic theory of gases, together with some ideas about energy and electric fields, to see why this is the case.

Two protons can only undergo fusion if they approach one another within about 5×10^{-14} m – the approximate range of the strong nuclear force.

When two protons approach one another, they gain electrostatic potential energy at the expense of their kinetic energy. If the approach is head-on, the kinetic energy becomes zero when the protons are at their smallest separation. We can use an energy conservation argument to relate the distance of closest approach to the initial kinetic energy of the protons and therefore to the temperature of a plasma of protons and electrons in which fusion can occur. In *Probing the Heart of Matter*, you used an equation for the electrostatic potential energy, E_e, of two particles, with charges Q and q, separated by a distance r:

$$E_e = \frac{qQ}{4\pi\varepsilon_0 r} \tag{59}$$

Question 53 takes you through a calculation of the temperature at which two protons can fuse.

QUESTION

 (a) Write down an expression for the total mean kinetic energy of two protons in a plasma of temperature T when they are widely separated.

(b) By referring to Equation 59 (and looking back at your work in *Probing the Heart of Matter*), write an equation that relates the protons' initial kinetic energy to their electrostatic potential energy when they are at their distance of closest approach.

(c) Use your equation from **(b)** to calculate the approximate temperature at which hydrogen fusion should be possible.

(d) Refer to Figure 5.67 and comment on your answer to **(c)**.

($k = 1.38 \times 10^{-23}$ J K^{-1}; proton charge $e = 1.60 \times 10^{-19}$ C; $\varepsilon_0 = 8.85 \times 10^{-12}$ C^{2}N^{-1} m^{-2}; $r = 5 \times 10^{-14}$ m)

Making stars

The stars we see in the sky are hot and dense, and yet they form from interstellar material (ISM) that is very tenuous. As you saw in Table 5.13, the temperature in the hottest parts of the ISM is not quite high enough for hydrogen fusion (which needs over 10^7 K), but the main problem is the density and pressure. You saw in Question 46 that the pressure inside the Sun is about 10^{14} Pa, while Question 51 showed that in the ISM it is no more than about 10^{-11} Pa – a difference of some 25 orders of magnitude. The following passage explains how star formation happens. Read it carefully, then answer Questions 53 to 55.

To form stars, an interstellar cloud must collapse under its own gravity to form a hot, dense concentration of matter. To see how this works, imagine a particle, such as a hydrogen atom or molecule, near the edge of an interstellar cloud (Figure 5.69). Newton's law of gravitation tells us that the particle will experience gravitational forces due to all the other matter around it. As there is a greater concentration of matter on one side of the particle, it accelerates towards the centre of the cloud – the particles are all drawn gradually inwards and so after many thousands of years all the material will eventually become concentrated within a small volume.

But if that were all that happens, then all interstellar material would be clumped together in stars, rather than existing as tenuous gas. As we know from the success of the kinetic theory of gases, the particles in a gas are moving around at high speeds. If they are moving fast enough, they can escape from the cloud, rather as a rocket fired from Earth at high enough speed can travel out to space instead of falling back to the ground; and so the cloud will disperse rather than collapse.

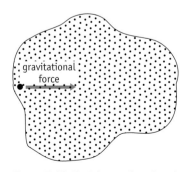

Figure 5.69 Particle near the edge of an interstellar cloud

Whether a cloud collapses under its own gravity to form stars, or disperses due to the motion of its particles, depends on the 'competition' between those two factors. A detailed analysis shows that only the coldest, densest regions of the ISM can collapse under their own gravity. But that then raises the problem of how such material can ever become hot enough for nuclear fusion to occur. It turns out that gravity and kinetic theory again play a part.

During the collapse of a cloud, particles can be thought of as gradually 'falling' towards the centre, rather like objects falling to Earth. As they fall, they accelerate. Put another way, they gain kinetic energy at the expense of gravitational potential energy. The increasing concentration of matter is enough to maintain the collapse, despite the increasing speeds of the particles.

As you can see from pictures of interstellar clouds (such as the Orion Nebula in Figure 5.61), they are not smooth and uniform. As a cloud collapses, it breaks up into fragments that continue to collapse separately.

By the time a fragment has collapsed into a small volume – several million years after the collapse first started – the particles are moving at very high speeds and are colliding with one another. As the temperature rises, the collisions become increasingly violent, so molecules are first disrupted into atoms, and then the electrons become dislodged from the atoms to make ions.

By the time the material has collapsed to the size of a star, it is a very dense plasma (an ionised gas) with a temperature of about 10^7 K. Now nuclear fusion begins to take place, and the release of energy increases the particles' kinetic energy and halts the collapse. A star is born.

This is a nice story, but do we have any evidence to support it? The processes are all much too slow for us to observe them happening during our lifetime, but one key piece of evidence comes from infrared astronomy. We know that all objects emit radiation at wavelengths that depend on their surface temperature, and so we would expect to detect infrared radiation from fragments that were collapsing and getting hotter on the way to becoming stars. And this is what astronomers find when they turn infrared-sensitive telescopes to regions such as the Orion Nebula – clusters of 'hotspots' that are just as we could expect to see if the cloud were in the process of collapsing, fragmenting and heating up (Figure 5.70).

⚙ ACTIVITY 29 FIRST GENERATION STAR
Sort a set of cards so that they correctly describe the formation of a star as described above.

Figure 5.70 Infrared image of the Orion Nebula

QUESTIONS

In answering these questions, you will need to refer to Table 5.13 and to your work in Section 4.1, as well as to the passage above.

Q 54 Estimate the mass of a typical molecular cloud, assuming that it is made entirely of hydrogen molecules, H_2, and therefore estimate the number of Sun-like stars that could form when such a cloud collapses. Quote your answer to the nearest order of magnitude. (Mass of H_2, $m = 3.34 \times 10^{-27}$ kg; mass of Sun, $M_{Sun} = 2 \times 10^{30}$ kg; 1 light year = 9.46×10^{15} m)

Q 55 **(a)** Estimate:
  (i) the acceleration of a hydrogen molecule at the edge of a typical molecular cloud (see Question 54)
  (ii) the time it would take the molecule to reach the centre of the cloud if it started from rest, travelled in a straight line, and its acceleration did not change
  (iii) the molecule's speed when it reached the centre
  (iv) the temperature of a gas or plasma where this was the rms speed of hydrogen molecules.

(b) Comment on your answer to **(a)(iv)**.
(G = 6.67×10^{-11} N m^2 kg^{-2}; 1 year ≈ 3×10^7 s; k = 1.38×10^{-23} J K^{-1})

Scientists who wish to publicise their research findings often produce a poster for display and discussion at conferences. The most eye-catching of these is likely to have the most immediate impact.

Make a conference poster showing the main stages in star formation from an interstellar cloud to the arrival of the star onto the main sequence. Use sketches or photographs and add plenty of information, including details of density, temperature and pressure in the interstellar medium. Include any relevant equations and example calculations to make your poster as informative as possible.

4.3 Evolution and end points – Summing up Part 4

This section continues the story of stars with an account of what happens after they are formed and, in doing so, reminds you of the physics that you have studied in Sections 4.1 and 4.3 (and also some of the physics from Parts 2 and 3 of this chapter). The questions and activities are intended to help you look back over your work.

Before going on to read more about stars, look back through your work in this part of the chapter and make sure you understand the meaning of all the terms printed in **bold**.

A star on the main sequence

In Section 4.2 you saw how a fragment of interstellar cloud could collapse under its own weight to form a small, dense ball of matter that was hot enough for nuclear fusion to begin. Once the star has settled down to a steady rate of fusion, its size, luminosity and temperature remain more or less constant until its supply of hydrogen fuel is exhausted. It has become a main-sequence star. This stage of its life takes a long time, which is why we observe so many stars on the main sequence of a Hertzsprung–Russell diagram (see Figure 5.71), which is based on observations of a very large number of stars. The numbers give the masses of the stars in units of the Sun's mass, $M_{Sun} \approx 2 \times 10^{30}$ kg.

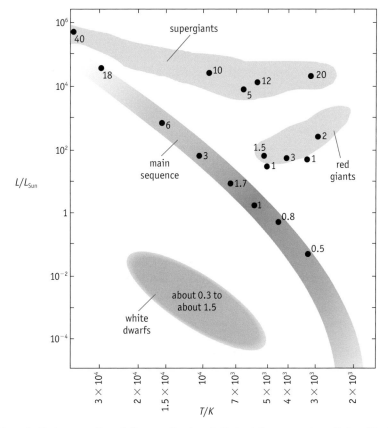

Figure 5.71 Schematic Hertzsprung–Russell diagram. Numbers indicate stellar masses in multiples of the Sun's mass

A star with the same mass as the Sun will remain on the main sequence for about 10^{10} years (see Question 26 in Section 3.2). The main-sequence lifetime of a star depends on its mass and the rate at which it converts hydrogen to helium. Perhaps surprisingly, low-mass stars live for much longer than the Sun. A star of $0.5\,M_{Sun}$ has $L \approx 0.1\,L_{Sun}$. It has only half the amount of fuel, but it is consuming it at one-tenth the rate found in the Sun, and so it will remain on the main sequence for five times as long – about 5×10^{10} years. The most massive stars are the shortest lived, because their nuclear reactions proceed more rapidly, giving them a very high luminosity, and so they exhaust their fuel supplies in a relatively short time. So if we observe any very luminous main-sequence stars, we know they must have formed quite recently – only a few million years ago.

QUESTIONS

Q 56 In two or three sentences, outline how the stars' masses in Figure 5.71 might have been determined.

Q 57 Use Figure 5.71 to find the approximate luminosity of a star of mass $18M_{Sun}$. Assuming that all stars fuse the same fraction of their total mass while on the main sequence, and assuming a main-sequence lifetime of 10^{10} years for the Sun, estimate the main-sequence lifetime of this star.

After the main sequence

As with the 'story' of star formation, astronomers have drawn on observations and on physics in order to piece together an account of what happens to a star after the hydrogen in its core has all undergone fusion to helium.

When hydrogen fusion ceases, the inner parts of the star cool down, which allows the star to collapse again under its own gravity. Just as happened when the star was forming, there is a gain of kinetic energy at the expense of gravitational potential energy, which raises the temperature. Now the material surrounding the helium core becomes hot enough for hydrogen fusion to helium to begin, so inside the star there is now a cool(ish) shrinking core surrounded by a layer where nuclear fusion is releasing energy. The material surrounding this fusion layer expands enormously, cooling as it does so. If the star's mass is less than that of the Sun, or only a few times greater, it is now a red giant – a star with a cool exterior, but very luminous because of its large size. A more massive star (say about five times the Sun's mass or more) becomes a **supergiant** – it is not only large but the exterior is still quite hot. Figure 5.71 shows where red giants and supergiants are located on an HR diagram.

As the core continues to shrink under its own gravity, the temperature rises. In stars at least as massive as the Sun, the core becomes hot enough for new fusion reactions to take place:

$${}^{4}_{2}\text{He} + {}^{4}_{2}\text{He} \rightarrow {}^{8}_{4}\text{Be} \qquad\qquad\qquad (\text{R}10)$$

$${}^{4}_{2}\text{He} + {}^{8}_{4}\text{Be} \rightarrow {}^{12}_{6}\text{C} + 2\gamma \qquad\qquad\qquad (\text{R}11)$$

(The beryllium nucleus is very unstable and decays in less than a second if it does not immediately undergo fusion with another helium nucleus.)

In stars up to about twice the mass of the Sun, the onset of helium fusion causes great upheavals in the outer parts of the star. The upheavals can cause the outer parts of a star to pulsate in and out with periods of only a few days, and astronomers observe a variable star as its luminosity changes. The outermost layers are eventually thrown off to form an expanding shell of glowing gas called a **planetary nebula** (though it has absolutely nothing to do with the planets). Figure 5.72 shows an example of such a shell – you can also see the star in its centre.

When the helium in the core has been converted to carbon, then the star again begins to collapse under its own weight and to heat up. In stars up to about the mass of the Sun, the temperature never gets high enough for any further fusion reactions that would stop the gravitational collapse, and so the star gradually shrinks to become a white dwarf – a small, hot, dense star about the size of the Earth that will eventually cool and fade from view.

Figure 5.72 Helix Nebula in Aquarius

⚙ ACTIVITY 32 END OF THE SUN

Figure 5.73 summarises the evolution of a Sun-like star by tracing the changes in its luminosity and surface temperature from just before it joins the main sequence right up to its end as a white dwarf. Table 5.14 summarises the physical processes that take place at each stage (though not in the right order). Use Figure 5.73 and the preceding text to help you complete the table – in the 'luminosity' column, state the nearest order of magnitude. In your answer, arrange the stages in the correct order.

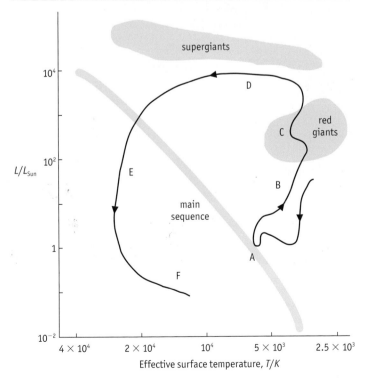

Figure 5.73 Evolution of a Sun-like star

Point on Figure 5.73	Stage	Physical process	Luminosity/ L_{Sun}
	evolution away from main sequence	fusion of H continues	
		all fusion stops; slow cooling	
	red giant		
		outer layers thrown off leaving hot core	
	variable star	violent pulsations of outer layers	
	main sequence		1

Table 5.14 The evolution of a Sun-like star

End points

In stars more massive than the Sun, there is now a pattern that repeats itself:

- the star contracts
- loss of gravitational energy leads to heating
- new fusion reactions produce more massive nuclei
- reactions cease
- the star contracts

STUDY NOTE
You can read more about the fate of massive stars on STA Additional Sheet 4.

... and so it goes on. The successive stages of fusion each happen more and more rapidly, and so a star can get through many stages of fusion in a much shorter time than it spends on the main sequence. Each stage of fusion can give rise to changes in the external appearance of the star: it may become a variable star as the outer parts repeatedly expand and shrink before being ejected to form a planetary nebula.

The ultimate fate of a star depends upon its initial mass, as is shown in Figure 5.74. A star several times the mass of the Sun has a cataclysmic and dramatic death, exploding in a tremendous explosion called a **supernova**. A supernova remnant is left behind, the best known of which is the Crab Nebula, which is the remains of a star that was seen to explode in AD 1054 (Figure 5.2). The explosion was so bright that it was visible in daytime, despite being about 6000 light years away.

Figure 5.74 A star's evolution depends on its mass. The mass increases going up the left-hand side of the diagram

Supernovae are the source of nearly all the elements heavier than hydrogen and helium. The Big Bang that began the Universe created only hydrogen, helium and traces of some other light elements. Heavier elements are created by nuclear fusion in stars, but most remain locked in stellar cores as they cool. It is only through the violent supernova explosions that elements are scattered into the interstellar matter. Here they cool down and might eventually condense to form another star, perhaps with planets in orbit around it, where carbon-based life (or perhaps some other life forms) can evolve.

The material left behind at the core of the supernova remnant might form a neutron star or a black hole. (A white dwarf cannot have a mass greater than 1.4 solar masses.) A **neutron star** forms if the material left behind is between about 1.4 and 2 solar masses. This is an object made mainly of neutrons and with the density of nuclear matter. The entire mass is contained within a sphere about 10 km in diameter.

When a neutron star rotates quickly, as in Figure 5.75, it sends out narrow beams of electromagnetic radiation. These objects are often known as **pulsars** because of the rapid sequence of pulses observed as the beams sweep past us (Figure 5.76).

STUDY NOTE

For more on neutron star rotation, see STA Additional Sheet 5.

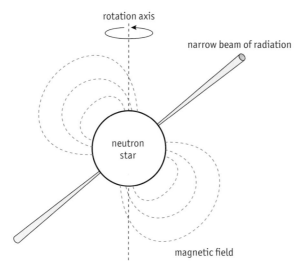

Figure 5.75 Schematic diagram of a pulsar

Figure 5.76 The pulsar in the Vela supernova remnant imaged using the Chandra X-ray observatory

If the remaining mass is more than about twice the mass of the Sun, then the strong gravitational forces crush even neutrons. The object becomes unimaginably small and compact, and it has such a strong gravitational field that all nearby matter will inexorably be drawn into it; nothing can escape – not even light. It is a **black hole**.

Black holes can be detected by their gravitational effect on surrounding matter; for example, if a black hole is a member of a binary star system. Supermassive black holes, with masses over 10^8 times the solar mass, have been detected in the centres of spiral galaxies, including the Milky Way. They probably formed in the early universe as part of the process of making galaxies, rather than being produced in supernovae.

STUDY NOTE

For more on black holes, see STA Additional Sheet 6.

ACTIVITY 33 FUSION IN STARS

The more massive the star, the more massive the nuclei it can produce by fusion, but even the most massive stars cannot produce nuclei more massive than iron in this way.

Write a paragraph explaining the physics behind this statement. Include the following terms: gravitational force, electrostatic force, binding energy, kinetic energy. Also include any other key terms from Parts 2 and 3 of this chapter that you think are relevant.

ACTIVITY 34 SETI

Imagine you are an astronomer involved in the SETI (Search for Extraterrestrial Intelligence) programme. You are bidding for time to use the Arecibo Radio Astronomy facility in Puerto Rico. Your aim is to try and pick up radio transmissions that might have been emitted by an alien civilisation on a planet orbiting another star. Prepare a bid to be given to the time-allocation committee that explains which type of stars you intend to observe and why.

You can find out more about SETI, planets and the possibility of life elsewhere in the Universe from the websites listed at **www.pearsonhotlinks.co.uk**.

QUESTIONS

Q 58
(a) Making the rash assumption that the core of a highly evolved star behaves like an ideal gas, calculate:

  (i) the mean kinetic energy of the nuclei

  (ii) the rms speed of ^{56}Fe nuclei at a temperature of 5×10^9 K.
  (Mass of ^{56}Fe nucleus, $m = 9.3 \times 10^{-26}$ kg; $k = 1.38 \times 10^{-23}$ J K^{-1})

(b) Repeat for an electron in the star's core and comment on your answer.
(Electron mass, $m_e = 9.11 \times 10^{-31}$ kg.)

Q 59
(a) Calculate the densities of:

  (i) a neutron star that has a radius of 5 km and a mass of 4×10^{30} kg (i.e. twice the mass of the Sun)

  (ii) a helium nucleus that has a radius of approximately 10^{-15} m and a mass of 6.7×10^{-27} kg.

  (iii) Thus comment on the statement that a neutron star 'has the density of nuclear matter'.

(b) **(i)** Calculate the gravitational field strength at the surface of the neutron star in **(a)**.

  (ii) To the nearest order of magnitude, express your answer as a multiple of the field strength at the Earth's surface. ($G = 6.67 \times 10^{-11}$ N m^2 kg^{-2}; $g_{\text{Earth}} = 9.81$ N kg^{-1})

Q 60
Pulsars have been detected with periods of about a millisecond. The material at the surface of a spinning neutron star must experience a centripetal force strong enough to keep it in orbit, and the only force available is the star's own gravity. If the gravitational force is not strong enough, then the star will fly apart.

(a) Show that the minimum possible rotation period, T, of a star of mass M and radius r is given by:

$$T \geq \sqrt{\frac{4\pi^2 r^3}{GM}}$$

(b) Find the minimum possible rotation period for the neutron star in Question 58 and then say whether a 'millisecond pulsar' could reasonably be a rotating neutron star.

Q 61
Theoretical physicists (and science-fiction writers) have speculated what might happen to space travellers who approached a black hole. One scenario is that as a person fell towards a black hole, they would suffer a process dubbed 'spaghettification', which this question illustrates.

Jo is very tall and she has a large head and large feet (see Figure 5.77). Her spacecraft has brought her close to a black hole of mass $M = 10^{26}$ kg.

(a) Calculate:

  (i) the gravitational force on Jo's feet, which are exactly 50 km from the black hole

  (ii) the gravitational force on her head, which is exactly 2.0 m further from the black hole.
  (Keep at least five figures in each answer, even though you might not feel that this is justified.)

(b) Explain how these forces will affect Jo.

mass of head = 10 kg

height $h = 2$ m

mass of feet = 10 kg

Figure 5.77 Diagram for Question 61

5 The story of our Universe

In this final part of the chapter, you will see how, since the early 20th century, scientists have been looking for answers to such incredible questions as: 'How did the Universe begin?' and 'How will it end?' You will see how some of these questions can be tackled using physics that you have been studying in Parts 2 and 3 of the chapter, and extend your knowledge and understanding of those areas of physics.

5.1 Taking on the biggest question in science

The birth of modern cosmology

Albert Einstein (1879–1955), the greatest scientist of the 20th century (Figure 5.78), hated school. But he was fascinated by science and maths, and read up on them at home. Having gained his degree in Switzerland (a mathematics and physics teaching diploma from the Zürich Polytechnic), Einstein hoped to obtain a teaching job there. After failing to obtain a post for two years he eventually started work in the patent office. In 1905, the same year he received his PhD, Einstein published four papers on the photoelectric effect, Brownian motion, special relativity, and the equivalence of mass and energy, which became classics and transformed some of our fundamental understanding. By 1908 he had been appointed to a lectureship at the University of Bern and had started work on a new theory of gravity. His thinking came to fruition when he published his General Theory of Relativity in 1915, having moved to Berlin to take up a professorship there. This work of great genius revolutionised our thinking about space and time.

Figure 5.78 The young Albert Einstein

In 1917 Einstein turned his attention to cosmology and tried to apply his equations of general relativity to the whole Universe. He didn't know much about astronomy, and had no experimental evidence on which to base his equations – truly pioneering work.

In order to simplify his calculations, Einstein applied the **cosmological principle**; namely that, on average, the Universe is homogeneous (looks the same from any point) and isotropic (looks the same in all directions). Matter, galaxies and clusters of galaxies are evenly scattered on a large scale. Without the cosmological principle, the equations of cosmology would be just too hard to solve, and scientists would not have made the astonishing progress they achieved in the 20th century. Mercifully, Einstein's assumption turns out to be true – the Universe does seem to be pretty much the same everywhere when viewed on a large enough scale.

STUDY NOTE

You will see some evidence for homogeneity in Section 5.2.

Like virtually everyone in his day, Einstein believed that the Universe was unchanging, that it was the same now as it was when it was created. But he realised that the gravitational attraction of all the matter in the Universe would cause it to be drawn together (rather like the collapsing cloud idea that you met in Section 4.2). He therefore added to his equations a term describing a force that would ensure that the Universe would indeed continue to look the same forever without contracting under gravity. He later called this 'the biggest blunder in my life', as Edwin Hubble made some observations that indicated that the Universe had not always been the same.

Figure 5.79 Edwin Hubble

Hubble and the galaxies

Edwin Hubble (1889–1953, Figure 5.79), a tall, athletic young American, had a tricky choice to make. He excelled in maths, astronomy and boxing, and had been offered the chance of fighting the then world heavyweight champion. Fortunately for the future of cosmology, he chose instead to come and study at Oxford University, England. But his discoveries had to wait until he'd served in the First World War. (He began as a private in the US army and ended the war a major.)

Eventually, at the age of 30, Hubble began his famous work at the new 100-inch telescope at Mount Wilson, California. His first discovery was a Cepheid variable star within a faint misty patch of light called the Andromeda Nebula (Figure 5.80). Astronomers had puzzled for centuries over what these nebulae were. Using observations of Cepheids in the Andromeda Nebula (as described in Section 2.6 of this chapter), Hubble was able to estimate the distance to the nebula, and showed that it was a congregation of stars – just like our Milky Way galaxy, but far beyond it.

Figure 5.80 Andromeda Nebula

For the first time, we knew something about the real scale of the Universe. The basic units of matter in the Universe are galaxies, not stars. And if that was not enough, Hubble was soon to make sense of a phenomenon that was to change our whole conception of the Universe even more. This concerned the line spectra of distant galaxies.

As mentioned in Part 3 of this chapter, the spectrum of light from the Sun and other stars contains many narrow absorption lines, produced when atoms in the cool outer parts of the star absorb radiation at their own characteristic wavelengths. When astronomers began looking at the spectra of whole galaxies beyond our own, they found the same general pattern of absorption lines in the light from all the stars superimposed, but they also noticed something rather surprising: for all but the closest galaxies, the dark lines seemed to be shifted from their normal positions towards longer wavelengths. This phenomenon became known as **redshift**, because the lines were shifted towards the red (low frequency, long wavelength) end of the spectrum.

Redshift is defined as the ratio of the change in wavelength $\Delta\lambda$ as a fraction of the wavelength, λ, measured in the laboratory and is usually given the symbol z. Provided the wavelength change is small, this is very nearly the same as the ratio of the change in frequency Δf to the 'lab' frequency f, and so:

$$z = \frac{\Delta\lambda}{\lambda} = \frac{\Delta f}{f} \tag{60}$$

Figure 5.81 shows the spectra of two galaxies with different redshifts.

Figure 5.81 Effect of redshift on wavelength. The upper spectrum shows the spectrum in the absence of any red shift.

You have met something like Equation 59 before – in the chapter *Technology in Space* and in Part 4 of this chapter. There you saw that the Doppler shift was given by:

$$\frac{\Delta\lambda}{\lambda} = \frac{\Delta f}{f} = \frac{v}{c} \qquad \text{(Equations 30 and 32)}$$

provided $v \ll c$, where v is the speed of the source of the radiation and c the speed of light in a vacuum. If we interpret the redshift in the light from galaxies as being due to the Doppler effect, we can relate redshift to the velocity, v, of the galaxy:

$$z = \frac{\Delta\lambda}{\lambda} = \frac{\Delta f}{f} = \frac{v}{c} \tag{60a}$$

STUDY NOTE

Although the relationship between z and v holds true, the galactic redshift is not really a Doppler shift (see later).

Using the largest telescope in the world, Hubble was able to measure the redshifts and distances of 24 galaxies. He noticed that the redshift was greater for more distant galaxies, and plotted his results on a graph of distance against 'recessional' speed. Figure 5.82 shows his results, taken from his paper published in 1929. Notice that the unit of distance here is the parsec, pc (see Section 2.3). Each black dot is an individual galaxy, and each circle is a group of galaxies. Figure 5.83 shows more recent data. You can see the linear relationship extends to very distant galaxies – note the huge range of values plotted, making it necessary to use a logarithmic scale. The distance unit here is the megaparsec, Mpc. 1 Mpc = 10^6 pc.

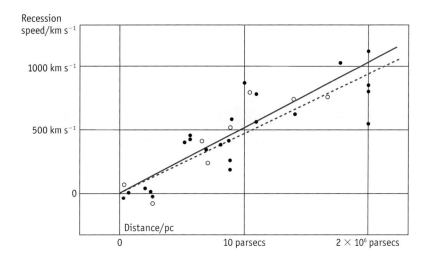

Figure 5.82 Hubble's graph of recession speed against distance

Figure 5.83 Graph of more recent redshift–distance data

STUDY NOTE

Notice the log scales in Figure 5.83. Zero does not feature on the graph, since that would require an infinite length of graph paper.

⚙ **ACTIVITY 35 REDSHIFT**

Use photographs and spectra of galaxies to determine their distances and to plot a graph of redshift against distance.

PRACTICAL SKILLS REFERENCE

Analysis and interpretation of data

See Practical skills notes 4.1–4.2

Hubble had found a simple link between distance, d, and redshift. If z is small ($z \ll 1$, $v \ll c$), which was the case for Hubble's original data:

$$z \approx \frac{H_0 d}{c} \tag{61}$$

or (using Equation 60):

$$v \approx H_0 d \tag{61a}$$

This is known as **Hubble's law**, and H_0 is called the **Hubble constant**.

STUDY NOTE

Equation 62 is included for completeness. You will not be expected to use it as part of this course.

Equation 60 implies that a redshift $z > 1$ is not possible, as then $v > c$. Redshifts greater than 1 are observed and can be accounted for by incorporating relativity, in which case the equation for Hubble's law is:

$$\frac{H_0 d}{c} = (z + 1)^2 - \frac{1}{(z + 1)^2} + 1 \qquad (62)$$

Equation 61 is an approximation of Equation 62, which works well for small redshifts.

The Hubble constant can be found from the gradients of graphs such as those in Figures 5.82 and 5.83. It is usually quoted in units of $\text{km s}^{-1} \text{Mpc}^{-1}$ (think of Equation 61 with v in km s^{-1} and d in Mpc). It is difficult to determine accurately, because it relies on the accurate measurement of d for very distant galaxies. Since 2000, estimated values have consistently been around $70 \, \text{km s}^{-1} \text{Mpc}^{-1}$, with the latest result (2013) from the Planck mission being $(67.80 \pm 0.77) \, \text{km s}^{-1} \text{Mpc}^{-1}$.

The expanding Universe

The most surprising thing about Hubble's discovery was its interpretation. It implied that the Universe was expanding. Activity 36 shows how Hubble's law implies an expanding Universe.

⚙ ACTIVITY 36 THE EXPANDING RUBBER UNIVERSE

Cut a wide elastic band to make a strip of rubber. With a pen, make three marks at 1 cm intervals, labelled 1, 2, and 3. These represent galaxies.

Draw an observer (a small bug?) on one of the galaxies.

Stretch the rubber band to simulate the expansion of the Universe (Figure 5.84).

Measure how far the other galaxies have moved away from the observer, and so compare the recession speeds of the two galaxies.

Repeat with the observer on one of the other galaxies.

galaxy 2

bug observer
on galaxy 1

galaxy 3

Figure 5.84 Stretching the Universe

Stretching the rubber band produces Hubble's law: speed is proportional to distance. Any uniformly stretching material would do the same. And the converse is true: if galaxies obey Hubble's law, you know that the 'material' that holds them – in this case the Universe – is uniformly stretching.

Hubble's observational evidence wasn't completely convincing. He had only been able to measure relatively nearby galaxies. It was a leap of faith to believe that Hubble's law applied to the rest of the galaxies in the Universe. But the leap turned out to be correct, as you can see from Figure 5.84. Hubble's law is now so well established that astronomers use it to estimate distances by measuring redshifts from galaxies that are too far away to have their distances measured by other methods.

In fact, Hubble's law had already been predicted. A Belgian physicist and priest, Georges Lemaître (Figure 5.85), had just written a theoretical paper that claimed that the Universe was expanding. The link that Hubble discovered followed directly from this. So, it seemed that Einstein's static view of the Universe had to be thrown away. The evidence pointed to a Universe that was expanding equally in all directions. Georges Lemaître is sometimes called the father of the Big Bang, as he predicted the expanding Universe before Hubble discovered it.

Thinking about Activity 36 may help you to understand a fundamental fact about the Universe. The Universe is expanding not because the things in it (galaxies) are moving outwards, but because space itself is expanding. The rubber band is like the space in the Universe. When it expands, it carries the dots (representing galaxies) with it.

Figure 5.85 Georges Lemaître

Try not to imagine the Big Bang as being like an ordinary explosion, where matter flies outwards into space in all directions. Instead, think of the Universe as a balloon, and imagine that the *surface* of the balloon contains every single thing in the Universe within it. So when the Universe expands, it's like blowing up the balloon. The Universe (or balloon) is not expanding into anything – it *is* everything. It's just that the surface of the balloon grows bigger. Or you can imagine that the expanding Universe is like raisin bread rising. In the edible Universe it is the expansion of the dough that moves the raisins apart (Figure 5.86). In the real Universe, it is the expansion of space that moves galaxies apart.

Figure 5.86 Raisin bread rising

When astronomers explain why galaxies appear redshifted, they are careful not to say that it's because of the Doppler effect. Why? Because the Doppler effect is a change in observed wavelength owing to a moving source. In the case of galaxies, it is the expansion of space itself that causes the shift in wavelength; galaxies are not moving through their local space. Figure 5.87 illustrates the difference. To mark the difference, the redshift of receding galaxies is properly called **cosmological redshift**. However, you can use the formula for the Doppler effect to calculate the cosmological redshift for receding galaxies.

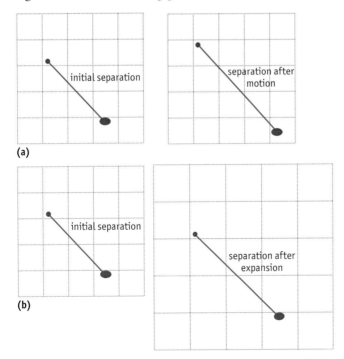

Figure 5.87 Galaxies increasing their separation because of (a) motion through space, and (b) the distance AB getting larger due to expansion of space (note the larger grid)

You might wonder why, if the Universe is expanding, the Earth is not getting further from the Sun and you are not getting bigger. The expansion of the Universe only applies to sufficiently distant objects such as galaxies. Particles that are relatively close together are much more affected by attractive forces, and stay together. For instance, the Sun and Earth are held together by gravity, and the particles in your body are kept in place by electrical forces.

Activity 36 helps answer the question: is our galaxy at the centre of the expanding Universe? Whichever ink mark you measure from, you will see that speed is proportional to distance. Likewise, an intelligent being on a planet in another galaxy would see all other galaxies speeding away from them too; we do not occupy a special place in the Universe.

It started with a bang

If we imagine travelling backwards in time, then galaxies would all be speeding towards each other, making the Universe a denser place in the past. If we continue going backwards, then we should reach a moment in time when the Universe was crammed together into a state of almost infinite density.

This reasoning suggests that the Universe began in a sort of explosion. It's hard to conceive of such a colossal event. The Universe began at an unimaginably high density and temperature, and all the matter and energy in the Universe today was released. Ever since, the matter and energy have been travelling outwards, thinning out and cooling. This is basically the **Big Bang** model of the origin of the Universe.

People sometimes ask what happened before the Big Bang. Strange as it may seem, there is no before the Big Bang. Time itself came into being at the moment of the Big Bang.

The UCLA (University of California at Los Angeles) has a useful website that addresses some frequently asked questions (FAQs) about cosmology, which you can access from **www.pearsonhotlinks.co.uk**.

The Big Bang follows logically from Einstein's theory of general relativity (and a few other assumptions), but it was not instantly accepted and had a rival theory.

QUESTIONS

 Hubble's law holds from any viewpoint in the Universe. Explain why this supports the cosmological principle.

 Explain the difference between Doppler redshift and cosmological redshift.

5.2 Was there a Big Bang?

Big Bang or Steady State?

The mid 20th century was a time of great rivalry between two conflicting cosmological theories. Some astronomers, including Sir Martin Rees (1942–, now Lord Rees of Ludlow and Astronomer Royal since 1995), thought that the Big Bang provided the best explanation for observations. Others, notably Sir Fred Hoyle (1915–2001), who described the Big Bang as 'a form of religious fundamentalism', favoured the alternative Steady State theory.

Fred Hoyle and colleagues developed their Steady State theory in the 1940s, based on a suggestion made by Sir James Jeans in 1920. Oddly enough, the idea for the Steady State theory came to Hoyle after seeing a film. The fact that the film ended with everything returning to how it was at the beginning made him wonder whether the Universe could be like that too (with the expansion of the Universe compensated by extra matter forming within it).

For a while, the Steady State theory presented a real challenge to the Big Bang explanation. It explained the expanding Universe by proposing that new matter is continually created in space, compensating for the spreading apart of individual galaxies. In this way, the Universe maintained its Steady State.

So what makes us think that Universe did begin with a bang? The two theories make different predictions about what the Universe looked like a long time ago. One way to find this out is simply to look far out into space. Because light travels with a finite speed, we observe distant galaxies as they were a long time ago, not as they are now.

The Steady State theory predicts that distant places in the Universe should look just the same as nearby places. However, the Big Bang theory predicts that distant places should be denser, since in the early Universe the same amount of matter occupied a smaller space.

In the mid-1950s, the British astronomer Sir Martin Ryle (1918–1984) and colleagues tried to settle the debate using a radio telescope to look at nearby and distant galaxies. They found that galaxies further away were much more concentrated than those near to us. This observation did not match the prediction of a steady state Universe.

Fred Hoyle was on the ropes, struck by a theory that he had originally named the 'Big Bang' as a term of abuse. But despite the evidence, the controversy continued. It was in the late 1960s that the Big Bang struck the knockout punch.

Knockout punch

The Big Bang theory made another specific prediction: the birth of the Universe must have produced colossal numbers of extremely high-energy photons from energetic collisions between sub-atomic particles. Some radiation would have been absorbed, but not all. What happened to it? Shouldn't we be able to see it today?

In the 1950s, Ukranian–American physicist George Gamow and colleagues realised that today's Universe should be bathed in **cosmic background radiation**, the remnant of the heat of creation, greatly cooled by the expansion of the Universe. (Remember, it was the whole Universe itself that emitted it, not just one part of it.) The radiation from the Big Bang has been travelling for about 14 billion years. The expansion of the Universe has literally stretched the waves, increasing the wavelength and reducing the energy of the photons (Figure 5.88). Gamow predicted that the peak wavelength of that radiation should now be in the microwave part of the spectrum, corresponding to the energy emitted by a black body at approximately 3 K – just three degrees above absolute zero. Scientists thought the radiation would be very difficult to detect – it would require extremely sensitive microwave equipment.

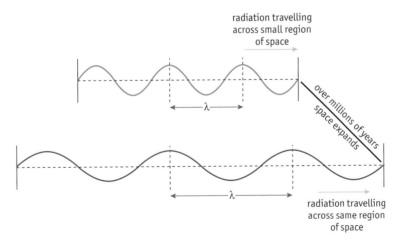

Figure 5.88 Wavelength of cosmic background radiation increases as space expands

However, in 1965, two American radio engineers, Arno Penzias and Robert Wilson (Figure 5.89), were testing a new horn-shaped microwave antenna. They had found an annoying constant hiss of static, which for months they thought was noise within the telescope. They tried many ways to get rid of the hiss, including taping up metal joints and clearing out pigeon droppings from inside the horn. But try as they might, they could not get rid of it.

Figure 5.89 Arno Penzias (left) and Robert Wilson

At Princeton University, 60 km away from Penzias and Wilson, cosmologist Jim Peebles and others were trying to work out how Gamow's predicted radiation might be detected. When Penzias and Wilson saw a preprint of Peebles' paper, they began to realise what they had found, and invited Peebles to see their antenna and data.

It was like finding a smoking gun – direct evidence that the Universe started off incredibly hot and dense. The Steady State theory practically vanished overnight. The discovery of cosmic background radiation also supported the cosmological principle. The intensity of the microwaves was the same in all directions.

Later measurements (including those by the European Space Agency's Planck satellite in 2013) have confirmed that the spectrum closely matches that of a black body at a temperature of 2.7 K (see Figure 5.90 and compare it with Figure 5.18 in Section 2.5).

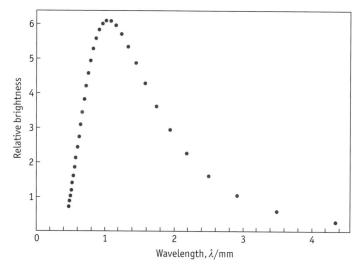

Figure 5.90 Spectrum of the microwave background radiation

A third line of evidence for the Big Bang is its successful prediction of the amount of helium in the Universe. Large amounts were produced in the first few minutes of the Big Bang (see Section 5.3). The model predicts that about 23% of the mass of matter in the early Universe was helium. Indeed, astronomers have been able to measure the proportion of helium in some of the oldest objects in the Universe – and they've found that it lies between 23% and 24%.

QUESTIONS

Q 64   Explain why the finding of the cosmic background radiation proved fatal to the Steady State theory.

Q 65   What will happen to the spectrum of the cosmic background radiation if the Universe expands forever?

⚙ ACTIVITY 37 COSMOLOGY CHAT SHOW

Imagine a chat show where the two guests are Sir Martin Rees and Sir Fred Hoyle. Each comes on in turn, answers questions from the host about their theory, and then there is a chance for them to 'discuss' together. In a small group write and act out the script for the show – or improvise what you think they might say.

Using Hubble's law

Redshift and distance

Measurements of redshift provide a way to estimate distances to distant galaxies and quasars (quasars are very luminous galaxies found at large distances, which indicates they were more common in the early Universe). It is relatively easy to measure an accuurate value for redshift (using the wavelengths of spectral lines) and then to calculate the distance using Equations 61 (for $z \ll 1$) or 62.

Astronomers often use redshifts, rather than actual distances, to indicate distances of galaxies and quasars – and not because they are too lazy to do the calculations. Any calculation of distance using redshift relies on having an accurate value of the Hubble constant, which in turn depends on having accurate measurements of distance using a standard candle method such as Cepheid variables or Type Ia supernovae, as discussed in Section 2.6. So, while a redshift measurement might have a very small uncertainty, the uncertainty in the calculated distance could be much greater.

How old is the Universe?

The Big Bang theory allows us to estimate the age of the Universe. By knowing how fast the Universe has been expanding, we can work out how long it must have taken to get to its current state.

The low-redshift approximation of Hubble's law gives us a simple way to estimate the age of the Universe:

$$v = H_0 d \qquad \text{(Equation 61a)}$$

Suppose a galaxy has been travelling at speed v relative to another galaxy (e.g. the Milky Way) for a time t since the Big Bang. The distance it has travelled is given by:

$$d = v t$$

where t is the age of the Universe. Substituting for d from Equation 52 gives:

$$\frac{v}{H_0} = vt$$

So:

$$t = \frac{1}{H_0} \qquad (63)$$

With H_0 expressed in its normal units of $\text{km s}^{-1}\,\text{Mpc}^{-1}$, its relation to the age of the Universe is not obvious. The following example shows how to deal with the units.

WORKED EXAMPLE

Q If $H_0 = 70\,\text{km s}^{-1}\,\text{Mpc}^{-1}$, what does Equation 63 imply is the age of the Universe?

A $1\,\text{Mpc} = 3.09 \times 10^{22}\,\text{m}$, and so $1\,\text{Mpc}^{-1} = (3.09 \times 10^{22}\,\text{m})^{-1}$

In SI units:

$$H_0 = \frac{70 \times 10^3\,\text{m s}^{-1}}{(3.09 \times 10^{22}\,\text{m})} = 2.27 \times 10^{-18}\,\text{s}^{-1}$$

Then:

$$t = \frac{1}{H_0} = \left(\frac{1}{2.27 \times 10^{-18}}\right)\text{s} = 4.41 \times 10^{17}\,\text{s}$$

$1\,\text{year} = 3.16 \times 10^7\,\text{s}$, so $t = 1.4 \times 10^{10}$ years (i.e. about 14 billion years)

Recent results (2015) have nudged our estimates of the Universe's age to 13.8 billion years. Cosmologists, however, keep an eye on the results of other research, looking for the oldest observable galaxies. Currently the oldest known galaxy appears to have formed some 700 million years after the Big Bang. Computer simulations suggest that galaxies might be able to form only 200 million years after the Big Bang, but if we find any galaxies that appear to be older than that, then something is going wrong with our understanding.

How big is the Universe?

Scientists find it difficult to think of the Universe as having an 'edge'. What would the edge of the universe be like and what would prevent you from going further? If the Universe had an edge, then the cosmological principle would not apply, as things would certainly be rather different near to the edge. However, not having an edge does not necessarily mean that the Universe is infinite; it could be finite and *closed*. The surface of a balloon is not infinite, but it does not have an edge. If you set off in one direction you will eventually find yourself back at your starting point. At the moment, we do not actually know if the Universe is finite or infinite, and we may never know.

However, astronomers often talk about the size of the universe and quote a value. They are referring to the size of the *visible* universe.

We can only see a small fraction of the Universe. If light has only been travelling for up to 13.8 billion years (since the Big Bang), we can only see parts of the Universe up to 13.8 billion light years away. Light from more distant parts has not reached us yet. When people talk about the 'size' of the Universe they mean the size of the part we can observe, which is related to its age.

The cosmic microwave background radiation we can detect is from about 300 000 years after the Big Bang, when the Universe became transparent to photons. This would correspond to a redshift of $z \approx 1000$. The highest confirmed redshift for a galaxy at the time of writing (spring 2015) is $z \approx 7.5$, with unconfirmed redshifts up to $z \approx 10$ for the furthest galaxies we can detect.

The big picture

Extensive surveys of galaxy redshift, such as the 2dF Sky Survey, have allowed us to measure the redshift of distant galaxies accurately and plot 3D maps of their distance from us, producing the largest scale maps of the Universe (Figure 5.91). These reveal some very unusual distributions, resulting in 'structures' such as enormous voids and the Sloan Great Wall. The Sloan Great Wall was for some time the largest known structure in the Universe. In 2013, results from a gamma ray burst mission suggested an even larger structure: the Hercules–Corona Borealis Great Wall.

You can read more about the large-scale distribution of galaxies using the Harvard weblink (see **www.pearsonhotlinks.co.uk**).

> **STUDY NOTE**
>
> 2dF stands for 'two-degree-Field' and refers to the angular size of the region they used for the survey.

Figure 5.91 2dF Sky Survey map

The Shapley Supercluster is the largest concentration of galaxies yet seen (tens of thousands that gravitationally interact, remaining a single unit). This was rediscovered by Dr Somak Raychaudhury at the University of Birmingham, and named after its original observer, the American astronomer Harlow Shapley (1885–1972).

On slightly smaller scales we can now clearly see some of these voids and superclusters (Figure 5.92). These superclusters are themselves composed of thousands of clusters of galaxies. Understanding the distribution of these features will help us to better explain how the Big Bang happened.

On even larger scale than that of the voids and superclusters, the cosmological principle (isotropy and homogeneity) still holds.

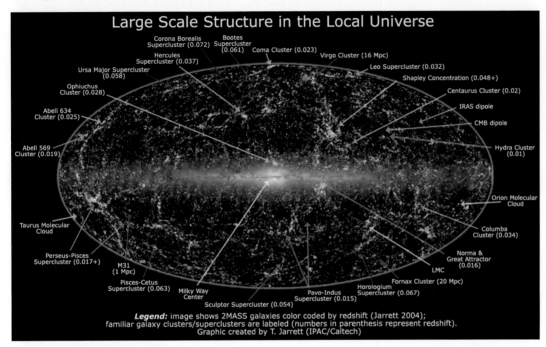

Figure 5.92 Voids and superclusters

QUESTIONS

 Q 66 One of the galaxies plotted in Figure 5.83 has $z = 0.0800$.
Use $H_0 = (67.80 \pm 0.77)\,\text{km s}^{-1}\,\text{Mpc}^{-1}$ to calculate:

(a) the galaxy's recession speed

(b) the distance to the galaxy together with its uncertainty. Express your answer to **(b)** in Mpc.

($c = 3.00 \times 10^8\,\text{m s}^{-1}$)

 Q 67 If $H_0 = 100\,\text{km s}^{-1}\,\text{Mpc}^{-1}$ what would be the approximate age of the Universe?

Q 68 A low-redshift quasar is believed to be 350 Mpc from our galaxy. If $H_0 = 70\,\text{km s}^{-1}\,\text{Mpc}^{-1}$, what are:

(a) the quasar's recessional speed in km s^{-1}?

(b) its redshift, z?

MATHS REFERENCE

Calculations with uncertainties

See Maths note 7.2

Combining uncertainties

See Maths note 7.5

Q 69 Radiation from a Cepheid variable star in the large Magellanic Cloud (a small galaxy close to the Milky Way) has an intensity $I_{LMC} = 9.5 \times 10^{-15}\,W\,m^{-2}$. The Hubble telescope observed a Cepheid with the same period in the galaxy known as M100 and found the intensity of its radiation to be $I_{M100} = 2.7 \times 10^{-19}\,W\,m^{-2}$. This is close to the smallest intensity that can be measured with current instruments.

The distance to the LMC has been determined by other methods; it is 65 kpc.

(a) According to the Hubble Cepheid measurements, what is the distance to the galaxy M100?

(b) What, roughly, would be the furthest distance at which Hubble (or a similar instrument) could measure the peak intensity from a Type Ia supernova ($L \approx 5 \times 10^7\,L_{Sun}$)? Express your answer in pc and in light years.

(c) Suggest how the supernova luminosity might originally have been measured.

(d) Comment on your answer to (b) in relation to the apparent age of the Universe.

(Sun's luminosity, $L_{Sun} = 2.9 \times 10^{26}\,W$; 1 pc $= 3.09 \times 10^{16}\,m$; 1 ly $= 9.46 \times 10^{15}\,m$)

5.3 Riding through time

It may surprise you to learn that scientists have experimental evidence that backs up theories of the early Universe to within a second of the Big Bang. Not from telescopes, but from particle physics. A particle accelerator, by accelerating particles to close to the speed of light, can recreate, in a tiny volume, the incredible temperatures of the birth of the Universe.

With the Big Bang model, and drawing on the results of particle-physics experiments, we can take a ride through time, charting the history of the Universe from the first instant of creation. So put on your virtual reality headset for a ride through the early history of the Universe – Figure 5.93 shows what you would see.

QUESTION

 Q 70 (a) Explain why looking at the light from distant galaxies is like looking back in time.

(b) We can't see through telescopes anything that happened earlier than 300 000 years after the Big Bang. Why not?

The 'ride through time' is also a 'ride' through some important areas of physics. The reactions between fundamental particles are like some of those you met in the chapter *Probing the Heart of Matter*, and the other key areas of physics are those you studied in Parts 2, 3 and 4 of this chapter: nuclear physics, kinetic theory and gravity. We will take a brief look at the role of each of these in turn.

> ⚙ **ACTIVITY 38 MATTER IN THE UNIVERSE**
> Draw a diagram to show the stages by which fundamental particles in the early Universe became the large structures that we see in the Universe today. Use the information in the 'ride' (Figure 5.93), and label each main stage with its temperature and its time from the Big Bang.

Temp:
10^{11} K dropping

Time:
0 yrs 0 hr
00 min 00.01 sec

Goggles tuned to:
gamma rays

Objects in focus:
quarks

Commentary:
'A myriad of tiny particles swarm around in the murky haze. Some are more massive than the particles that exist today. But almost as soon as they appear, they are gone, never to be seen again. The temperature is dropping rapidly, and the particles are losing speed a little.
  Here are quarks, the building blocks of matter. As they collide with one another, they fuse together to make two kinds of bigger particles, called protons and neutrons. There are about equal numbers of each.' The haze of photons swarm around the universe being absorbed and radiated by free charges.

(a)

Temp:
10^8 K dropping

Time:
0 yrs 0 hr 10 min

Goggles tuned to:
X-rays

Objects in view:
protons, neutrons nuclei of helium, and lithium

Commentary:
'It's still hazy. The newly created protons and neutrons will later go on to form atoms. But wait! The neutrons are disappearing – there's only one left for every 7 protons. Just in time, the particles are slowing down enough that when a neutron collides with a proton, they stick together, preventing more neutrons from disappearing. So we can see the first atomic nuclei.'

(b)

Temp:
10^4 K dropping

Time:
10 000 yrs

Goggles tuned to:
visible light

Objects in view:
electrons

Commentary:
'The universe continues to cool. These electrons are no longer just bouncing off everything they collide with. They're moving slowly enough that they get caught up when they meet oppositely charged protons and nuclei of helium.
  At last, 300 000 years after the Big Bang, isolated charges are bound into atoms and the haze clears. The Universe is now revealed in all its full glory, as photons of light can travel freely, without colliding with electrons.
  The Universe now is made up mainly of hydrogen and helium atoms. There about 12 hydrogen atoms for each helium atom.'

(c)

Temp:
10^4 K dropping

Time:
10 000 yrs

Goggles tuned to:
microwave

Objects in view:
pink areas – hotter than average
blue areas – colder than average

Commentary:
'Without moving forward in time, we switch to a microwave view. This is no animation. This image came from a microwave detector sent into space in 2001. The radiation you can see began its journey nearly 14 billion years before it was detected. Most of the image is just "noise" from the detector, but careful analysis shows slight variations in the temperature and density. The denser parts are the seeds of galaxies that will form millions of years later.'

(d)

Temp:
1000 K dropping

Time:
1 million yrs

Goggles tuned to:
visible

Objects in view:
dust, gas, clouds

Commentary:
'The Universe continues to cool and expand. Bits of dust and gas clump together, pulled by their gravity. The first structures start to appear in the Universe – big and small.
  As time goes on they grow.'

(e)

Temp:
100 K dropping

Time:
30 million yrs

Goggles tuned to:
infrared

Objects in view:
galaxies, stars

Commentary:
'Now the Universe is a very cold place. You can see vast galaxies have formed. Inside them things are heating up dramatically. The blazing pinpoints of light that are breaking out all over the sky are the first stars beginning to shine.'

(f)

Figure 5.93 (a)–(f) Riding through time

Making nuclei

An important piece of evidence supporting the Big Bang theory is that it predicts the ratio of the mass of hydrogen to helium in the Universe exactly as scientists have measured it: 75% hydrogen, 25% helium. How were scientists able to make such a prediction?

According to the Big Bang model, the very early Universe was extremely hot and dense. Did you notice how in the virtual-reality ride, as the temperature dropped, particles began to move more slowly? This illustrates the relationship between the temperature of a substance and the kinetic energy of its particles, which you met in Section 4.2.

Figure 5.94 shows how the temperature changed with time as the Universe expanded, as calculated using the Big Bang model. Particle physicists can work out what reactions took place, and which particles could survive at each stage. At first the Universe would consist of fundamental particles (quarks and leptons) and high-energy photons. As the temperature dropped, quarks could become bound into hadrons (e.g. protons and neutrons) without being shaken apart in violent collisions. Then, as the Universe cooled further, larger particles could begin to form in reactions such as this:

$$_0^1\text{n} + {}_1^1\text{H} \rightarrow {}_1^2\text{H} + \gamma \tag{R12}$$

which forms the isotope $_1^2\text{H}$, known as 'heavy hydrogen' or deuterium; its nucleus $_1^2\text{H}$ is sometimes known as a deuteron. Fusion of $_1^2\text{H}$ and $_1^1\text{H}$ produces another hydrogen isotope, $_1^3\text{H}$, called tritium, and then a tritium and deuterium nucleus can fuse to make a nucleus of helium $_2^4\text{He}$.

Detailed calculations show that there is time only to convert less than one-twelfth of the protons into helium, and to produce minute amounts of lithium, before the Universe becomes too cold for further fusion to take place.

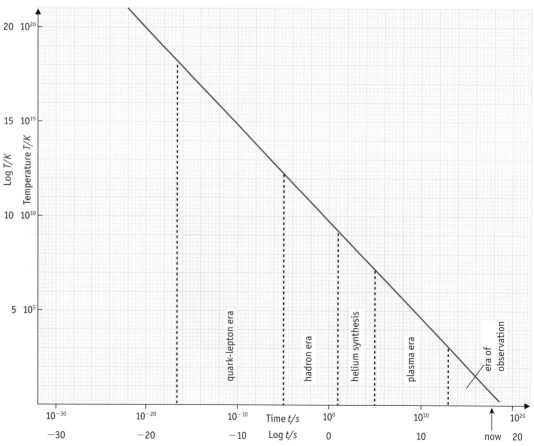

Figure 5.94 Temperature of the Universe falls with time

QUESTIONS

Q 71 Write a balanced nuclear equation showing the fusion of ^3_1H and ^2_1H to make ^4_2He. Which other particle must be produced in this reaction?

Q 72 Calculate the mean kinetic energy and rms speed of helium nuclei when the temperature is $1.0 \times 10^8\,\text{K}$. (Mass of ^4He nucleus, $m = 6.68 \times 10^{-27}\,\text{kg}$; $k = 1.38 \times 10^{-23}\,\text{JK}^{-1}$)

Q 73 Given that the mass of a helium nucleus is four times that of a hydrogen nucleus, show that a ratio of 12 hydrogen atoms to one helium atom is consistent with a mass ratio of about 75% H to 25% He.

Q 74 In stars, the production of helium is followed by further fusion to produce carbon. Explain why no carbon would have been produced in the early Universe.

Gravity takes over

Electric forces gave rise to atoms in the early Universe. Then, over large distances, gravity holds sway, and is the driving force of the evolution of the Universe. In the early Universe, the fluctuations in density and temperature were tiny. But without these fluctuations, gravity would not have been able to clump matter together. Gravity pulls vast galaxies together. Gravity keeps planets orbiting their suns (Figure 5.95a). Yet with one finger, you can overcome the gravitational pull from the whole Earth (Figure 5.95b).

Gravity is an incredibly weak force. Compare it with another familiar force: the electrostatic force, which is responsible for the force when any two objects are in contact, for friction, and for surface tension.

(a)

(b)

Figure 5.95 (a) Gravity holds vast objects together, but (b) it can easily be overcome

QUESTION

Q 75 Compare the magnitudes of the electrostatic force and the gravitational force between two electrons.

(Data: electron mass, $m = 9.11 \times 10^{-31}\,\text{kg}$; electron charge $e = -1.60 \times 10^{-19}\,\text{C}$; $G = 6.67 \times 10^{-11}\,\text{N m}^2\,\text{kg}^{-1}$; $\varepsilon_0 = 8.85 \times 10^{-12}\,\text{F m}^{-1}$)

Your answer to Question 75 should show that the electric force between electrons is an incredible 10^{42} times stronger than gravity, regardless of the separation of the particles.

So why is gravity so important on a big scale? Matter tends to be made of pairs of oppositely charged particles whose charge cancels, leaving no net electrostatic force. Furthermore, the amounts of matter in big structures such as stars and galaxies is huge, and so their gravitational effect is noticeable, causing atoms and molecules to gather together to form stars, holding stars together in galaxies and binding whole galaxies together into clusters.

We've seen how the Big Bang can give us a glimpse of the ancient past. It can also take us on a trip to the future, as you will see in Section 5.4.

5.4 Into the future

An open and shut case

What do cosmologists put bets on? Not horses or football matches. They're much more likely to be putting money on a number known as omega, Ω. Why? Because, wrapped up in this single constant, is the fate of the Universe.

If $\Omega < 1$ the Universe will carry on expanding forever, a so-called **open Universe** (a fate sometimes described as a 'big chill' or a 'big yawn'). There is a possibility that the expansion might accelerate rather than slow down.

If $\Omega > 1$ the Universe will eventually stop expanding, contract, and end up in a **Big Crunch**. This is a **closed Universe**.

You can see these possible futures illustrated in Figure 5.96. If the Universe is closed, then instead of ending after its collapse, it could begin a new expansion – a 'bounce'. In which case, maybe we are living in one of the bounces, and our Universe is actually much older than 10–20 billion years. However, scientists have found no plausible mechanism to produce these bounces and the theory is currently out of fashion.

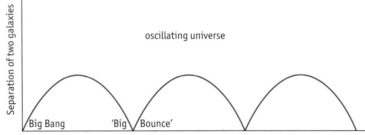

STUDY NOTE

The critical Universe and the accelerating Universe are discussed later in this section.

Figure 5.96 Graphs showing open and closed Universes

Why should the Universe ever stop expanding? Because the gravity between all the matter is trying to pull it together, and slow the expansion. Gravity and expansion are in direct competition. The winner dictates the fate of the Universe.

We can help explain the closed and open Universe by imagining what happens to a rocket fired away from the Earth (Figure 5.97). The rocket in Figure 5.97(a) has been fired at fairly low velocity. As it climbs, it loses kinetic energy and it gains gravitational potential energy. Eventually, the rocket's kinetic energy will decrease to zero, and it then falls back to Earth. This is very much like what happens in a closed Universe, where gravity wins (Figure 5.96). The rocket in Figure 5.97(b) has a higher velocity at launch. Even when it has reached a very great distance from Earth, it still has some kinetic energy and will keep on travelling out into space. This is similar to what happens in an open Universe, where the expansion wins over gravity – or, more correctly, kinetic energy wins over gravitational potential energy.

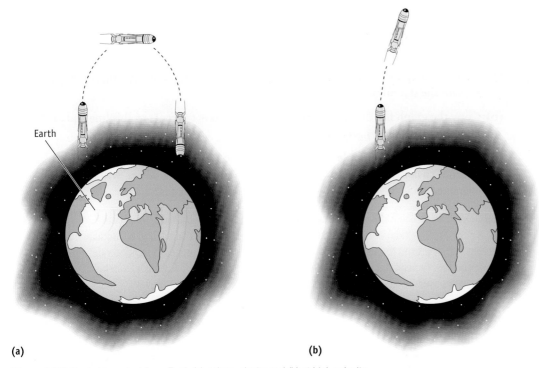

(a)   **(b)**

Figure 5.97 Rocket launched from Earth (a) at low velocity and (b) at high velocity

The argument is very like the one that determines whether a cloud will collapse and form stars or disperse, owing to the motion of its particles. The main difference is that particles in a gas cloud are moving randomly, whereas galaxies are all moving apart from one another.

An open Universe will carry on expanding forever. A closed Universe will one day (a long time in the future) stop expanding and start contracting again. A closed Universe also has a limited size. That doesn't mean, as you might think, that there is anything outside it. It sort of bends around on itself, like the surface of a sphere does, so if you could walk across the Universe, you'd eventually end up where you started. However, it's very hard to picture the shape of the Universe; you need mathematics.

An open Universe, on the other hand, has unlimited size. It extends infinitely in all directions. It might be a bit confusing to think of something that's already infinite expanding even more. Yet all it means is that the distance between any two galaxies is increasing.

It's interesting to speculate what would happen if the Universe did collapse. Most of the time during the collapse, things would be pretty similar to conditions during the expansion phase we're in now. Planets and stars would be born, live and die. Similarly, galaxies would evolve and black holes form from big dying stars. The main difference is that spectra from distant galaxies would be blueshifted (a cosmological blueshift), and the cosmic background radiation would get slowly warmer and warmer.

Later on, clusters of galaxies would start to merge, and the galaxies they're made of would also combine. Stars in the galaxies would collide, forming black holes, and other stars and planets would evaporate, as the temperature of space started really heating up.

Ultimately, the Universe could just end in the state it probably began in, as a 'quantum vacuum' (where particles pop in and out of existence). Or maybe it would bounce back into life again, reincarnated. However, cosmologists do not think this is likely.

What about an open Universe? One day (in about 120 000 billion years), all the stars would eventually run out of fuel, and they will stop shining. Then the Universe would become dark, populated by planets and black dwarfs. Its temperature would be just over absolute zero.

Galaxies would gradually get sucked into the black holes at their centres, and would be further apart than the whole of the observable Universe extends today.

QUESTIONS

Q 76 Another way of saying that the expansion of the Universe is slowing down is to say that *the rate of change of expansion is decreasing*.

(a) Copy the graph for the open Universe (Figure 5.96). By drawing tangents to the line, explain the statement in *italics*.

(b) Copy the graph for the accelerating Universe. Draw tangents to this line and explain how the graph shows acceleration.

(c) In Section 5.2, the Hubble constant was used to estimate the age of the Universe, assuming a constant expansion. If the expansion is slowing down, will the values estimated in Section 5.2 be overestimates or underestimates?

✿ ACTIVITY 39 BACK TO THE FUTURE?

Plan a virtual-reality 'ride into the future'. You could write your plan as a story, or produce a series of cartoons showing what the user will see at each stage, called a 'storyboard', to describe what the user will experience. You could add commentary and sound effects, too. Think about whether the 'rider' will be able to choose to experience an open or a closed Universe, and how they will make that choice.

Big chill or big crunch?

You might think that cosmologists are wasting their time betting on the fate of the Universe. Surely we cannot know the result in advance. Wrong. The value of omega, which determines the outcome, is already fixed. It is as though it were in a sealed envelope waiting to be opened. The challenge is to determine its value.

Omega is a ratio of densities:

$$\Omega = \frac{\rho}{\rho_c} \tag{64}$$

where ρ is the actual average density of the Universe and ρ_c is called the **critical density**. If ρ is greater than this critical value (i.e. $\Omega > 1$), then there is enough mass in the Universe for gravity to defeat expansion and the Universe to be closed. However, if $\rho < \rho_c$, the Universe is less dense than the critical value, expansion will win and the Universe will be open.

The way to calculate critical density is from the Hubble constant, as critical density is a measure of how much mass would be needed to halt the current expansion. Estimates of the critical density lie around $10^{-26}\,\text{kg m}^{-3}$. This is an extremely small value, and it is uncertain because of the difficulties of measuring the Hubble constant. From this value, the Universe must on average contain at least the mass of one poppy seed in each volume the size of the Earth for gravity to win, collapsing the Universe in on itself one day. Does the Universe contain this density of matter?

Scientists have estimated ρ, the actual density of the Universe, by examining huge volumes of space and calculating the mass of stars and galaxies needed to produce the observed amount of light. They have also calculated the density by estimating how the motions of galaxies are affected by local concentrations of mass. Unfortunately, all these estimates depend on knowing the distances to the galaxies, and we have seen that this is hard to calculate accurately. The highest estimates turn out to be only about $2 \times 10^{-27}\,\text{kg m}^{-3}$, in which case:

$$\Omega = \frac{10^{-27}\,\text{kg m}^{-3}}{10^{-26}\,\text{kg m}^{-3}} = 0.1$$

From these results, it looks as though the Universe is open, destined to end its days with matter spread ever wider, in a cold, lonely emptiness. But despite this, many cosmologists are betting on what seems an unlikely third way – that the density is exactly the critical value, making omega precisely one. This third possibility is called a **critical Universe** (or, sometimes, a flat Universe), as shown in Figure 5.96.

STUDY NOTE

STA Additional Sheet 7 shows how to calculate the critical density using the Hubble constant.

Why do many cosmologists think that gravity and expansion are so perfectly balanced? The evidence for omega being one is partly that we know it is close to one. It sounds absurd, but the actual density and the critical density vary with the age of the Universe. It is obvious that the density of the Universe changes with time: as the Universe expands its volume increases, but the amount of matter remains the same. The critical density also changes but in a different way, as the strength of gravity also changes as the matter in the universe spreads out. The upshot is that the two densities change with time in a different way, and so their ratio, Ω, also changes over time.

This scaling of Ω with time never allows a closed universe to become open. If $\Omega > 1$ at any time, then it is always greater than 1. Equally, if $\Omega < 1$ then it remains that way always. In between the two extremes, if $\Omega = 1$ then in that unique case, it remains 1 at all times.

Now, if we plug $\Omega = 0.1$ (the current estimate) into the equation that scales Ω with the age of the Universe, then it comes out to be almost exactly 1 at the Big Bang – in fact, it only differs from being 1 in the 59th decimal place.

Cosmologists find it hard to believe that Ω could have been so close to 1 at the beginning, without actually being 1. That would mean that Ω should be 1 right now and our estimations of the density would have to be (very!) wrong – perhaps there are some exotic types of matter and energy that we don't yet know about.

There is also reason to believe that a Universe with $\Omega = 1$ is 'easy' to create, on the grounds of energy. Finding the Universe so close to critical today means the kinetic energy and gravitational potential energies are closely balanced. The potential energies of all the matter in the Universe (due to being in a gravitational field) are negative. The positive kinetic energy balances exactly with the negative potential energy, making the total energy in the Universe zero. Some scientists have suggested that creating a Universe with zero total energy is easier.

In the chapter *Probing the Heart of Matter*, you read that physicist Alan Guth had a spectacular realisation that led him to propose that the Universe had undergone a rapid inflation about 10^{-35} seconds after the Big Bang. One consequence of this inflation is that the Universe would achieve a density exactly equal to the critical density.

The riddle of dark matter

We can only see a few per cent of the matter needed to make $\Omega = 1$. The rest is lurking unseen, invisible to our telescopes. The hidden stuff is called **dark matter** and there is also evidence for something called dark energy which we will discuss later. According to theoretical calculations, the relative amounts of dark matter and dark energy have changed over time (Figure 5.98).

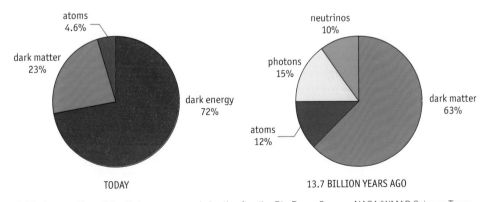

Figure 5.98 Composition of the Universe now and shortly after the Big Bang. Source: NASA/WMAP Science Team

Just because we cannot see dark matter does not mean it is not there. When you turn out the light in a room, most objects disappear – only the luminous objects remain visible.

In fact we already know that dark matter exists. Though we cannot see it, astronomers have measured its gravitational effect on galaxies and stars, and believe there must be at least ten times more dark matter than 'ordinary' matter.

Some of the strongest evidence comes from the rotation of galaxies. Just as you saw in Part 3 that astronomers can determine the masses of stars from observing their effect on the orbits of planets or on one another, they can also determine the mass of a galaxy from the motions of stars near its edge. In the early 1970s, American astronomer Vera Rubin (1928–) did just this for the Andromeda Galaxy and some other spiral galaxies, and was surprised to find that the mass thus calculated was about ten times the mass of the stars and interstellar material needed to account for the output of radiation.

Other evidence comes from whole clusters of galaxies, in which the galaxies are seen to move around at random – rather like the kinetic-theory model of particles in a gas. Some years before Rubin's work, Swiss–American astronomer Fritz Zwicky (1898–1974) measured the random speeds of galaxies in clusters and calculated the mass that would need to be present in order to stop the clusters dispersing. (This is another example of the sort of argument that we used in Part 4 when considering whether a gas cloud would disperse or collapse.) Here, too, there seemed to be about ten times as much mass in the clusters as could be accounted for by the radiating material.

⚙ ACTIVITY 40 MASS AND MOTION

Use computer simulations to explore:

- how the rotation of a galaxy is affected by the distribution of its mass
- how the mass in a cluster of galaxies influences the galaxies' random speeds.

The website **www.pearsonhotlinks.co.uk** provides links to some websites that provide suitable applets.

The baffling question is: 'What is dark matter?' Simply, it is matter that we cannot detect by either its emission or its absorption of radiation (the dust that shows up as dark streaks on photographs of nebulae is not 'dark matter', as we detect its absorption of starlight). There are two possibilities. Dark matter could be:

- baryonic matter, made from baryons (protons and neutrons) and their accompanying electrons – in other words, ordinary, everyday matter in a form that renders it undetectable by the emission or absorption of radiation

- non-baryonic matter – which could include all manner of exotic particles not yet detected; we often refer to these as WIMPS (weakly-interacting massive particles).

Baryonic matter could be in the form of planets or dim stars. Earth-like planets are not going to contribute a lot of mass to the Universe, but perhaps Jupiter-sized planets could. Other candidates include brown dwarfs – these are failed stars that were not quite massive enough to start fusion and become luminous. But we do not know.

However, we know that nearly all the dark matter must be non-baryonic. The Big Bang theory makes a clear prediction for the total amount of ordinary, atom-forming matter in the Universe. It says there should only be enough baryonic matter to make $\Omega = 0.1$. If Ω really is one, then the remaining 90% of the Universe has to be made of stranger forms of matter.

Current research

More evidence for dark matter comes from gravitational lensing. Einstein realised that, as well as matter, light is affected by gravity. So a beam of light would curve as it passes a large concentration of matter. When light from a very distant galaxy (billions of light years away) makes the long journey to Earth, it is sometimes deflected on its journey to Earth by matter en route. The matter acts like a lens, distorting and splitting the image into two. Nearly one hundred gravitational lenses have been discovered so far.

Figure 5.99 shows some examples of gravitational lensing. In Figure 5.99(a) the four images around the central image show the same galaxy, and in 5.99(b) the arc of light is an image of a galaxy that has been magnified and distorted. Figure 5.100 shows how a cluster of galaxies can produce gravitational lensing of light from a very distant object. Notice how the rays behave, as if they were passing through a converging lens. By studying images suc as those

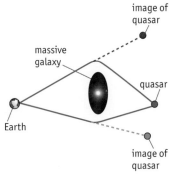

Figure 5.100 Schematic diagram of gravitational lensing

Figure 5.99 Images produced by gravitational lensing

in Figure 5.99, physicists can then reconstruct what the lens would be, its distribution in space and, importantly, its total mass. Often the mass calculated from its gravitational effects significantly exceeds that of the visible matter in the galaxy, indicating the presence of dark matter.

Other physicists are taking a completely different approach in the search for dark matter particles and are using sensitive instruments in an attempt to detect individual dark matter particles. Weakly-interacting massive particles (WIMPs) are generally heavy particles (thousands of times the mass of a proton) that only interact weakly with other matter, and are some of the best candidates for dark matter. WIMPs can glide through the gaps between electrons and nuclei with ease, and could be streaming right through you as you read this.

Dark energy – a repulsive idea

In the late 1990s, two separate groups of scientists studied the explosions of Type Ia supernovae in several very distant galaxies whose distances were known from their redshifts. Their aim was to determine the rate at which the expansion of the Universe was slowing down. They found that the supernovae were much fainter than expected, and their surprising interpretation of the results was that the Universe is actually expanding faster, and must have been speeding up for several billion years.

This has since been confirmed by other observations, including two detailed surveys of the sky – the *2-degree field* (2dF survey) and the *Sloan Digital Sky Survey* (SDSS). It means that the Universe isn't just open; it is accelerating (Figure 5.101).

To produce this effect, it is believed that three-quarters of the Universe must be composed of gravitationally repulsive **dark energy**, a radical and controversial idea in the world of physics. Several have suggested that the idea of dark energy is remarkably similar to Einstein's cosmological constant (see Section 5.1), and indeed, it may be that we need to revive the idea.

STUDY NOTE

Type Ia supernovae were discussed in Section 4.2.

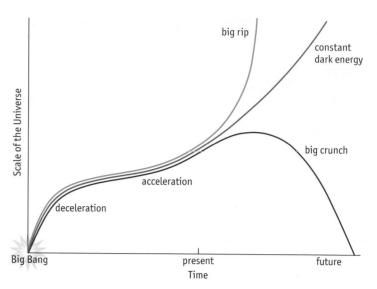

Figure 5.101 Expansion of the Universe is accelerating

As the Universe expanded, it is believed that the balance between the attractive pull of dark matter and the repulsive push of dark energy changed, with the latter becoming more dominant 6 to 8 billion years ago (Figure 5.102).

It is difficult to explain dark energy using conventional physics. It produces effects similar to a negative pressure, thus its repulsion. It may be that dark matter and dark energy are connected in some way that we do not yet understand. It appears that dark matter may be more 'clumpy' in its distribution than dark energy.

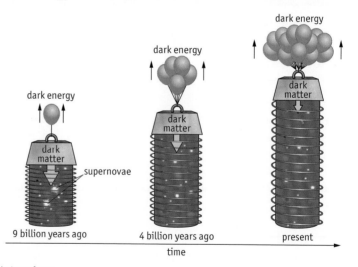

Figure 5.102 Cosmic tug of war

One alarming consequence of dark energy is the possibility of a *Big Rip*, tens of billions of years in the future, as suggested by Robert Caldwell of Dartmouth College. The accelerated expansion of the fabric of space–time by dark energy could cause matter to lose its attractive grip, and structures such as galaxies, then stars, planets and eventually atoms themselves, would break apart. Could this then be the end of the Universe?

At the time of writing (2015) NASA is developing the Wide Field Infrared Survey telescope for a projected launch in 2020. The mission will use independent techniques to study dark energy, as well as joining the search for exoplanets.

ACTIVITY 41 THE FATE OF THE UNIVERSE

To explore how dark energy might affect the fate of the Universe, use an activity developed by the Villiers Park Educational Trust. This can be accessed from **www.pearsonhotlinks.co.uk**.

5.5 Summing up Part 5

In this part of the chapter you have read about some of the biggest questions currently being addressed by astronomers and cosmologists, and you have seen how they relate to the physics you have studied in the whole of this chapter and elsewhere in the course.

Activity 42 is intended to help you look back over this part of the chapter. You will find further questions in Part 6.

ACTIVITY 42 THE COSMOLOGISTS

Put yourself in the position of one of the great astronomers or cosmologists of the 20th century. You need to give a five-minute presentation about your ideas at a scientific conference. You could choose someone featured in this chapter or someone that you have researched yourself.

Do the presentation as a group. One person in the group could introduce the scientists, for example explaining to the audience when the scientists worked and a few details about their lives.

Include a brief account of your theory or experiment, and say why it was important. To help you communicate your message, use props or audiovisuals (such as a clear diagram on an overhead projector). You can find further information from books or websites – some useful websites can be accessed from **www.pearsonhotlinks.co.uk**.

6 Link the learning: Universal physics

6.1 Big ideas

In this chapter, you have studied four main areas of physics: black-body radiation, nuclear physics, gravitation and kinetic theory. You have also touched on several areas of physics from elsewhere in the course, including radiation intensity, circular motion, the Doppler effect, electrostatic force, and energy conservation. You have used many of the key ideas in more than one part of the chapter, and seen how they have helped, and are still helping, astronomers and cosmologists to find out more about stars, galaxies and the Universe itself.

> ### ⚙ ACTIVITY 43 UNIVERSAL PHYSICS
>
> For each of the major areas of physics covered in this chapter (black-body radiation, nuclear physics, gravitation and kinetic theory), make a summary chart in the form of a concept map.
>
> Use a separate sticky note for each key term, diagram or equation, and arrange the stickers on a large piece of paper, moving them around so as to make the connections between them as clear as possible. Then stick them down firmly, and add extra notes to emphasise the links between the points.

6.2 Questions on the whole chapter

Q 76 A keen UFO spotter claims to have seen a 'mysterious object, about as bright as the brightest stars – maybe a bit brighter – and moving'. A cynic remarks that it was probably an aeroplane.

An Airbus landing-light has a power of 600 W. The brightest star in the night sky, Sirius, has a luminosity of 1.2×10^{28} W and is 8.3×10^{16} m from Earth. By carrying out a suitable order-of-magnitude calculation, decide whether an aeroplane light would typically be seen from a distance at which it could look about as bright as Sirius. Show the steps in your working, and state clearly any assumptions and approximations you have made.

Q 77 The Sun can be approximated to a black body with a temperature of 6000 K, whose luminosity is 4×10^{26} W and whose spectrum peaks at 500 nm. If its temperature fell to 3000 K, what would be the new luminosity and peak wavelength?

Q 78 Refer to Table 5.4 (Section 3.1) to help you answer this question.

(a) Complete the following reactions that take place in some giant stars:

    (i) $^{4}_{2}\text{H} + {}^{12}_{6}\text{C} \rightarrow \;.... + \gamma$

    (ii) $^{1}_{0}\text{n} + {}^{38}_{16}\text{S} \rightarrow {}^{39}_{16}\text{S} \rightarrow {}^{39}_{17}\text{Cl} + \;....$

(b) In a supernova explosion, neutron absorption produces a long-lived isotope of indium, $^{115}_{49}\text{In}$. This can absorb another neutron before decaying to give ^{116}In, which is itself unstable, decaying by beta-minus decay into an isotope of tin (Sn). Write equations for these two reactions, starting with ^{115}In.

Q 79 Nuclear power stations use nuclear fission to provide energy for electricity generation. When a nucleus of ^{235}U absorbs a neutron, it becomes unstable and undergoes fission, producing two lighter nuclei and some more neutrons. A typical reaction is:

$$^{235}_{92}\text{U} + ^{1}_{0}\text{n} \rightarrow ^{141}_{56}\text{Ba} + ^{92}_{36}\text{Kr} + \text{neutrons}$$

The neutrons released can produce more fission reactions if they are captured by further ^{235}U nuclei. A reactor in a power station contains neutron-absorbing materials designed so that the chain reaction proceeds at a steady rate. The energy released per unit mass of nuclear fuel is several orders of magnitude greater than that produced by burning chemical fuels such as oil or gas. After use, the spent fuel can be chemically treated to extract any useful materials. The remaining waste is highly radioactive.

(a) (i) How many neutrons are produced in the reaction above? Explain your reasoning.
(ii) Explain why a nuclear fission reaction releases energy.

(b) The uranium fuel used in reactors consists of a mixture of isotopes. $^{238}_{92}$U, the most common isotope, does not undergo fission when it absorbs a neutron. Instead, it continues to absorb neutrons and heavier nuclei are produced. These nuclei are radioactive, and contribute to the hazards associated with nuclear waste. For example:

$$^{238}_{92}\text{U} + ^{1}_{0}\text{n} \rightarrow ^{239}_{92}\text{U} \rightarrow ^{239}_{93}\text{Np} + \beta^{-} + \bar{\nu}_{e}$$

Further absorption reactions produce $^{239}_{94}$Pu (plutonium), one of several nuclei that are particularly hazardous both because of the radiation they emit and because of their complex biological effects if they enter the body. $^{239}_{94}$Pu has a half-life of 24 400 years and decays by alpha emission.

(i) What fraction of the $^{239}_{94}$Pu nuclei in a sample of waste will remain after 1000 years?
(ii) By what percentage will the activity due to $^{239}_{94}$Pu have decreased in this time?
(iii) If you started with a sample of pure $^{239}_{94}$Pu, the activity of the sample after 1000 years would in fact be greater than indicated by your answer above. Suggest a reason for this.

Q 80 Geophysicists study volcanoes in order to understand how they work and to predict eruptions. In some types of volcano, magma (molten rock) from deep inside the Earth moves upwards through underground fractures (large cracks forming gaps in the solid rock) and emerges at the Earth's surface. As the fractures fill with magma, the gravitational field at the surface changes, giving warning of an eruption.

To measure small changes in gravitational field strength, geophysicists use a gravimeter, which is essentially a very sensitive spring balance. The gravimeter is so sensitive that precise measurements of height need to be made at the same time as the gravity measurements.

(Data: near Earth's surface g = 9.81 N kg^{-1}; G = 6.67 × 10^{-11} N m^2 kg^{-2})

(a) (i) What is meant by the term *gravitational field strength*?
(ii) If a piece of magma moves upwards towards the Earth's surface, filling a gap in the rocks, what would happen to the gravitational field strength at the surface?
(iii) One type of gravimeter used in volcanic studies can detect changes in gravitational field of one part in 100 million at the Earth's surface. What is the order of magnitude of the smallest change in gravitational field that it could detect?

(b) Calculate the strength of the gravitational field due to a cube of magma, density 4000 kg m^{-3}, measuring 100 m along each side, with its centre of mass 1.5 km below the surface.

(c) Explain why it is necessary to measure the height when monitoring changes in gravitational field due to underground magma movements.

Q 81 During 1996 and 1997, the Galileo spacecraft investigated the planet Jupiter and its moons. One such moon is Io. Measurements of Io's orbital motion enable Jupiter's mass to be determined.

(a) The radius of Io's orbit around Jupiter, r, is 4.22×10^8 m and its period of rotation, T, is 1.53×10^5 s. What is Io's orbital speed v?

(b) (i) Write down an expression in terms of Io's mass, m_{Io}, for the centripetal force, F_{cent}, needed to keep it in orbit.

(ii) Using m_{Ju} to denote Jupiter's mass, write down an expression for the gravitational force of attraction, F_{grav}, between Jupiter and Io.

(c) Using your answer to **(b)**, obtain an expression for m_{Ju} in terms of v, G and r.

(d) Calculate a value for the mass of Jupiter ($G = 6.67 \times 10^{-11}$ N m^2 kg^{-2}).

Q 82 (a) A galaxy is moving away from Earth at 4.58×10^5 m s^{-1}. It emits light at a frequency 6.00×10^{14} Hz. Calculate the observed frequency of this light on Earth.
($c = 3.00 \times 10^8$ m s^{-1})

(b) Hydrogen atoms emit and absorb electromagnetic radiation with a wavelength of 21.2 cm. This '21 cm line' is used by radio astronomers to study the Doppler shifts of radiation from the ISM and rotating galaxies. The receivers must be tuned to precisely the right frequency.

(i) What is the emitted frequency of this radiation?

(ii) If the hydrogen in a galaxy is moving at 7370 km s^{-1} away from Earth, what is the difference between the observed frequency and the emitted frequency?

Q 83 Our galaxy, the Milky Way, is one of a group of galaxies. The light from our neighbour in the group, the Andromeda galaxy, is slightly blueshifted. Suggest a reason why it is not redshifted as other galaxies are.

Q 84 The Kepler space telescope has discovered many exoplanets. Around one particular star, Kepler-444, which is 117 light years away, five Earth-sized planets have been identified.

(Data for this question: Earth–Sun distance = 1.50×10^{11} m; mass of Earth = 6.0×10^{24} kg)

(a) The basis of finding an exoplanet is a change in the light curve of the star. Describe briefly how the presence of five different exoplanets would appear on such a light curve.

(b) The closest planet to Kepler-444 has an orbit period of 6 days and is a distance of a tenth of that from the Earth to the Sun. What is the mass of Kepler-444?

(c) At what speed will the exoplanet be orbiting Kepler-444?

(d) If the exoplanet has an average density 5.7×10^3 kg m^{-3}, and has one-third the mass of the Earth, what would be the gravitational field strength at its surface?

(e) What is the exoplanet's escape velocity (that is, the minimum vertical velocity that would enable a spacecraft to leave the exoplanet's surface)?

(f) Suppose it takes two minutes for the spacecraft to accelerate from rest to the escape velocity. Estimate the spacecraft's acceleration and comment on your answer.

(g) Suppose the spacecraft has a mass of 10 tonnes. What is the charge in gravitational potential energy as it travels to a distance 2000 km from the exoplanet's surface?

Q 85   Creationists claim that the Universe was created 6000 years ago. If this were true, calculate the value that the Hubble constant would have:

(a) in years^{-1}

(b) in s^{-1}

(c) in km s^{-1} Mpc^{-1} (assuming uniform expansion).

(1 year = 3.16×10^7 s; 1 Mpc = 3.09×10^{22} m)

Q 86   Figure 5.103 shows the spectrum of light received from the quasar known as 3C273 (a quasar is an extremely luminous type of galaxy). The four peaks marked are hydrogen lines that have been shifted in wavelength. In a laboratory, these lines have wavelengths 410 nm, 434 nm, 486 nm and 656 nm.

Figure 5.103 Spectrum of 3C273

(a) (i) What is the name given to this shift in wavelength?
     (ii) Without doing any calculations, what does the spectrum of 3C273 allow you to deduce about its motion?

(b) Using any one of the lines marked in the diagram, calculate the speed of 3C273 relative to the Earth.

(c) Assuming that the Hubble constant H_0 = 67 km s^{-1} Mpc^{-1}, calculate the distance to 3C273. Express your answer in Mpc.

Q 87   The planet Jupiter orbits the Sun at a distance of 7.78×10^{11} m and has a radius of 7.14×10^7 m. It is composed mainly of hydrogen and helium, but it is believed that its central core has a similar composition to the rocks of the Earth. The outer parts of Jupiter are at a temperature of about 130 K, which is slightly higher than that which is calculated on the basis of its absorption of solar energy alone, suggesting that there is some internal heat source. The Earth is 1.49×10^{11} m from the Sun.

(a) (i) Calculate the maximum and minimum angular diameter, in radians, of Jupiter when observed from the Earth.
     (ii) Express the smaller value as a percentage of the larger one.

(b) Calculate the ratio $I_{\text{Jupiter}}:I_{\text{Earth}}$, where I is the intensity of solar radiation at each planet. Express the ratio as a percentage.

(c) Suggest a possible mechanism by which Jupiter might be heated internally.

6.3 Achievements

Now you have studied this chapter you should be able to achieve the outcomes listed in Table 5.15.

TABLE 5.15 ACHIEVEMENTS FOR THE CHAPTER REACH FOR THE STARS

	Statement from Examination Specification	Section(s) in this chapter
1	know and understand the distinction between base and derived quantities and their SI units	2.5, 4.1
2	be able to demonstrate their knowledge of practical skills and techniques for both familiar and unfamiliar experiments	2.2, 2.4, 3.1, 4.2
3	be able to estimate values for physical quantities and use their estimate to solve problems	2.7, 4.2
4	understand the limitations of physical measurement and apply these limitations to practical situations	2.2, 2.4, 3.1, 4.2
5	be able to communicate information and ideas in appropriate ways using appropriate terminology	3.1, 4.1, 4.2, 5.3
7	understand the role of the scientific community in validating new knowledge and ensuring integrity	5.1, 5.2, 5.4
146	understand the concept of internal energy as the random distribution of potential and kinetic energy amongst molecules	4.2 (and see BLD)
147	explain the concept of absolute zero and how the average kinetic energy of molecules is related to the absolute temperature	4.2
148	be able to derive and use the equation $pV = \frac{1}{3}Nm\langle c^2 \rangle$ using the kinetic theory model	4.2
149	be able to use the expression $pV = NkT$ for an ideal gas	4.2
150	Core Practical 14: investigate the relationship between the pressure and volume of a gas at fixed temperature	4.2
151	be able to derive and use the expression $\frac{1}{2}m\langle c^2 \rangle = \frac{3}{2}kT$	4.2
152	understand what is meant by a *black-body radiator* and be able to interpret radiation curves for such a radiator	2.5
153	be able to use the expression the Stefan–Boltzmann law equation $L = \sigma A T^4$ for black-body radiators	2.5
154	be able to use Wien's law equation $\lambda_{max}T = 2.90 \times 10^{-3}\,\text{m K}$ for black-body radiators	2.5
155	be able to use the equation intensity of radiation $I = \dfrac{P}{A}$	2.4
156	be able to use the equation intensity of radiation $I = \dfrac{L}{4\pi d^2}$, where L is luminosity and d is distance from the source	2.4
157	understand how astronomical distances can be determined using trigonometric parallax	2.3
158	understand how astronomical distances can be determined using measurements of intensity received from standard candles (objects of known luminosity)	2.6
159	be able to sketch and interpret a simple Hertzsprung–Russell diagram which relates stellar luminosity to surface temperature	2.5
160	understand how to relate the Hertzsprung–Russell diagram to the life cycle of stars	4.3
161	understand how the movement of a source of waves relative to an observer/detector gives rise to a shift in frequency (Doppler effect)	4.1

continued

TABLE 5.15 ACHIEVEMENTS FOR THE CHAPTER REACH FOR THE STARS *continued*

Statement from Examination Specification	Section(s) in this chapter
162 be able to use the equations for redshift, $z = \dfrac{\Delta\lambda}{\lambda} \approx \dfrac{\Delta f}{f} \approx \dfrac{v}{c}$ for a source of electromagnetic radiation moving relative to an observer, and $v = H_0 d$ for objects at cosmological distances	4.1, 5.1, 5.2
163 understand the controversy over the age and ultimate fate of the Universe associated with the value of the Hubble constant and the possible existence of dark matter	5.1, 5.2, 5.4
164 understand the concept of *nuclear binding energy*, and be able to use the equation $\Delta E = c^2 \Delta m$ in calculations of nuclear mass (including *mass deficit*) and energy	3.2
166 understand the processes of nuclear fusion and fission with reference to the binding energy per nucleon curve	3.2, 5.3
167 understand the mechanism of nuclear fusion and the need for very high densities of matter and very high temperatures to bring about and maintain nuclear fusion	3.2, 4.2
168 understand that there is background radiation and how to take appropriate account of it in calculations	3.1
169 understand the relationships between the nature, penetration, ionising ability and range in different materials of nuclear radiations (alpha, beta and gamma)	3.1
170 be able to write and interpret nuclear equations given the relevant particle symbols	3.1, 3.2 and see PRO
171 Core Practical 15: Investigate the absorption of gamma radiation by lead	3.1
172 understand the spontaneous and random nature of nuclear decay	3.1
173 be able to determine the half-lives of radioactive isotopes graphically and be able to use the equations for radioactive decay: activity $A = \dfrac{dN}{dt} = -\lambda N$, $\lambda = \ln\dfrac{2}{t_{\frac{1}{2}}}$, $N = N_0\, e^{-\lambda t}$ and $A = A_0\, e^{-\lambda t}$, and derive and use the corresponding log equations	3.1
174 understand that a gravitational field (force field) is defined as a region where a mass experiences a force	4.1
175 understand that gravitational field strength is defined as $g = \dfrac{F}{m}$ and be able to use this equation	4.1
176 be able to use the equation $F = \dfrac{Gm_1 m_2}{r^2}$ (Newton's law of universal gravitation)	4.1
177 be able to derive and use the equation $g = \dfrac{Gm}{r^2}$ for the gravitational field due to a point mass	4.1, 4.2
178 be able to use the equation $V_{\text{grav}} = \dfrac{-Gm}{r}$ for a radial gravitational field	4.1
179 be able to compare electric fields with gravitational fields	4.1
180 be to apply the gravitational field equations to orbital motion	4.1

Answers

Q 1

(a) time, t = distance/speed so:

$$t = \frac{1.50 \times 10^{11}\,\text{m}}{3.00 \times 10^{8}\,\text{m s}^{-1}} = 500\,\text{s}$$

(b) Distance to Sun = 500 light seconds

$$= \frac{500\,\text{s}}{3.16 \times 10^{7}\,\text{s yr}^{-1}}$$

$$= 1.58 \times 10^{-5}\,\text{light years}$$

(c) $\dfrac{d_{\text{PC}}}{d_{\text{Sun}}} = \dfrac{4.24\,\text{light years}}{1.58 \times 10^{-5}\,\text{light years}}$

$$= 2.68 \times 10^{5}$$

To the nearest order of magnitude, the ratio is 10^{5}

(d) (i) $(250 \times 60)\,\text{s} \times 3.00 \times 10^{8}\,\text{m s}^{-1} = 4.50 \times 10^{12}\,\text{m}$

(ii) $(100\,000 \times 3.16 \times 10^{7})\,\text{s} \times 3.00 \times 10^{8}\,\text{m s}^{-1}$
$$= 9.48 \times 10^{20}\,\text{m}$$

Q 2

We are told that the Sun is about 400 times further away than the Moon, and that the Sun and Moon have very similar angular diameters.

We can therefore deduce that:

$$d_{\text{Sun}} \approx 400\,d_{\text{Moon}} \approx 400 \times 3.85 \times 10^{8}\,\text{m}$$

From Equation 1, $\alpha_{\text{Sun}} = \dfrac{D_{\text{Sun}}}{d_{\text{Sun}}}$

$$= \frac{1.4 \times 10^{9}\,\text{m}}{400 \times 3.85 \times 10^{8}\,\text{m}}$$

$$= 9 \times 10^{-3}\,\text{radians}$$

We have already said (and used the fact) that $\alpha_{\text{Sun}} = \alpha_{\text{Moon}}$, so, with no further calculation, $\alpha_{\text{Moon}} = 9.0 \times 10^{-3}\,\text{radians}$.

(Alternatively, we could have started by saying

$D_{\text{Moon}} = \dfrac{D_{\text{Sun}}}{400}$. This would give exactly the same answers.)

Q 3

Volume of sphere, $V = \dfrac{4\pi r^{3}}{3}$

Volume of Earth, $V_{\text{Earth}} = \dfrac{4\pi \times (6.4 \times 10^{6}\,\text{m})^{3}}{3}$

$$- 1.1 \times 10^{21}\,\text{m}^{3}$$

Sun's radius $= \dfrac{D_{\text{Sun}}}{2} = \dfrac{1.4 \times 10^{9}\,\text{m}}{2}$

$$= 7.0 \times 10^{8}\,\text{m}$$

$$V_{\text{Sun}} = \frac{4\pi \times (7.0 \times 10^{8}\,\text{m})^{3}}{3} = 1.4 \times 10^{27}\,\text{m}^{3}$$

$$\frac{V_{\text{Sun}}}{V_{\text{Earth}}} = \frac{1.4 \times 10^{27}\,\text{m}^{3}}{1.1 \times 10^{21}\,\text{m}^{3}} = 1.3 \times 10^{6}$$

So in terms of volume, the Sun is about a million times bigger than the Earth.

Q 4

(a) 1 light year $= 3.00 \times 10^{8}\,\text{m s}^{-1} \times 3.16 \times 10^{7}\,\text{s}$

$$= 9.48 \times 10^{15}\,\text{m}$$

$$d = 4.351 \times 9.48 \times 10^{15}\,\text{m} = 4.125 \times 10^{16}\,\text{m}$$

$$\phi = \frac{r}{d} = \frac{1.49 \times 10^{11}\,\text{m}}{4.125 \times 10^{16}\,\text{m}}$$

$$= 3.61 \times 10^{-6}\,\text{rad}$$

$$= 3.61 \times 10^{-6}\,\text{rad} \times 2.06 \times 10^{5}\,\text{arcsec rad}^{-1}$$

$$= 0.744\,\text{arcsec}$$

(b) $\phi = 1\,\text{arcsec} = \dfrac{1}{2.06 \times 10^{5}\,\text{rad}}$, so:

$$d = \frac{r}{\phi} = 1.49 \times 10^{11}\,\text{m} \times 2.06 \times 10^{5}$$

$$= 3.1 \times 10^{16}\,\text{m, which is as stated.}$$

Q 5

$\phi = 0.002\,\text{arcsec} = \dfrac{0.002\,\text{arcsec}}{2.06 \times 10^{5}\,\text{arcsec rad}^{-1}}$

$$= 9.71 \times 10^{-9}\,\text{rad}$$

From Equation 2, $d = \dfrac{r}{\phi/\text{rad}}$

$$= \frac{1.49 \times 10^{11}\,\text{m}}{9.71 \times 10^{-9}\,\text{rad}}$$

$$= 1.53 \times 10^{19}\,\text{m}$$

From Question 4, 1 light year $= 9.46 \times 10^{15}\,\text{m}$, so:

$$r = \frac{1.53 \times 10^{19}\,\text{m}}{9.46 \times 10^{15}\,\text{m ly}^{-1}}$$

$$= 1.62 \times 10^{3}\,\text{light years}$$

$1\,\text{pc} = 3.1 \times 10^{16}\,\text{m so:}$

$$r = \frac{1.53 \times 10^{19}\,\text{m}}{3.1 \times 10^{16}\,\text{m pc}^{-1}} = 4.9 \times 10^{2}\,\text{pc}$$

Q 6

(a) Using Equation 3:

$$\frac{I_{\text{body}}}{I_{\text{Earth}}} = \left(\frac{L}{4\pi d^{2}_{\text{body}}}\right) \div \frac{L}{4\pi d^{2}_{\text{Earth}}}$$

$$= \left(\frac{d_{\text{Earth}}}{d_{\text{body}}}\right)^{2}$$

For Mercury:

$$\frac{I_{\text{Mercury}}}{I_{\text{Earth}}} = \left(\frac{149.6 \times 10^{9}\,\text{m}}{57.9 \times 10^{9}\,\text{m}}\right)^{2}$$

$$= \left(\frac{149.6}{57.9}\right)^{2} = 6.7$$

so $I_{\text{Mercury}} = 6.7\,I_{\text{Earth}}$

Notice that the units of m and the factor 10^9 cancel in the final calculation. Using a similar approach for the other bodies:

$$\frac{I_{Venus}}{I_{Earth}} = 1.9; \quad \frac{I_{Mars}}{I_{Earth}} = 0.43; \quad \frac{I_{Pluto}}{I_{Earth}} = 6.4 \times 10^{-4}$$

(b) In general we can expect that the bodies with the higher solar intensity (Mercury and Venus) will be hotter than Earth and have brighter sunlight. Mars is cooler, and Pluto is only very dimly lit by the Sun and very cold.

(c) You might have thought of some of the following effects. The curvature of the body reduces the incident energy intensity near the poles. The body's surface may reflect radiation back into space rather than absorbing it. It may reflect or absorb radiation so the surface does not get so warm. Any atmosphere may trap radiation (the greenhouse effect). (This causes Venus to be hotter than Mercury, even though it is further from the Sun.) There may be heating due to radioactive materials. Geothermal effects may warm up parts of the body (volcanic activity, hot springs, etc.).

Q 7 **(a)** From Equation 3, $d^2 = \dfrac{L}{4\pi I}$

Assuming that Sirius and the Sun have the same luminosity, L, then:

$$\frac{d^2_{Sir}}{d^2_{Sun}} = \left(\frac{L}{4\pi I_{Sir}}\right) \div \left(\frac{L}{4\pi I_{Sun}}\right) = \frac{I_{Sun}}{I_{Sir}}$$

Using Huygens's measurements:

$$\frac{d^2_{Sir}}{d^2_{Sun}} = 26\,644^2$$

$$\frac{d^2_{Sir}}{d^2_{Sun}} = 26\,644^2 \approx 2.7 \times 10^4$$

(b) True $\dfrac{d^2_{Sir}}{d^2_{Sun}} = \dfrac{8 \text{ light years}}{1.58 \times 10^{-5} \text{ light years}} = 5.1 \times 10^5$

(c) The ratio calculated by Huygens is too small by a factor of about 20. He was wrong to assume that Sirius has the same luminosity as the Sun; we now know that it is about 26 times more luminous. Also, his memory of the brightness of Sirius could not be very reliable, so the five significant figures he quotes for the ratio of diameters is incredibly optimistic.

Q 8 Brown dwarf (coolest), red giant, white dwarf, blue supergiant (hottest)

Q 9 From Equation 6, $\sigma = \dfrac{L}{4\pi r^2 T^4}$, so its SI units are $W\,m^{-2}\,K^{-4}$

K and m are base units.

$1\,W = 1\,J\,s^{-1}$

$1\,J = 1\,Nm$

$\quad = 1\,kg\,m\,s^{-2}\,m$

so:

$1\,W = 1\,kg\,m^2\,s^{-3}$

and so the SI base units of σ are $kg\,m^2\,s^{-3}\,m^{-2}\,K^{-4} = kg\,s^{-3}\,K^{-4}$.

Q 10 From Equation 6, $r = \sqrt{\dfrac{L}{4\pi\sigma T^4}}$

(a) For Spica A:

$$r = \sqrt{\frac{8 \times 10^{29}\,W}{4\pi \times 5.67 \times 10^{-8}\,W\,m^{-2}\,K^{-4} \times (3 \times 10^4\,K)^4}}$$

$$= 1.2 \times 10^9\,m$$

$$\approx 1.7 \times r_{Sun}$$

(b) For Barnard's Star:

$$r = \sqrt{\frac{4 \times 10^{23}\,W}{4\pi \times 5.67 \times 10^{-8}\,W\,m^{-2}\,K^{-4} \times (2.7 \times 10^3\,K)^4}}$$

$$= 1.0 \times 10^8\,m$$

$$\approx 0.15 \times r_{Sun}$$

Q 11 **(a)** Blue-white

(b) Orange-red

Q 12 **(a)** **(i)** The hotter the star, the greater its luminosity.

(ii) The main sequence is not quite a straight line, so L and T cannot be related by a simple power law (see *Maths Note 8.7: Using log graphs*). But if you draw a straight line going approximately through the main sequence, then you can see that an increase of a factor 10 in temperature (e.g. from 2500 K to 25 000 K) corresponds to an increase in luminosity of about a factor 10^6 (from $10^{-2} L_{Sun}$ to $10^4 L_{Sun}$), which suggests a relationship something like $L \propto T^6$.

(b) Comparing a(ii) with Stefan's law ($L \propto T^4$) suggests that the assumption about size was indeed very rash. It suggests that hot main sequence stars are bigger than the cool ones; their greater luminosities are due to their larger surface areas as well as their higher temperatures.

(c) It is very difficult to say anything at all. Although size *can* be deduced from measurements of luminosity and temperature, to deduce mass you would need to know a star's density, or else determine its mass by some completely independent means (which you will meet in Section 4).

(d) Again, it is very difficult to say anything conclusive. One might be tempted to speculate that stars are cool and faint as they are formed, and gradually heat up and become more luminous as a result of nuclear fusion in their interiors. Or perhaps they start off hot and luminous and gradually cool down. Similar suggestions were made when the HR diagram was first produced, but they are entirely speculative and, we now believe, totally wrong. See Part 4 of this chapter.

Assuming that the stars are all at different stages in their evolution, the most we can say is that some combinations of luminosity and temperature must be more stable than others, indicating long-lasting phases in a star's life, since there are large numbers of stars on the main sequence, quite a few red giants, and some parts of the HR diagram that contain no stars.

Q 13 Rearranging Equation 3, $d^2 = \dfrac{L}{4\pi I}$

For Sirius:

$$d = \sqrt{\frac{9.9 \times 10^{27}\,\text{W}}{4\pi \times 1.17 \times 10^{-7}\,\text{W}\,\text{m}^{-2}}}$$

$$= 8.21 \times 10^{16}\,\text{m}$$

Similarly for the other stars – see Table 5.16.

Star	Luminosity, L/W	Intensity, $I/\text{W m}^{-2}$	Distance, d/m
Sirius	9.9×10^{27}	1.17×10^{-7}	8.21×10^{16}
Rigel	3.8×10^{31}	5.37×10^{-7}	2.37×10^{18}
Betelgeuse	5.3×10^{30}	1.11×10^{-8}	6.16×10^{18}
Proxima Centauri	2.8×10^{23}	1.09×10^{-11}	4.52×10^{16}
Sun	3.83×10^{26}	1.37×10^{3}	1.49×10^{11}

Table 5.16 Answers to Question 13

Q 14 (a) Wien's law (Equation 4):

Betelgeuse: $\lambda_{max} = \dfrac{c}{f_{max}}$

$$= \frac{3.00 \times 10^{8}\,\text{ms}^{-1}}{4.3 \times 10^{14}\,\text{Hz}}$$

$$= 7.0 \times 10^{-7}\,\text{m}$$

$$T = \frac{2.89 \times 10^{-3}\,\text{m K}}{\lambda_{max}}$$

$$\approx \frac{2.89 \times 10^{-3}\,\text{m K}}{7.0 \times 10^{-7}\,\text{m}}$$

$$\approx 4.1 \times 10^{3}\,\text{K}$$

Rigel: $\lambda_{max} = \dfrac{c}{f_{max}}$

$$= \frac{3.00 \times 10^{8}\,\text{m s}^{-1}}{10 \times 10^{14}\,\text{Hz}}$$

$$= 3.0 \times 10^{-7}\,\text{m}$$

$$T = \frac{2.89 \times 10^{-3}\,\text{m K}}{\lambda_{max}}$$

$$\approx \frac{2.89 \times 10^{-3}\,\text{m K}}{3.0 \times 10^{-7}\,\text{m}}$$

$$\approx 9.6 \times 10^{3}\,\text{K}$$

(b) Any hot object will emit radiation across a broad range of wavelengths, so even though Rigel appears blue, it will emit over a range of blue wavelengths. The quoted frequencies might not correspond to the peak wavelengths. A detailed spectral analysis would be needed. Also, stars are not perfect black-body emitters (there are absorption lines due to the cooler outer atmospheres).

Q 15 We don't have any material that was alive when the Earth formed. Also, while carbon dating is good for dating objects from thousands up to millions of years old, in very ancient objects there would be too little carbon-14 to detect.

Q 16 Uranium-235: $^{235}_{92}\text{U}$, contains 143 neutrons.

Uranium-238: $^{238}_{92}\text{U}$, 146 neutrons.

Thorium-232: $^{232}_{90}\text{Th}$, 142 neutrons.

Potassium-40: $^{40}_{19}\text{K}$, 21 neutrons.

Q 17

(a) This is beta-minus decay. $^{87}_{37}Rb \rightarrow \, ^{87}_{38}Sr + \, ^{0}_{-1}e + \bar{v}_e$

(b) (i) $^{211}_{83}Bi \xrightarrow{\alpha} \, ^{4}_{2}\alpha + \, ^{207}_{81}Tl$

$^{211}_{83}Bi \xrightarrow{\beta} \, ^{0}_{-1}\beta + \, ^{211}_{84}Po$

$^{211}_{84}Po \xrightarrow{\alpha} \, ^{4}_{2}\alpha + \, ^{207}_{82}Pb$

(ii) Therefore $^{207}_{81}Tl \rightarrow \, ^{0}_{-1}\beta + \, ^{207}_{82}Pb$
The mode of decay from Tl to Pb must be beta-minus .

Q 18

(a) (i) Taking natural logs of Equation 12:

$\ln(A) = \ln(A_0) - \lambda t$

$\lambda t = \ln(A_0) - \ln(A) = \ln(A_0/A)$

$\lambda = \dfrac{\ln(A_0/A)}{t}$

$= \dfrac{\ln(1.0 \times 10^{11} Bq/2.0 \times 10^{10} Bq)}{10 \times 60 \times 60\, s}$

$= \dfrac{\ln 5}{36\,000} s^{-1}$

$= 4.47 \times 10^{-5}\, s^{-1}$

(ii) $t_{\frac{1}{2}} = \dfrac{\ln 2}{\lambda} = 15\,504\, s = 4.3\, h$

(b) Using dice, $\lambda = \dfrac{1}{6}$ per throw:

$t_{\frac{1}{2}} = 4.16$ throws

Using coins, $\lambda = \dfrac{1}{2}$ per toss:

$t_{\frac{1}{2}} = 1.39$ tosses

Q 19 If n half-lives have elapsed, then the amount of parent isotope is $\dfrac{1}{2^n} \times$ the initial amount.

(Table 5.5 listed only whole-number values of n, but this rule works for n being any number.)

Q 20 Using Equation 13:

For ^{238}U, $t_{\frac{1}{2}} = \dfrac{\ln(2)}{\lambda} = \dfrac{\ln(2)}{1.552 \times 10^{-10}\, yr^{-1}}$

$= 4.466 \times 10^9\, yr = 4466 \times 10^6\, yr$

Similarly ^{40}K has $t_{\frac{1}{2}} = 1193 \times 10^6\, yr$

For ^{235}U, $\lambda = \dfrac{\ln(2)}{t_{\frac{1}{2}}} = \dfrac{\ln(2)}{(704 \times 10^6\, yr)}$

$= 9.846 \times 10^{-10}\, yr^{-1}$

Similarly ^{87}Rb has $\lambda = 1.420 \times 10^{-11}\, yr^{-1}$

$= 0.1420 \times 10^{-10}\, yr^{-1}$

Q 21 24 400 million years is half of the half-life,

so $n = 0.5$, and the fraction remaining is $\dfrac{1}{2^{0.5}} = 0.71$

or 71%.

Q 22 1/16 of the parent nuclei must have survived (and 15/16 changed into lead), so the rock must be 4 half-lives old, i.e. its age must be 4×704 million years $= 2816$ million years.

Q 23

(a) 4600 million years

(b) The meteorites could have formed in a volcanic explosion in which they were hurled at very high speed from the Martian surface (exceeding the escape velocity), eventually landing on Earth.

(c) The rocks from Mars are much younger than those formed on Earth, suggesting that they crystallised from molten rock long after Mars formed. This is consistent with their crystallising from the molten rock ejected from a volcano.

Q 24 $\dfrac{3.84 \times 10^{26}\, W}{746\, W\, hp^{-1}} \approx 5 \times 10^{23}\, hp$

Q 25

(a) Total energy emitted in 1000 years:

$\Delta E = 3.84 \times 10^{26}\, J\, s^{-1} \times 1000 \times 3.16 \times 10^7\, s$

$= 1.21 \times 10^{37}\, J$

(b) (i) $\Delta m = \dfrac{\Delta E}{c^2} = \dfrac{1.21 \times 10^{37}\, J}{(3.00 \times 10^8\, m\, s^{-1})^2}$

$= 1.35 \times 10^{20}\, kg$

(ii) Using m to represent mass of Sun:

$\dfrac{\Delta m}{m} = \dfrac{1.35 \times 10^{20}\, kg}{1.99 \times 10^{30}\, kg} = 6.78 \times 10^{-11}$

$= 6.78 \times 10^{-9}\%$

(c) In 10^{10} years, the percentage mass loss will be $10^7 \times 6.78 \times 10^{-9}\% = 6.78 \times 10^{-2}\%$. (This is a very small fraction of the Sun's mass.)

Q 26 **(a)** $4{}^{1}_{1}\text{H} + 2{}^{0}_{-1}\text{e}^{-} \rightarrow {}^{4}_{2}\text{He} + 2\nu_{\text{e}} + 6\gamma$

(b) Initial mass $= 4 \times 1.673 \times 10^{-27}\,\text{kg} + 2 \times 9.11 \times 10^{-31}\,\text{kg}$

$= 6.694 \times 10^{-27}\,\text{kg}$

Final mass $= 6.645 \times 10^{-27}\,\text{kg}$

Mass deficit $\Delta m = 6.694 \times 10^{-27}\,\text{kg} - 6.645 \times 10^{-27}\,\text{kg}$

$= 4.9 \times 10^{-29}\,\text{kg}$

(c) $\Delta E = c^{2}\Delta m = (3.00 \times 10^{8}\,\text{m s}^{-1})^{2} \times 4.9 \times 10^{-29}\,\text{kg}$

$= 4.41 \times 10^{-12}\,\text{J}$

(d) The Sun emits energy at a rate of $3.84 \times 10^{26}\,\text{J s}^{-1}$. To maintain this output, n nuclei of ${}^{4}_{2}\text{He}$ must be produced per second, where:

$$n = \frac{3.84 \times 10^{26}\,\text{J s}^{-1}}{4.41 \times 10^{-12}\,\text{J}} = 8.71 \times 10^{37}\,\text{s}^{-1}$$

Each ${}^{4}_{2}\text{He}$ is made from $4\,{}^{1}_{1}\text{H}$, so the number of hydrogen nuclei consumed per second is:

$4n = 4 \times 8.71 \times 10^{37}\,\text{s}^{-1} = 3.48 \times 10^{38}\,\text{s}^{-1}$

(e) There are various ways to tackle this part, which all give the same answer. Here is one way.

Initial mass of hydrogen in Sun is:

$m_0 = 0.75 \times 1.99 \times 10^{30}\,\text{kg}$

$= 1.49 \times 10^{30}\,\text{kg}$

Mass of hydrogen that will be converted to helium over the Sun's lifetime is:

$m = 0.15 \times 1.49 \times 10^{30}\,\text{kg} = 2.24 \times 10^{29}\,\text{kg}$

Number of hydrogen nuclei n_0 contained in this mass is:

$$n_0 = \frac{m}{m_{\text{H}}} = \frac{2.24 \times 10^{29}\,\text{kg}}{1.673 \times 10^{-27}\,\text{kg}} = 1.34 \times 10^{56}$$

If hydrogen nuclei are consumed at a rate of $3.48 \times 10^{38}\,\text{s}^{-1}$, then the time, t, for which the Sun can continue to emit energy is:

$$t = \frac{1.34 \times 10^{56}}{3.48 \times 10^{38}\,\text{s}^{-1}} = 3.85 \times 10^{17}\,\text{s}$$

$\approx 1.2 \times 10^{10}$ years

Q 27 See Figure 5.104. The curve is like an upside-down binding-energy curve (Figure 5.42). Think of the thought experiment in which nuclei are assembled from their constituent nucleons. In forming the most tightly bound nuclei, the nucleons lose the greatest amount of 'nuclear potential energy' – they lose energy, therefore they also lose mass. You have also seen, in the worked example in the text and in Question 26, that nuclei formed in fission and fusion reactions have less mass than the nuclei that initially took part in the reactions.

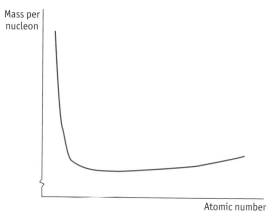

Figure 5.104 Answer to Question 27

Q 28 **(a)** Argon has atomic number 18. To conserve charge, the emitted particle must have a proton number of +1. The particle must therefore be a positron – this is an example of beta-plus decay. Since a positron has a nucleon number of zero, the argon isotope must have a nucleon number 40 – the same as the parent isotope. And just as beta-minus emission is accompanied by the emission of an anti-neutrino, beta-plus is accompanied by a neutrino. The reaction is thus:

$${}^{40}_{19}\text{K} \rightarrow {}^{40}_{18}\text{Ar} + {}^{0}_{+1}\text{e} + \nu_{\text{e}}$$

Alternatively a captured electron could have joined with a proton in the nucleus to make a neutron. (This is in fact what happens.)

(b) One way to do this calculation is to use the same method as the Worked example on page 267, using Equation 9.

$$t = \frac{\ln(N_0/N)}{\lambda}$$

The two isotopes have the same mass number, so the initial mass of potassium is $m_0 = 3.8\,\mu\text{g}$ and the numbers of nuclei of each isotope are in proportion to their masses, i.e.

$$\frac{N_0}{N} = \frac{m_0}{m} = \frac{3.8\,\mu\text{g}}{3.3\,\mu\text{g}} = 1.15$$

From Table 5.5, $\lambda = 5.810 \times 10^{-10}\,\mathrm{yr}^{-1}$

$$t = \frac{\ln(1.15)}{5.810 \times 10^{-10}\,\mathrm{yr}^{-1}} = 2.4 \times 10^{8}\,\mathrm{yr}$$

(Alternatively, using the method of Question 19 and the half-life you calculated in Question 20 gives the same answer.)

(c) Mass of ^{40}K nucleus $\approx 40 \times$ mass of proton (from Table 5.8)

$$= 40 \times 1.673 \times 10^{-27}\,\mathrm{kg}$$

$$= 6.692 \times 10^{-26}\,\mathrm{kg}$$

no. of ^{40}K nuclei, $N = \dfrac{5.0 \times 10^{-10}\,\mathrm{kg}}{6.692 \times 10^{-26}\,\mathrm{kg}}$

$$= 7.5 \times 10^{15}$$

Equation 8, activity $A = \dfrac{\mathrm{d}N}{\mathrm{d}t} = -\lambda N$

so $A = 7.5 \times 10^{15} \times 5.810 \times 10^{-10}\,\mathrm{yr}^{-1}$

$$= 4.36 \times 10^{6}\,\mathrm{yr}^{-1} = \frac{4.36 \times 10^{6}\,\mathrm{yr}^{-1}}{3.16 \times 10^{7}\,\mathrm{s}\,\mathrm{yr}^{-1}}$$

$$\approx 0.14\,\mathrm{s}^{-1} = 0.14\,\mathrm{Bq}$$

Q 29 Rearranging Equation 15:

$G = \dfrac{Fr^2}{m_1 m_2}$, so it has SI units $\mathrm{N\,m^2\,kg^{-2}}$

The units m and kg are SI base units.

$1\,\mathrm{N} = 1\,\mathrm{kg\,m\,s^{-2}}$, so the SI base units of G are $\mathrm{kg\,m\,s^{-2}} \times \mathrm{m^2\,kg^{-2}} = \mathrm{m^3\,kg^{-1}\,s^{-2}}$

Q 30 (a) Using Equation 21 for Mars:

$$g = \frac{GM}{r^2}$$

$$= \frac{6.67 \times 10^{-11}\,\mathrm{N\,m^2\,kg^{-2}} \times 6.42 \times 10^{23}\,\mathrm{kg}}{(3.39 \times 10^{6}\,\mathrm{m})^2}$$

$$= 3.73\,\mathrm{N\,kg^{-1}}$$

(b) You need to multiply your mass (in kg) by the field strength; eg if your mass is 60.0 kg:

$$W = mg = 3.73\,\mathrm{N\,kg^{-1}} \times 60.0\,\mathrm{kg} = 224\,\mathrm{N}$$

Q 31 Use Equation 21, $g = \dfrac{GM}{r^2}$, and substitute for the mass M:

$$M = \rho V = \frac{4\pi r^3 \rho}{3} \text{ so}$$

$$g = \frac{4\pi r^3 \rho G}{3r^2} = \frac{4\pi r \rho G}{3}$$

Q 32 From Equation 15 with

r = comet radius + height above surface

$$= 26\,\mathrm{km} = 26 \times 10^{3}\,\mathrm{m}$$

$$F = \frac{6.67 \times 10^{-11}\,\mathrm{N\,m^2\,kg^{-2}} \times 3.1 \times 10^{12}\,\mathrm{kg} \times 100\,\mathrm{kg}}{(26 \times 10^{3}\,\mathrm{m})^2}$$

$$= 3.1 \times 10^{-5}\,\mathrm{N}$$

Q 33 From Equation 21 with $M = 3.1 \times 10^{12}\,\mathrm{kg}$

(a) With $r = 2\,\mathrm{km} = 2 \times 10^{3}\,\mathrm{m}$,
$g = 5.2 \times 10^{-5}\,\mathrm{N\,kg^{-1}} = 5.3 \times 10^{-6}\,g_{\mathrm{Earth}}$

(b) With $r = 26\,\mathrm{km} = 26 \times 10^{3}\,\mathrm{m}$,
$g = 3.1 \times 10^{-7}\,\mathrm{N\,kg^{-1}} = 3.1 \times 10^{-8}\,g_{\mathrm{Earth}}$

Q 34 The lander had a mass 100 kg and was travelling at $20\,\mathrm{km\,s^{-1}}$. To reduce its momentum to zero would need a large impulse (force × time); to reduce its kinetic energy to zero requires a lot of work to be done (force × distance). In this situation the gravitational force is very small so the lander could bounce to a great height.

Q 35 Using Equation 25 with $r = 22\,\mathrm{km} = 22 \times 10^{3}\,\mathrm{m}$

$$V_{\mathrm{grav}} = -\frac{6.67 \times 10^{-11}\,\mathrm{N\,m^2\,kg^{-2}} \times 3.1 \times 10^{12}\,\mathrm{kg}}{22 \times 10^{3}\,\mathrm{m}}$$

$$= -9.4 \times 10^{-3}\,\mathrm{J\,kg^{-1}}$$

(Notice that the complete answer includes the negative sign.)

Q 36 Using Equation 23a with the same notation as in the Worked example:

$$GmM\left(\frac{1}{r_1} - \frac{1}{r_2}\right) = \frac{1}{2}mv^2$$

$$\frac{GM}{r_2} = \frac{GM}{r_1} - \frac{1}{2}v^2$$

$$\frac{GM}{r_1} = \frac{6.67 \times 10^{-11}\,\mathrm{N\,m^2\,kg^{-2}} \times 3.1 \times 10^{12}\,\mathrm{kg}}{2 \times 10^{3}\,\mathrm{m}}$$

$$= 0.103\,\mathrm{J\,kg^{-1}}\ (= 0.103\,\mathrm{m^2\,s^{-2}})$$

$$\frac{1}{2}v^2 = \frac{1}{2} \times (0.10\,\mathrm{m\,s^{-1}})^2 = 0.005\,\mathrm{m^2\,s^{-2}}$$

$$\frac{GM}{r_2} = (0.103 - 0.005)\,\mathrm{m^2\,s^{-2}} = 0.098\,\mathrm{m^2\,s^{-2}}$$

$$r_2 = \frac{6.67 \times 10^{-11}\,\mathrm{N\,m^2\,kg^{-2}} \times 3.1 \times 10^{12}\,\mathrm{kg}}{0.098\,\mathrm{m^2\,s^{-2}}} = 2.11 \times 10^{3}\,\mathrm{m}$$

Substracting the radius of the comet ($2 \times 10^{3}\,\mathrm{m}$), the bounce height would be about 110 m.

Q 37 They all orbit around the centre of mass of the Earth. A geostationary orbit can only ever be above the equator. The orbit must be in the same plane as the rotation of the Earth. If it were to be at a higher or lower latitude, then the centre of the orbit would not be around the centre of mass of the Earth, which would not make it a possible orbit.

Q 38 For a geostationary satellite $T = 24$ hours $= 86\,400\,s$.

Rearranging Equation 29, with r as the orbital radius:

$$r = \sqrt[3]{\frac{GMT^2}{4\pi^2}}$$

$$= \sqrt[3]{\frac{6.67 \times 10^{-11}\,\mathrm{N\,m^2\,kg^{-2}} \times 6.0 \times 10^{24}\,\mathrm{kg} \times (86\,400\,\mathrm{s})^2}{4\pi^2}}$$

$$= 4.23 \times 10^7\,\mathrm{m}$$

Height above Earth's surface $= 4.23 \times 10^7\,\mathrm{m} - 6.4 \times 10^6\,\mathrm{m}$
$$\approx 3.6 \times 10^7\,\mathrm{m}$$

Q 39 **(a)** $v = \dfrac{2\pi r}{T} = \dfrac{2\pi \times 4.23 \times 10^7\,\mathrm{m}}{86\,400\,\mathrm{s}} = 3.08 \times 10^3\,\mathrm{m\,s^{-1}}$

(b) Equation 21:

$$g = \frac{GM}{r^2}$$

$$= \frac{6.67 \times 10^{-11}\,\mathrm{N\,m^2\,kg^{-2}} \times 6.0 \times 10^{24}\,\mathrm{kg}}{(4.23 \times 10^7\,\mathrm{m})^2}$$

$$= 0.22\,\mathrm{N\,kg^{-1}}$$

Q 40 **(a)** The smaller star must have the brighter surface, since when it passes behind its companion there is a large dip in the light curve, whereas when it blocks off light from the larger star the dip is smaller.

(b) The period is the time for one complete orbit, which could be shown as the time from the middle of one large dip in the light curve to the middle of the next large dip.

Q 41 **(a)** See Figure 5.105(a). The intensity of the radiation halves each time one star passes in front of the other.

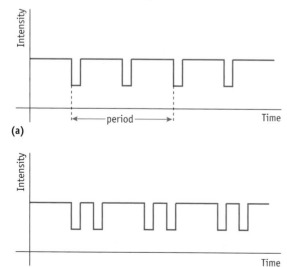

Figure 5.105 Answer to Question 41

(b) See Figure 5.105(b)

Q 42 Using Equation 30 and 31: $\dfrac{v}{c} \approx \dfrac{(f_{em} - f_{rec})}{f_{em}}$

Substituting $f = \dfrac{c}{\lambda}$:

$$\frac{v}{c} = \left(\frac{c}{\lambda_{em}} - \frac{c}{\lambda_{rec}}\right) \times \left(\frac{\lambda_{em}}{c}\right)$$

On right-hand side, cancel c and combine the terms inside the bracket:

$$\frac{v}{c} = \frac{(\lambda_{rec} - \lambda_{em})}{\lambda_{em}\lambda_{rec}} \times \lambda_{em}$$

Then cancel λ_{em} to get:

$$\frac{v}{c} = \frac{(\lambda_{rec} - \lambda_{em})}{\lambda_{rec}}$$

and use $\lambda_{rec} \approx \lambda_{em}$ to get

$$\frac{v}{c} = \frac{(\lambda_{rec} - \lambda_{em})}{\lambda_{em}}$$

Q 43 The received wavelength is shorter than the emitted wavelength, so the gases were approaching. Using Equations 32 and 33:

$$\frac{v}{c} = \frac{\Delta\lambda}{\lambda_{em}} = \frac{(\lambda_{rec} - \lambda_{em})}{\lambda_{em}}$$

$$= \frac{1.070\,\mu\mathrm{m} - 1.083\,\mu\mathrm{m}}{1.083\,\mu\mathrm{m}}$$

$$= -0.012$$

(The negative sign means the gases are approaching.)

$v = 0.012\,c = 3.6 \times 10^3\,\mathrm{km\,s^{-1}}$

Q 44 Using Equation 32,

$$v = \frac{c\Delta\lambda}{\lambda_{em}}$$

$$= \frac{3.00 \times 10^8\,\mathrm{m\,s^{-1}} \times 0.041\,\mathrm{nm}}{656.28\,\mathrm{nm}}$$

$$= 1.87 \times 10^4\,\mathrm{m\,s^{-1}}$$

Assuming circular motion, $v = \dfrac{2\pi r}{T}$ so:

$$r = \frac{vT}{2\pi}$$

$$= \frac{1.87 \times 10^4\,\mathrm{m\,s^{-1}} \times 8.6 \times 3.16 \times 10^7\,\mathrm{s}}{2\pi}$$

$$= 8.1 \times 10^{11}\,\mathrm{m}$$

Then using Equation 28:

$$M = \frac{v^2 r}{G}$$

$$= \frac{(1.87 \times 10^4\,\text{m s}^{-1})^2 \times 8.1 \times 10^{11}\,\text{m}}{6.67 \times 10^{-11}\,\text{N m}^2\,\text{kg}^{-2}}$$

$$= 4.2 \times 10^{30}\,\text{kg}$$

(You would get the same answer if you used Equation 29.)

Q 45 pV has SI units $\text{N m}^{-2} \times \text{m}^3 = \text{N m}$, and $1\,\text{N m} = 1\,\text{J}$.

From Equation 39, $k = \dfrac{pV}{NT}$. N is a number (no units),

T is in kelvin (K) so k has units J K^{-1}.

Q 46 From Equation 39, $p = \dfrac{NkT}{V}$.

For a sphere, $V = \dfrac{4\pi r^3}{3}$, so $p = \dfrac{3NkT}{4\pi r^3}$

No. of ions $= \dfrac{M}{m}$

$$= \frac{2 \times 10^{30}\,\text{kg}}{1.67 \times 10^{-27}\,\text{kg}} \approx 1.2 \times 10^{57}$$

Total number of particles (ions + electrons)
$N = 2 \times$ no. of ions $\approx 2.4 \times 10^{57}$

$$p = \frac{3 \times 2.4 \times 10^{57} \times 1.38 \times 10^{-23}\,\text{J K}^{-1} \times 1 \times 10^7\,\text{K}}{4\pi \times (7 \times 10^8\,\text{m})^3}$$

$$\approx 2 \times 10^{14}\,\text{Pa}$$

Q 47 (a) Mean speed $= 460\,\text{m s}^{-1}$

(b) rms speed $= 471\,\text{m s}^{-1}$

Q 48 From Equation 51, $\langle c^2 \rangle = \dfrac{3p}{\rho}$, so $\sqrt{\langle c^2 \rangle} = \sqrt{\dfrac{3p}{\rho}}$

$$= \sqrt{\frac{3 \times 10^5\,\text{Pa}}{1\,\text{kg m}^{-3}}}$$

$$\approx 500\,\text{m s}^{-1}$$

Q 49 From Equation 52, at any given temperature:

$\sqrt{\langle c^2 \rangle} \propto \sqrt{\dfrac{1}{m}}$, so the less massive molecules will be moving

faster. Since $\dfrac{m_{\text{ox}}}{m_{\text{hyd}}} = 16$, $\sqrt{\dfrac{m_{\text{ox}}}{m_{\text{hyd}}}} = 4$, and so the rms speed
of hydrogen molecules will be four times that of oxygen
molecules at the same temperature.

Q 50 $\sqrt{\langle c^2 \rangle} = \sqrt{\dfrac{3kT}{m}}$

$$= \sqrt{\frac{3 \times 1.38 \times 10^{-23}\,\text{J K}^{-1} \times 1 \times 10^7\,\text{K}}{1.67 \times 10^{-27}\,\text{kg}}}$$

$$= 5 \times 10^5\,\text{m s}^{-1}$$

Q 51 (a) (i) $\sqrt{\langle c^2 \rangle} = \sqrt{\dfrac{3kT}{m}}$

$$= \sqrt{\frac{3 \times 1.38 \times 10^{-23}\,\text{J K}^{-1} \times 30\,\text{K}}{(2 \times 1.67 \times 10^{-27}\,\text{kg})}}$$

$$\approx 600\,\text{m s}^{-1}$$

(ii) Since $c \propto \sqrt{T}$, and T is one-tenth that in
Question 50,

$$c = \frac{5 \times 10^5\,\text{m s}^{-1}}{\sqrt{10}} = 1.6 \times 10^5\,\text{m s}^{-1}$$

(b) (i) Supernova remnant:

$$p = \frac{N}{V} \times kT$$

$$= 10^5\,\text{m}^{-3} \times 1.38 \times 10^{-23}\,\text{J K}^{-1} \times 10^6\,\text{K}$$

$$\approx 1.4 \times 10^{-12}\,\text{Pa} = 14 \times 10^{-13}\,\text{Pa}$$

Similarly for ionised hydrogen region:
$$p = 10^8\,\text{m}^{-3} \times 1.38 \times 10^{-23}\,\text{J K}^{-1} \times 8000\,\text{K}$$

$$\approx 1.1 \times 10^{-11}\,\text{Pa} = 110 \times 10^{-13}\,\text{Pa}$$

(ii) Despite the huge differences in the temperatures
and number densities in the interstellar gas, it
is remarkable that the pressure varies only by
about two orders of magnitude.

Q 52 (a) From Equation 56, $\Delta T = \dfrac{2\Delta E}{3Nk}$:

$$N = \frac{M}{m}$$

$$\Delta T = \frac{2m\Delta E}{3Mk}$$

$$= \frac{2 \times 6.68 \times 10^{-27}\,\text{kg} \times 1000\,\text{J}}{3 \times 1.00\,\text{kg} \times 1.38 \times 10^{-23}\,\text{J K}^{-1}}$$

$$= 0.32\,\text{K}$$

Equivalently, use Equations 57 and 58 to reach the same
answer. (Equation 58 gives $C = 3.10 \times 10^3\,\text{J kg K}^{-1}$.)

(b) As an argon atom has ten times the mass of a helium
atom, there will be only one-tenth the number of
atoms in the 1 kg sample. Each atom will receive
(on average) ten times as much energy, and so the
temperature rise will be ten times that for helium,
i.e. 3.2 K.

Another way to look at this is to notice that, from
Equation 58, the specific heat capacity is inversely
proportional to atomic mass. Argon therefore has a
specific heat capacity one-tenth that of helium, so it
requires only one-tenth the amount of energy for a
given temperature rise.

(c) If we were dealing with a monatomic gas, the reasoning would be the same as in **(b)**; i.e. the specific heat capacity would be 1/4.5 that of helium. But water molecules have rotational and vibrational energy, and so a sample of water vapour needs additional energy in order to increase its temperature. Its specific heat capacity is therefore greater than that of a monatomic gas made from particles of the same mass.

Q 53

(a) Frrom Equation 53: $E_k = \dfrac{2 \times 3kT}{2} = 3kT$

(b) Increase in E_e = decrease in E_k

When $E_k = 0$, then $E_e = 3kT$, so:

$$3kT = \dfrac{e^2}{4\pi\varepsilon_0 r}$$

(c) From **(b)**:

$$T = \dfrac{e^2}{12\pi k\varepsilon_0 r}$$

$$= \dfrac{(1.60 \times 10^{-19}\,\text{C})^2}{(12\pi \times 1.38 \times 10^{-23}\,\text{J K}^{-1}}$$
$$\times 8.85 \times 10^{-12}\,\text{C}^2\,\text{N}^{-1}\,\text{m}^{-2} \times 5 \times 10^{-14}\,\text{m})$$

$$\approx 1 \times 10^8\,\text{K}$$

(d) The temperature calculated in **(c)** is much greater than the 'fusion temperature' of 10^7 K quoted in the text. The calculation in **(c)** used the *average* kinetic energy at temperature T. Figure 5.67 shows that a considerable fraction of particles have above-average kinetic energy, so at 10^7 K there will be *some* protons with enough energy to undergo fusion, even though not all will be able to do so.

Q 54

Your answer will depend on what you have assumed for the size of the cloud. For a rough estimate, volume, $V \approx d^3$ where d is the diameter. If $d \approx 10$ light years $\approx 1 \times 10^{17}$ m, then $V \approx 10^{51}$ m³. Such a cloud would have about 10^{10} molecules per m³ so would contain roughly 10^{61} molecules of H_2, so its mass would be:

$$M_{\text{cloud}} \approx 10^{61} \times 3.34 \times 10^{-27}\,\text{kg} \approx 3 \times 10^{34}\,\text{kg}$$

$$\dfrac{M_{\text{cloud}}}{M_{\text{Sun}}} \approx \dfrac{3 \times 10^{34}\,\text{kg}}{2 \times 10^{30}\,\text{kg}} \approx 10^4\,\text{stars}$$

(Some star clusters are found that do contain many thousand stars, so this is reasonable.)

Q 55

(a) (i) Since gravitational acceleration is equal to the field strength, use Equation 21 $\left(g = \dfrac{GM}{r^2}\right)$ and your values from Question 54.

With our values, $M = M_{\text{cloud}} = 3 \times 10^{34}$ kg, $r = \dfrac{d}{2} = 5 \times 10^{16}$ m, we get:

$$g = \dfrac{6.67 \times 10^{-11}\,\text{N m}^2\,\text{kg}^{-2} \times 3 \times 10^{34}\,\text{kg}}{(5 \times 10^{16}\,\text{m})^2}$$

$$\approx 8 \times 10^{-10}\,\text{m s}^{-2}$$

(ii) $s = ut + \dfrac{1}{2}at^2$, with $u = 0$, $a = g$ and $s = r$,

$$t = \sqrt{\dfrac{2r}{g}} = \sqrt{\dfrac{10^{17}\,\text{m}}{8 \times 10^{-10}\,\text{m s}^{-2}}} = 1 \times 10^{13}\,\text{s}$$

$$\approx 3 \times 10^5\,\text{years}$$

(iii) $v = u + at$, $u = 0$ so

$$v = gt \approx 8 \times 10^{-10}\,\text{m s}^{-2} \times 10^{13}\,\text{s} \approx 8000\,\text{m s}^{-1}$$

(iv) From Equation 52, $T = \dfrac{m\langle c^2\rangle}{3k}$

$$\approx \dfrac{3.34 \times 10^{-27}\,\text{kg} \times (8000\,\text{m s}^{-1})^2}{3 \times 1.38 \times 10^{-23}\,\text{J K}^{-1}}$$

$$\approx 5000\,\text{K}$$

(b) The temperature in **(a) (iv)** is not high enough for fusion. However, it illustrates that gravitational collapse does lead to a substantial rise in temperature. In practice, the gravitational acceleration does *not* remain constant throughout the collapse – as the fragment shrinks to a smaller radius, the field strength at its surface increases, and so particles will 'fall' with an ever-increasing acceleration, leading to a much greater temperature rise than estimated here.

Q 56

Many stars are found in binary systems. Doppler shifts in the stars' spectral lines enable their orbital speeds and periods, and hence the sizes of the orbits, to be determined. The stellar mass can then be calculated using equations of gravitation.

Q 57

Luminosity $\approx 7 \times 10^4\,L_{\text{Sun}}$ (note that the axis has a log scale).

Lifetime $\propto$ amount of fuel $\propto$ mass of star

and

Lifetime $\propto \dfrac{1}{\text{rate at which fuel is used}} \propto \dfrac{\text{luminosity of Sun}}{\text{luminosity of star}}$

so

Lifetime of star,

$$t = \text{lifetime of Sun} \times \dfrac{\text{mass of star}}{\text{mass of Sun}} \times \dfrac{\text{luminosity of Sun}}{\text{luminosity of star}}$$

$$t = \dfrac{18 \times 10^{10}\,\text{years}}{7 \times 10^4} = 2.5 \times 10^6\,\text{years}$$

Q 58 **(a)** From Equation 53, $= \langle E_k \rangle = \dfrac{3kT}{2} = \dfrac{m\langle c^2 \rangle}{2}$

(i) $\langle E_k \rangle = \dfrac{3 \times 1.38 \times 10^{-23}\,\mathrm{J\,K^{-1}} \times 5 \times 10^9\,\mathrm{K}}{2}$

$= 1 \times 10^{-13}\,\mathrm{J}$

(ii) $\sqrt{\langle c^2 \rangle} = \sqrt{\dfrac{3kT}{m}}$

$= \sqrt{\dfrac{3 \times 1.38 \times 10^{-23}\,\mathrm{J\,K^{-1}} \times 5 \times 10^9\,\mathrm{K}}{9.3 \times 10^{-26}\,\mathrm{kg}}}$

$= 1.5 \times 10^6\,\mathrm{m\,s^{-1}}$

(b) **(i)** $\langle E_k \rangle = 1 \times 10^{-13}\,\mathrm{J}$ (i.e. the same as the iron nuclei, because the mean kinetic energy is independent of mass).

(ii) $\sqrt{\langle c^2 \rangle} = \sqrt{\dfrac{3kT}{m}}$

$= \sqrt{\dfrac{3 \times 1.38 \times 10^{-23}\,\mathrm{J\,K^{-1}} \times 5 \times 10^9\,\mathrm{K}}{9.11 \times 10^{-31}\,\mathrm{kg}}}$

$= 4.8 \times 10^8\,\mathrm{m\,s^{-1}}$

This is greater than the speed of light, but no particle can travel this fast. (The material is so hot that electrons do travel close to the speed of light. To describe their behaviour properly we need the equations of special relativity instead of the 'classical' equations used here.)

Q 59 **(a)** $\rho = \dfrac{M}{V}$, $V = \dfrac{4\pi r^3}{3}$, so $\rho = \dfrac{3M}{4\pi r^3}$

(i) Neutron star: $\rho = \dfrac{3 \times 4 \times 10^{30}\,\mathrm{kg}}{4\pi \times (5 \times 10^3\,\mathrm{m})^3}$

$= 7.6 \times 10^{18}\,\mathrm{kg\,m^{-3}}$

(ii) He nucleus: $\rho = \dfrac{3 \times 6.7 \times 10^{-27}\,\mathrm{kg}}{4\pi \times (10^{-15}\,\mathrm{m})^3}$

$= 1.6 \times 10^{18}\,\mathrm{kg\,m^{-3}}$

(iii) The two densities are the same order of magnitude so the statement is justified.

(b) **(i)** From Equation 21,

$g = \dfrac{GM}{r^2}$

$= \dfrac{6.67 \times 10^{-11}\,\mathrm{N\,m^2\,kg^{-2}} \times 4 \times 10^{30}\,\mathrm{kg}}{(5 \times 10^3\,\mathrm{m})^2}$

$= 1.1 \times 10^{13}\,\mathrm{N\,kg^{-1}}$

(ii) $\dfrac{g}{g_{\mathrm{Earth}}} = \dfrac{1.1 \times 10^{13}\,\mathrm{N\,kg^{-1}}}{9.81\,\mathrm{N\,kg^{-1}}}$

$\approx 10^{12}$

Q 60 **(a)** Consider a particle mass m that is part of the star's surface at its equator, which moves in a circle of radius r. The gravitational force must be at least strong enough to provide the centripetal force so, from Equations 15 and 26:

$\dfrac{GMm}{r^2} \geq \dfrac{mv^2}{r}$

We can cancel m and multiply by r, and since

$v = \dfrac{2\pi r}{T}$, we can write:

$\dfrac{GM}{r} \geq \left(\dfrac{2\pi r}{T}\right)^2$

i.e. $\dfrac{GM}{r} \geq \dfrac{4\pi^2 r^2}{T^2}$ and hence:

$T \geq \sqrt{\dfrac{4\pi^2 r^2}{GM}}$

(b) $T \geq \sqrt{\dfrac{4\pi^2 \times (5 \times 10^3\,\mathrm{m})^3}{6.67 \times 10^{-11}\,\mathrm{N\,m^2\,kg^{-2}} \times 4 \times 10^{30}\,\mathrm{kg}}}$

$= 1.4 \times 10^{-4}\,\mathrm{s} \approx 0.1\,\mathrm{ms}$

So a period of a few ms would be reasonable, but a neutron star of this size and mass with a period less than 0.1 ms would spin so fast that it would fly apart.

Q 61 **(a)** **(i)** Using Equation 15, $F_{\mathrm{feet}} = \dfrac{GMm}{r^2}$

$= \dfrac{6.67 \times 10^{-11}\,\mathrm{N\,m^2\,kg^{-2}} \times 10^{26}\,\mathrm{kg} \times 10.0\,\mathrm{kg}}{(5.0 \times 10^4\,\mathrm{m})^2}$

$= 2.6680 \times 10^7\,\mathrm{N}$

(ii) Similarly for the head with
$r = 5.0002 \times 10^4\,\mathrm{m}$
$F_{\mathrm{head}} = 2.6678 \times 10^7\,\mathrm{N}$

(b) The difference between the forces is $0.0002 \times 10^7\,\mathrm{N} = 2 \times 10^3\,\mathrm{N}$. A difference of 2 kN will cause some serious stretching — imagine hanging by your hands with 200 kg suspended from your feet. Now you know why Jo is so tall.

Q 62 The cosmological principle states that all large volumes of space have the same properties. They should then be expanding at the same rate, which implies Hubble's law holds everywhere.

Q 63 Doppler shift is a change in wavelength (and frequency) resulting from movement of the source of waves away from the observer. Cosmological redshift is a similar change, but occurs as a result of space itself expanding.

Q 64 The existence of the cosmic background radiation suggests the Universe was a hotter, denser place than it is now, whereas the Steady State theory predicts the Universe has always been as it is now.

Q 65   As the Universe expands, the peak wavelength of the radiation will become longer and longer and the temperature will move ever closer to absolute zero.

Q 66   (a) Equation 60a:

$$v = cz = 0.0800 \times 3.00 \times 10^8 \, \text{m s} = 2.40 \times 10^7 \, \text{m s}^{-1}.$$

(b) Use Equation 61a, with speed in units of km s^{-1}:

$$v = 2.40 \times 10^7 \, \text{m s}^{-1} = 2.40 \times 10^4 \, \text{km s}^{-1}$$

$$d = \frac{v}{H_0}$$

'best value' of distance

$$d = \frac{2.40 \times 10^4 \, \text{km s}^{-1}}{67.80 \, \text{km s}^{-1} \, \text{Mpc}^{-1}}$$

$$= 354 \, \text{Mpc}$$

EITHER:

Maximum distance

$$d_{max} = \frac{2.40 \times 10^4 \, \text{km s}^{-1}}{(67.80 - 0.77) \, \text{km s}^{-1} \, \text{Mpc}^{-1}}$$

$$= 358 \, \text{Mpc}$$

Minimum distance

$$d_{min} = \frac{2.40 \times 10^4 \, \text{km s}^{-1}}{(67.80 + 0.77) \, \text{km s}^{-1} \, \text{Mpc}^{-1}}$$

$$= 350 \, \text{Mpc}$$

so $d = 354 \, \text{Mpc} \pm 4 \, \text{Mpc}$

OR:

Fractional or percentage uncertainty in

$$H_0 = \frac{0.77}{67.80} = 0.011 = 1.1\%$$

Fractional or percentage uncertainty in $d = 0.011 = 1.1\%$

Uncertainty in $d = 0.011 \times 354 \, \text{Mpc} = 3.89 \, \text{Mpc} \approx 4 \, \text{Mpc}$

Q 67   If H_0 has about 1.5 times the value used in the worked example, then the age, t, will be about two-thirds that calculated, i.e. about 10 billion years

Q 68   (a) From Equation 61a:

$$v = H_0 d = 70 \, \text{km s}^{-1} \, \text{Mpc}^{-1} \times 350 \, \text{Mpc}$$

$$= 2.5 \times 10^4 \, \text{km s}^{-1}$$

(b) As $v \ll c$ we can use Equation 60a:

$$z = \frac{v}{c} = \frac{2.5 \times 10^7 \, \text{m s}^{-1}}{3.00 \times 10^8 \, \text{m s}^{-1}} = 8.3 \times 10^{-2}$$

Q 69   (a) Using Equation 3, a Cepheid of this period has:

$$L = 4\pi d^2_{LMC} I_{LMC} = 4\pi d^2_{M100} I_{M100}$$

Therefore:

$$d^2_{M100} I_{M100} = d^2_{LMC} I_{LMC}$$

$$d^2_{M100} = d^2_{LMC} \frac{I_{LMC}}{I_{M100}}$$

$$d_{M100} = d_{LMC} \sqrt{\frac{I_{LMC}^2}{I_{M100}}}$$

$$= 65 \, \text{kpc} \times \sqrt{\frac{9.5 \times 10^{-15} \, \text{W m}^{-2}}{2.7 \times 10^{-19} \, \text{W m}^{-2}}}$$

$$= 1.2 \times 10^7 \, \text{pc} = 12 \, \text{Mpc}$$

As an intermediate step you could use Equation 3 to find the star's luminosity (4.5×10^{29} W) but this cancels out in the final calculation.

(b) From the text, the supernova has peak luminosity $L \approx 5 \times 10^7 \times 2.9 \times 10^{26}$ W.

The smallest intensity that can be measured is close to $I = 2.7 \times 10^{-19} \, \text{W m}^{-2}$.

Using Equation 3, the maximum distance at which the supernova can be measured is given approximately by:

$$d = \sqrt{\frac{L}{4\pi I}} = \sqrt{\frac{5 \times 10^7 \times 2.9 \times 10^{26} \, \text{W}}{4\pi \times 2.7 \times 10^{-19} \, \text{W m}^{-2}}}$$

$$\approx 6.5 \times 10^{25} \, \text{m} \approx 2 \times 10^9 \, \text{pc} \, (= 2 \, \text{Gpc})$$

$$\approx 7 \times 10^9 \, \text{ly}$$

(c) Supernovae of this type would need to be observed in galaxies whose distances were already known e.g. from observations of Cepheids. Measurements of the intensity of their radiation would then allow their luminosity to be found using Equation 3.

(d) Light observed from such a distant object has been en route for 7 billion years. It must have been emitted when the Universe was only about half its present age.

Q 70 **(a)** Light travels at finite speed. So light reaching us from a large distance has taken time to get here. We therefore see galaxies as they were when the light was emitted, not as they are right now (see the answer to Question 69d).

(b) In the first 300 000 years after the Big Bang, light could not travel freely through the Universe. Photons were continuously colliding with other particles.

Q 71 $^3_1\text{H} + ^2_1\text{H} \rightarrow ^4_2\text{He} + ^1_0\text{n}$

The neutron must be produced in order to conserve charge (proton number) and nucleon number (baryon number).

Q 72 Using Equation 53, $\langle E_k \rangle = \dfrac{3kT}{2} = \dfrac{m\langle c^2 \rangle}{2}$

$\langle E_k \rangle = \dfrac{3 \times 1.38 \times 10^{-23}\,\text{J K}^{-1} \times 1.0 \times 10^8\,\text{K}}{2}$

$\quad = 2.1 \times 10^{-15}\,\text{J}$

$\sqrt{\langle c \rangle^2} = \sqrt{\dfrac{3kT}{m}} = \sqrt{\dfrac{2\langle E_k \rangle}{m}}$

$\quad = \sqrt{\dfrac{2 \times 2.1 \times 10^{-15}\,\text{J}}{6.68 \times 10^{-27}\,\text{kg}}}$

$\quad = 7.9 \times 10^5\,\text{m s}^{-1}$

Q 73 If there are $12n$ hydrogen atoms (^1_1H) to n helium atoms (^4_2He), then:

mass of H : mass of He = 12 : 4 = 75% : 25%

Q 74 He nuclei carry more charge than H nuclei, so their mutual electrostatic repulsion is greater and hence they need to be moving with more kinetic energy in order to get close enough to fuse. Kinetic energy depends on temperature, so producing carbon requires a higher temperature than producing helium. In the expanding Universe, unlike inside stars, temperature is falling so the next stage of fusion cannot take place.

Q 75 From Equations 15 and 19:

$\dfrac{F_e}{F_g} = \dfrac{e^2}{4\pi\varepsilon_0 G m^2} = 4.16 \times 10^{42}$

(Note that you do not need to know the separation r, since the magnitudes of both forces are proportional to $1/r^2$.)

Q 76 **(a)** See Figure 5.106(a). The tangents to the curve show that the slope, which represents the rate of expansion, is decreasing.

(b) See Figure 5.106(b). Here the tangents to the curve show that the slope is increasing, i.e. the rate of expansion is increasing.

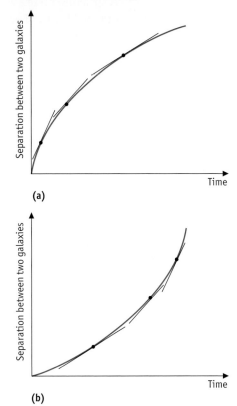

Figure 5.106 Answers to Question 76

(c) If the expansion is slowing down, the actual age of the Universe will be less than estimated in Section 5.2. As the Universe must have been expanding more rapidly in the past, it will have reached its present size in a shorter time than if the expansion had always been at the rate which is deduced from the present value of the Hubble constant. (In the very distant past, H_0 would have been larger, as the expansion was more rapid.)

PRACTICAL SKILLS

Developing practical skills in SHAP

Throughout the SHAP course there are opportunities to develop your practical skills. A structured framework for developing practical skills through an investigative approach is outlined below. The SHAP Teacher and Technician Resource Pack provides a practical skills self-evaluation sheet which you can use to record and reflect on the skills you develop within practical activities. You will not necessarily address all the skills in every practical activity; rather, an activity might focus on one or more specific aspects of practical work. You will need draw on these practical skills when completing the written assessment at the end of your AS or A level course. Evidence of the development of practical skills will also enable you to obtain a practical endorsement alongside your A level grade.

1 Scientific questions and information research

1.1 State the scientific problem to be investigated

Before you start any practical task, you should have a clear idea of what you are setting out to do. Any report of your work should begin by saying what this is. It is helpful to express the task as an **hypothesis** (an 'educated guess') to be tested, for example as a question or as a design brief:

- Is the refractive index of glass independent of the angle of incidence?

- How does the frequency of a vibrating string vary with tension?

- Design and carry out an experiment to find the Young modulus of copper.

- Calibrate a thermistor and potential divider circuit for use as a thermostat.

Sometimes you might start with only a general idea of a topic for investigation (for instance, you might be interested in absorption of radiation). You will need to research some information and discuss your ideas with your teacher before stating your research question or design brief (using this example, you might investigate how the intensity of gamma radiation varies with thickness of absorber).

1.2 Research relevant information

When preparing to carry out your task, you might need to research some or all of the following aspects:

- applications and implications: try to relate your task to a 'real life' problem or situation and find some information about it;

- background science, including relevant equations;

- methods that other people have used to tackle a similar task (What problems did they encounter? How were they overcome? What were the advantages and disadvantages of their approach?);

- outcomes of other people's investigations into similar questions.

For your research, you should consult *at least three* information sources (online or offline).

Wikipedia might seem an obvious starting point, but it is not always reliable so do not treat it as the only information source; consult some of the other references given in the articles.

If using an Internet search engine, do not just consult the first 'hit'. Scroll down and have a look at some of the others.

Write a summary of what you find out *in your own words*; do not just copy and paste.

1.3 Give details of your sources

A written report should include details of your sources. Wherever you refer to a printed source, give the surname(s) of the author(s) in brackets followed by the year of publication e.g. (Smith, 2013). When you refer to a website, give the name of the person or organisation, the url and the date you accessed it, e.g. (Institute of Physics, www.iop.org,17/06/14)

List details of all your sources in alphabetical order in a **bibliography** at the end of the report. To help other people locate the same sources, there are conventions for presenting this information:

Books

● Surname(s), initials (Year of publication in brackets), *Title of book in italic*, Publisher, Place of publication, Chapter number, Page number(s).

Journals and magazines

● Surname(s), initials (Date of publication in brackets), *Title of journal/magazine in italic*, **Volume number (and issue number if relevant) in bold**, Page number(s).

Websites

● Organisation/author/editor, (Year) *Title in italic*, (Date of last update if known), URL, Date when you visited the site.

Comment on your sources

You should comment critically on the reliability of each of your information sources. Questions to ask yourself include:

● Who produced this source? What authority do they have?

● Why was it produced? Is there a vested interest or likely bias?

● How recent is the information?

In a word-processed research summary, you could insert a footnote on the first occasion that you mention each source and use it to record your comments on the source.

2 Planning and experimental design

2.1 Identify the variables to be measured and controlled

At an early stage in your planning and design, decide what you are going to measure in order to answer your research question or fulfil your design brief.

Identify the relevant independent and dependent variables. Think about the range of values that you will use for the independent variable(s) and the range that you might expect to find for the dependent variable(s).

Consider variables that need to be controlled or otherwise accounted for. For example, if you are investigating how the length of a vibrating string affects its frequency, how can you ensure that you change *only* the length and that other factors that might be important (such as tension) are kept constant? If you think that external factors you cannot control (such as room temperature) might affect your results, how will you account for them?

2.2 Describe the experimental apparatus and methodology

The report of your work should include a clear, labelled diagram of the apparatus that you intend to use. If you have been given a basic arrangement to start with, explain how you will adapt and improve it. Add notes and labels to your diagram to show what you will be measuring.

Sometimes you need to carry out a trial experiment to get a rough idea of what will be involved and what measurements to make. Include a brief report of any such experiment and say how it has helped you to plan your main investigation.

Explain why you will be using specific items of apparatus in order to carry out the task. For example, if you have chosen to use a micrometer, say why you are using that rather than callipers or a ruler.

Describe and explain the method that you will be using. Include a summary of standard procedures (e.g. for handling radioactive sources) and any techniques that you have used in order to address a particular problem (e.g. avoiding parallax when reading a scale). It can be helpful to write yourself a set of instructions that lists what needs to be done. This is particularly useful if there will be several steps that must be carried out in a certain order.

In some tasks, you might be given instructions to follow for all or part of the practical work. If so, make sure you know and understand what will be involved before you start setting up apparatus or making measurements.

2.3 Identify sources of systematic and random error

As part of your planning, you need to think about sources of error, and how errors could be minimised through appropriate apparatus and methodology.

Systematic errors affect each reading in the same way (e.g. zero errors on a micrometer).

MATHS REFERENCE

Experimental measurements

See Maths note 7.1

Random errors arise because there is always some uncertainty in reading an instrument (due to the size of scale divisions) and sometimes repeated attempts to make the same measurement give different results.

For each measurement, consider what might contribute to its uncertainty and estimate a ± range within which you are confident the measurement will lie. Table 1 shows a way to record this.

Measurement	Source of error/ uncertainty	Systematic or random?	Estimate of size of error	Comment

Table 1 Table for recording sources of error

A particular measurement might have both a systematic and random error, so could appear twice. Consider how you might modify your experimental design to reduce sources of error, paying most attention to factors that are likely to have the greatest effect.

2.4 Identify safety issues, discuss how to reduce risk and complete a risk assessment

When carrying out any practical work your own safety is important, and so is the safety of other people. You should also consider how to avoid damage to apparatus.

Think about the design of the apparatus and how everyone can be safe in its vicinity. List any safety issues in your experiment and note how you will deal with each one. Table 2 shows a way to record this.

Safety issue	How it will be minimised

Table 2 Table for recording safety issues

Consider how you might modify your experimental design to address safety issues.

3 Carrying out practical work safely and ethically

3.1 Carry out the experiment following correct procedures and with appropriate safety precautions

Having planned your work carefully and considered safety issues, you should be in a good position to carry out your practical work in a safe, ethical and well-organised manner, following your own plan and any instructions that you might have been given. 'Ethical' in this context means that you should behave considerately towards people and other living things and the environment. This is important if, for example, you are working outdoors, or if you involve other people in your investigation (e.g. if you ask people to carry out a sporting activity so that you can make measurements).

If unexpected ethical or safety issues arise, deal with them sensibly, taking advice where needed, and make a note of them to include in the final report.

3.2 Make measurements and record data in an appropriate format using suitable precision

Record all measurements, including repeated ones, as soon as they are taken and with appropriate precision (i.e. a suitable number of significant figures) and units.

Make a blank table for recording your measurements, and include columns for quantities that you will calculate. For example, if you are investigating how a sample extends under a load, you could start with the headings in Table 3.

Added mass m/kg	Load $F = mg$/N	Length x/m	Extension $\Delta x = (x - x_0)$/m

Table 3 Partial results table for investigating extension produced by a load

Set up your results table in such a way that you can use it to record *all* your measurements as soon as you make them. Include repeated measurements, and use the units of the measuring instrument even though you might later need to convert to some other unit. Table 4 shows how you might do this if measuring the diameter of a wire in mm in order to work out its cross-sectional area in m^2.

Diameter measurements d/mm	Average diameter d/mm	Average diameter d/m	Cross sectional area $A = \pi(d/2)^2/m^2$

Table 4 Table for recording diameter measurements and calculations

As you make your measurements, note possible systematic and random errors, especially where they differ from original estimates.

4 Analysis and interpretation of data

4.1 Process, analyse and display data using appropriate mathematical, ICT and statistical techniques

The way you analyse your data will depend on the quantity and nature of data that you have collected, and on what you are trying to find out.

- State the errors/uncertainties in measured and calculated values wherever possible.
- Record a suitable number of significant figures in measured and calculated values.
- Use standard scientific notation to represent very large and very small numbers.
- Look for opportunities to develop and demonstrate your mathematical and ICT skills.

If you have a lot of data and need to carry out several repeat calculations, use a spreadsheet to record and manipulate your data. Use settings that give correct numbers of significant figures.

Use statistical tests to explore correlations between variables. However, in most physics tasks, just finding a correlation is not enough. If two variables are correlated, look for a mathematical relationship between them that can be expressed as an equation.

Think how best to present your data and results. In many physics tasks, this will involve drawing one or more graphs or charts.

4.2 Plot a graph and use it to derive further information

In some cases, a pie chart or bar chart might be suitable, but more often you will need to plot one variable against another and produce a line graph,

MATHS REFERENCE

Units
See Maths note 2.1
Significant figures
See Maths notes 7.1, 7.2, 7.3, 7.4

MATHS REFERENCE

Significant figures
See Maths note 7.2

MATHS REFERENCE

Relationships and graphs
See Maths notes 5.1–5.5
Measurement and uncertainty
See Maths notes 7.1–7.7

When using ICT to plot line graphs, use settings that give sensible scales and labelling, graph gridlines and sharply marked data points.

Use your estimates of experimental uncertainty to add error bars and error boxes to your line graphs, and draw the 'best' trend line that passes through all the boxes. On linear graphs, draw the steepest and least-steep straight lines that go through all the error boxes; these lines can be used to estimate the uncertainty in the results.

A straight-line graph indicates a linear relationship between variables, and further information can often be deduced from the graph's gradient and intercept.

If the relationship is non-linear, the graph is a curve whose shape can indicate the mathematical relationship involved. By manipulating the variables mathematically, you can sometimes produce numbers that give a straight-line graph, which in turn can be used to deduce further information.

5 Conclusion and evaluation

5.1 State a conclusion based on experimental evidence

State a clear conclusion to your work, summarising what you have found out.

- If you set out to test a hypothesis, say whether or not your findings support it.
- If your aim was to find a mathematical relationship between variables, write an equation to express that relationship.
- If you were determining the value of a quantity, state your final result.
- If you were working to a design brief, summarise the outcome of your work.

State the uncertainty in any numerical result. If you have combined several measurements to calculate a value, consider how their uncertainties contribute to the overall uncertainty in the final result. If you have derived your result from a graph, use the graph to estimate the uncertainty in the result. Then write your result, complete with units, in the form:

result = best experimental value ± uncertainty

> **MATHS REFERENCE**
>
> Combining uncertainties
> See Maths note 7.5
> Uncertainties and graphs
> See Maths note 7.7

5.2 Use appropriate scientific knowledge to explain your conclusion and comment on its validity

A conclusion is **valid** if it is based on sound reasoning using data obtained from a well-designed experiment. When discussing the validity of your conclusion, reflect on the extent to which your practical and research work enabled you to meet your original aim.

In the commentary on your conclusion, draw on your scientific knowledge to explain the outcome of your work. Refer to physics that you have studied in this course and to any additional research that you carried out, as well as bringing in relevant scientific knowledge from elsewhere (e.g. GCSE work, other AS/A level courses, general knowledge, etc.).

Comment on whether the outcome of your work was as you expected. If it was not, try to explain why not.

Comment on the **accuracy** of your result. If your investigation involved determining the value of a quantity, comment on how close your result is to the accepted value (if known). Consider whether your experimental apparatus and method might be expected to produce a result that is higher, or lower, than the true value.

Comment on the **precision** of your result. Discuss the sources of experimental uncertainty in any measurements that you made, and consider which made the greatest contribution to the uncertainty in your overall results.

Where possible, comment on how the outcome of your work relates to an application of physics.

Finally, analyse the strengths and weaknesses of your apparatus and method, and make detailed suggestions for one or two improvements.

MATHS NOTES

0 Signs and symbols

0.1 Equations and comparisons

In physics, we are often interested in whether two quantities are exactly equal, or almost equal, or whether one is greater than the other. Table 1 lists the signs used for expressing such relationships.

Sign	Meaning	Notes
$=$	is equal to	
$\equiv$	is exactly the same as	used to emphasise the point that two expressions are two ways of writing exactly the same thing (as opposed to two different things being the same size)
$\neq$	is not equal to	
$\approx$	is approximately equal to	
$\sim$	is the same order of magnitude as	
$<$	is less than	the smaller quantity is written at the narrow end of the symbol
$>$	is greater than	
$\leqslant$	is less than or equal to	
$\geqslant$	is greater than or equal to	
$\ll$	is much less than	
$\gg$	is much greater than	

Table 1 Signs for equations and comparisons

MATHS REFERENCE

Order of magnitude

See Maths note 7.4

0.2 The delta symbol

The symbol Δ (the capital Greek letter delta) is used to mean 'a small amount of' or 'a change in'. Notice that Δ does *not* represent a number, so resist the temptation to cancel Δ, e.g. if it appears on the top and bottom of an expression.

For example, the symbol Δt represents a time interval and is often used when describing rates of flow or rates of change. For example, if an amount of charge ΔQ flows past a point in a time interval Δt, then the current I can be written

$$I = \frac{\Delta Q}{\Delta t}$$

The delta symbol is also used to denote an experimental uncertainty. For example, if a distance x is measured as 23 mm but could be out by 1 mm in either direction, then the uncertainty in the measurement is $\Delta x = 1$ mm. The measurement is written as $x \pm \Delta x$, i.e. 23 mm $\pm$ 1 mm.

0.3 Summation

The symbol Σ (the capital Greek letter sigma) means 'the sum of'. Note that Σ does *not* represent a number so resist the temptation to cancel it, e.g. if it appears on both sides of an equation.

The symbol Σ is usually used with other symbols representing the items to be added together. For example, the mass of a sample of gas is Σm, where m represents the mass of an individual gas molecule and the sum includes all the molecules in the sample.

When dealing with vectors, the sum is a vector sum i.e. it takes account of direction as well as magnitude. ΣF, meaning the sum of forces in a given situation, can be represented by a vector diagram.

1 Index notation

An **index** (plural **indices**) or **power** is the superscript number which, when a positive whole number, means squared, cubed, etc. For example:

$$5^2 = 5 \times 5 = 25$$

$$7^3 = 7 \times 7 \times 7 = 343$$

$$0.6^4 = 0.6 \times 0.6 \times 0.6 \times 0.6 = 0.1296$$

1.1 Index notation and powers of 10

Table 2 shows 'powers of 10'. The number in any row is found by dividing the number in the row above by 10.

$100\,000 =$	$10 \times 10 \times 10 \times 10 \times 10 =$	10^5
$10\,000 =$	$10 \times 10 \times 10 \times 10 =$	10^4
$1\,000 =$	$10 \times 10 \times 10 =$	10^3
$100 =$	$10 \times 10 =$	10^2
$10 =$	$10 =$	10^1
$1 =$	$1 =$	10^0
$0.1 =$	$\dfrac{1}{10} =$	10^{-1}
$0.01 =$	$\dfrac{1}{10 \times 10} = \dfrac{1}{10^2} =$	10^{-2}
$0.001 =$	$\dfrac{1}{10 \times 10 \times 10} = \dfrac{1}{10^2}$	10^{-3}

Table 2 Positive and negative powers of 10

MATHS REFERENCE

Units and physical quantitites

See Maths note 2.1

Extending the pattern gives a meaning to zero and negative indices. If you replace all the 10 s in Table 2 by any other number that you choose, you should be able to convince yourself that

$$x^0 = 1 \qquad \text{for } any \text{ value of } x.$$

1.2 Standard form

To represent very large and very small numbers, we generally use **standard form**, also called **scientific notation**.

A number written in standard form consists of a number with a single digit (not zero) before the decimal point, multiplied by a power of 10.

Large numbers

$5\,620\,000$ (five million six hundred and twenty thousand) becomes 5.62×10^6.

$407\,300$ (four hundred and seven thousand, three hundred) becomes 4.073×10^5.

Small numbers

$0.5680 = 5.680 \times 0.1 = 5.68 \times 10^{-1}$

$0.000\,702\,3 = 7.023 \times 0.0001 = 7.023 \times 10^{-4}$

1.3 Combining powers

Powers of the same number

When multiplying two numbers expressed as 'powers' of the same number, the powers add:

$$10^2 \times 10^3 = (10 \times 10) \times (10 \times 10 \times 10) = 10^5$$

i.e. $\quad 10^2 \times 10^3 = 10^{(2+3)}$

$$6^2 \times 6^2 = (6 \times 6) \times (6 \times 6) = 6^4$$

When dividing, the powers subtract:

$$10^6 \div 10^2 = (10 \times 10 \times 10 \times 10 \times 10 \times 10) \div (10 \times 10) = 10^4$$

i.e. $\quad 10^6 \div 10^2 = 10^{(6-2)}$

The rules still work when negative powers are involved:

$$10^5 \times 10^{-2} = 10^5 \times \left(\frac{1}{10^2}\right) = 10^5 \div 10^2 = 10^3$$

i.e. $\quad 10^5 \times 10^{-2} = 10^{(5-2)}$

$$x^4 \times x^{-3} = x^{(4-3)} = x$$

$$4^3 \div 4^{-2} = 4^3 \div \left(\frac{1}{4^2}\right) = 4^3 \times 4^2 = 4^5$$

i.e. $\quad 4^3 \div 4^{-2} = 4^{(3--2)} = 4^{(3+2)}$

MATHS REFERENCE

Reciprocals

See Maths note 3.3

Powers of different numbers

When dealing with a mixture of numbers of different type, collect together all numbers of the same type and combine their powers by adding or subtracting:

$$2 \times 10^4 \times 3 \times 10^5 = (2 \times 3) \times (10^4 \times 10^5) = 6 \times 10^9$$

$$1.38 \times 10^{-23} \times 2.3 \times 10^3 = 1.38 \times 2.3 \times 10^{(-23+3)} = 3.174 \times 10^{-20}$$

$$3y^2 \times 7y^5 = 21y^7$$

$$5z^2 \times 3z^{-2} = 15z^0 = 15$$

1.4 Manipulating powers on a calculator

Powers of 10

Think of the EXP or EE key as 'times 10 to the power of'.

To enter 7.54×10^9: enter 7.54, press EXP and enter 9. (Notice that you do *not* type in 10 – if you do, you will multiply your number by 10, making it 10 times too big.)

Your calculator might use its own shorthand to display this number as 7.54 09, or 7.54^9, or 7.54 EE 9 (or similar). But you should always *write* it as 7.54×10^9.

Negative powers of 10

To enter a negative index, use the ± or +/– key (*not* the 'minus' key, because that will subtract the next number from the one you have just entered).

To enter 1.38×10^{-23}: enter 1.38, press EXP, enter 23 and press ±.

Squares, etc.

To square a number, use the x^2 key. For example, to work out 1.3^2, enter 1.3 and press x^2 to get 1.69.

Pressing x^2 again squares the answer, i.e. calculates your original number to the power of 4. Pressing x^2 three times altogether gives you your original number to the power of 8, and so on – each time you press x^2, you double the power.

Other powers

Use the y^x key to raise one number to the power of a second number. y is the first number you enter, and x the second.

To calculate 2.5^3: enter 2.5, press y^x, enter 3, press =.

Other negative powers

As with powers of 10, use the ± or +/– key to enter negative numbers.

To calculate 2.5^{-3}: enter 2.5, press y^x, enter 3, press ±, press =.

1.5 Powers that are not whole numbers

The square root of a number x can be written as $x^{\frac{1}{2}}$ or $x^{1/2}$:

$$x^{\frac{1}{2}} \times x^{\frac{1}{2}} = x^{\left(\frac{1}{2}+\frac{1}{2}\right)} = x^1 = x$$

so $x^{\frac{1}{2}} = \sqrt{x}$.

Similarly, $x^{\frac{1}{3}} = \sqrt[3]{x}$ (the cube root of x); $x^{\frac{1}{4}} = \sqrt[4]{x}$ and so on.

Other fractional powers can also be interpreted in terms of roots, for example:

$$x^{\frac{3}{2}} = \sqrt{(x)^3} \text{ (the square root of } x\text{-cubed)}$$
$$= (\sqrt{x})^3 \text{ (the cube of the square root of } x\text{)}$$

and

$$x^{-\frac{1}{2}} = \frac{1}{x^{\frac{1}{2}}} = \frac{1}{\sqrt{x}}$$

Fractional powers can also be written using decimal numbers, for example:

$$x^{\frac{1}{2}} = x^{0.5}$$
$$x^{\frac{3}{2}} = x^{1.5}$$

Powers that are neither simple fractions nor whole numbers are less easy to interpret, but they still exist and can be calculated (e.g. using the y^x key of a calculator. For example:

$$10^{0.333} = 2.153$$
$$10^{0.6021} = 4.000$$
$$5.6^{\pi} = 224.1$$
$$9.34^{-0.83} = 0.1565$$

(All these answers are given to four significant figures.)

MATHS REFERENCE

Significant figures

See Maths note 7.2

2 Units

The SI system of units (Système Internationale d'Unités) has been established by international agreement. In your study of physics you will use mainly SI units. The basic SI units are listed in Table 3. Notice that, when a unit is named after a person, the unit symbol has a capital but the *name* of the unit does not.

Quantity	SI unit	Notes
Mass	kilogram, kg	
Time	second, s	
Length	metre, m	
Electric current	ampere, A	used to define the unit of charge, the coulomb
Temperature	kelvin, K	
Luminous intensity	candela, cd	not used in this course, but included here for completeness
Amount of substance	mole, mol	

Table 3 The basic SI units

2.1 Units and physical quantities; graphs and tables

A physical quantity consists of a number and a unit. Without the unit, the quantity is incomplete. When a symbol represents a physical quantity, it represents the *complete* quantity – units and all. For example, suppose v represents speed, and a particular speed is found to be $5\,\mathrm{m\,s^{-1}}$. You should write

$$v = 5\,\mathrm{m\,s^{-1}}$$

(*not* just $v = 5$ and *not* $v\,(\mathrm{m\,s^{-1}}) = 5$).

Units can be manipulated just like numbers and other symbols. When labelling axes of graphs, and when listing physical quantities in tables, it is conventional to divide each quantity by its unit to get a pure number.

For example, you can divide both sides of the expression for v above by $\mathrm{m\,s^{-1}}$ and write

$$v/(\mathrm{m\,s^{-1}}) = 5$$

If you are plotting values of v on a graph, or listing them in a table, you should label the graph axis, or the table column, as $v/\mathrm{m\,s^{-1}}$.

Large and small numbers

Suppose you were dealing with speeds that were all several million metres per second:

$$v = 2 \times 10^6\,\mathrm{m\,s^{-1}}, \qquad v = 7 \times 10^6\,\mathrm{m\,s^{-1}}, \text{ etc.}$$

To make the numbers more manageable, you could use the same rule as above to write $v/(10^6\,\mathrm{m\,s^{-1}}) = 2$, etc., and label your graph and table as shown in Figure 1.

t/s	$v/10^6\,\mathrm{m\,s^{-1}}$
1	2
2	7

Figure 1 Labelling graphs and tables

MATHS REFERENCE

Index notation and powers of 10

See Maths note 1.1

2.2 Manipulating units; index notation and units

In calculations, the units should be manipulated as well as the numbers. This can help you keep track of what you are doing as well as being correct – so it is a good habit to get into.

Indices can be used with units and with algebraic symbols. For example;

$$4^{-1} = \frac{1}{4} = 0.25, \qquad x^{-2} = \frac{1}{x^2}$$

Units such as coulombs per second, or joules per coulomb, can be written either as C/s and J/C or using index notation: $\mathrm{C\,s^{-1}}$ and $\mathrm{J\,C^{-1}}$. Similarly, metres per second, in calculations of unit of speed, can be written as m/s or $\mathrm{m\,s^{-1}}$. For example;

$$70\,\mathrm{m} \div 20\,\mathrm{s} = 3.5\,\mathrm{m\,s^{-1}}$$

Using the index notation helps prevent table headings and graph labels having too many oblique strokes.

When multiplying numbers, units or symbols, collect together all those of the same type and add their indices. For example:

$$2\,\mathrm{C\,s^{-1}} \times 4\,\mathrm{s} = 8\,\mathrm{C}$$

$$10\,\mathrm{m\,s^{-1}} \div 5\,\mathrm{s} = 2\,\mathrm{m\,s^{-2}}$$

2.3 Derived units

Table 4 shows how SI units are combined to give units of various quantities. Some common combinations are given 'shorthand' names.

Quantity	Unit name	Symbol	Equivalent
Speed			$m\,s^{-1}$
Acceleration			$m\,s^{-2}$
Force	newton	N	$1\,N = 1\,kg\,m\,s^{-2}$
Gravitational field strength			$1\,N\,kg^{-1} = 1\,m\,s^{-2}$
Energy, work	joule	J	$1\,J = 1\,N\,m = 1\,kg\,m^2\,s^{-2}$
Power	watt	W	$1\,W = 1\,J\,s^{-1}\ (= 1\,kg\,m^2\,s^{-3})$
Frequency	hertz	Hz	$1\,Hz = 1\,s^{-1}$
Electric charge	coulomb	C	$1\,C = 1\,A\,s$ $1\,A = 1\,C\,s^{-1}$
Potential difference, emf	volt	V	$1\,V = 1\,J\,C^{-1}\ (= 1\,kg\,m^2\,C^{-1}\,s^{-2})$
Electrical resistance	ohm	Ω	$1\,\Omega = 1\,V\,A^{-1}\ (= 1\,kg\,m^2\,C^{-2}\,s^{-1})$

Table 4 Some common derived SI units

2.4 SI prefixes

When dealing with quantities that are large or small, we often use prefixes as an alternative to standard form. For example, a distance of $1.3 \times 10^4\,m$ could be written as 13 km, and a distance of 0.0037 m could be written as 3.7 mm. The official SI prefixes go up and down in steps of 10^3. Table 5 lists the SI prefixes that you are likely to encounter in your study of physics.

Prefix	Symbol	Equivalent in powers of 10
Tera	T	10^{12}
Giga	G	10^9
Mega	M	10^6
Kilo	k	10^3
Centi	c	10^{-2}
Milli	m	10^{-3}
Micro	µ	10^{-6}
Nano	n	10^{-9}
Pico	p	10^{-12}
Femto	f	10^{-15}

Table 5 SI prefixes

When dealing with conversions involving prefixes, it is wise to write down each step using appropriate powers of 10, *and include the units at each stage*. For example, suppose light of a certain colour has a wavelength of 468 nm and you want to use standard form to write the wavelength in metres:

$$468\,nm = 468 \times 10^{-9}\,m$$
$$= 4.68 \times 10^2 \times 10^{-9}\,m$$
$$= 4.68 \times 10^{-7}\,m$$

Suppose the tension in a rope is $1.35 \times 10^5\,\text{N}$ and you want to express it in kN:

$$1\,\text{kN} = 10^3\,\text{N}, \text{ so } 1\,\text{N} = \frac{1}{10^3}\,\text{kN} = 10^{-3}\,\text{kN}$$

$$1.35 \times 10^5\,\text{N} = 1.35 \times 10^5 \times 10^{-3}\,\text{kN}$$

$$= 1.35 \times 10^2\,\text{kN}$$

$$= 135\,\text{kN}$$

Suppose an electric current is $4.56 \times 10^{-4}\,\text{A}$ and you want to express it in μA:

$$1\,\mu\text{A} = 10^{-6}\,\text{A}, \text{ so } 1\,\text{A} = \frac{1}{10^6}\,\mu\text{A} = 10^6\,\mu\text{A}$$

$$4.56 \times 10^{-4}\,\text{A} = 4.56 \times 10^{-4} \times 10^6\,\mu\text{A}$$

$$= 4.56 \times 10^2\,\mu\text{A}$$

$$= 456\,\mu\text{A}$$

2.5 Dimensions

The **dimensions** of a quantity show how it is related to the basic quantities listed in Table 3. Symbols M, L and T are used to represent the dimensions of mass, length and time.

For example, volume is calculated from length × breadth × height so has dimension of length^3 or L^3; speed is found from distance ÷ time so has dimensions of L/T or LT^{-1}. The dimensions of force are those of mass × acceleration: MLT^{-2}.

Square brackets are used to denote the dimensions of a quantity, e.g.

$$[\text{velocity}] = \text{LT}^{-1}$$

$$[\text{force}] = [\text{mass}] \times [\text{acceleration}] = \text{MLT}^{-2}$$

Dimensions are more fundamental than units. You might, for example, choose to express a speed in miles per hour rather than SI units of m s^{-1}, but the dimensions are still LT^{-1}, i.e. length (miles) ÷ time (hours).

Any equation must be dimensionally consistent, that is, the dimensions of the left-hand side must be the same as those of the right-hand side. This can help you check whether a particular equation is correct, and can also enable you to derive relationships between quantities.

3 Arithmetic and algebra

3.1 Fractions, decimals and percentages

A fraction is really a division sum, e.g.

$$\frac{4}{5} = 4 \div 5; \qquad \frac{7}{3} = 7 \div 3$$

You can express a fraction as a decimal number by doing the division on a calculator.

When fractions are multiplied together, you can often simplify the arithmetic by using the fact that the multiplication and division can be carried out in any order, e.g.

$$\frac{7}{5} \times \frac{3}{14} = \frac{7 \times 3}{5 \times 14}$$

and cancelling any common factors, e.g.

$$\frac{7 \times 3}{5 \times 14} = \frac{3}{5 \times 2} = \frac{3}{10} = 0.3$$

You can think of the **percentage** sign, %, as being made up of a 1, 0, 0 to remind you that it is a fraction of 100 parts. To calculate a percentage from a number expressed as a fraction or a decimal, you multiply by 100:

$$\frac{1}{2} = 0.5 \text{ and } 100 \times 0.5 = 50 \text{ so } \frac{1}{2} = 50\% \text{ (or 50/100)}$$

$$\frac{1}{4} = 0.25 \text{ and } 100 \times 0.25 = 25 \text{ so } \frac{1}{4} = 25\% \text{ (or 25/100)}$$

$$\frac{7}{8} = 0.875 \text{ and } 100 \times 0.875 = 87.5 \text{ so } \frac{7}{8} = 87.5\%$$

For example, if a solar array produces an output power of 600 W from an input power of 4 kW (4000 W), its efficiency is:

$$\frac{600 \text{ W}}{4000 \text{ W}} = 0.15 = 15\%.$$

To find a percentage of a quantity, you *multiply* the quantity by the percentage expressed as an ordinary fraction or decimal number. For example, to find 15% of 60, multiply 60 by 15/100 (or by 0.15):

$$\frac{15}{100} \times 60 = \frac{90}{10} = 9$$

or

$$0.15 \times 60 = 9$$

3.2 Brackets and common factors

To evaluate an expression such as

$$6(2+3-4+5), \quad \frac{12+8}{4} \quad \text{or} \quad I(R_1 + R_2 + R_3)$$

you usually deal first with the additions and subtractions inside the bracket and then multiply or divide the result by the number or symbol outside. Alternatively, you can carry out several separate multiplications or divisions on each number or symbol inside the bracket in turn, then do the additions or subtractions. For example:

either $6(2+3-4+5) = 6 \times 6 = 36$

or $6(2+3-4+5) = 12 + 18 - 24 + 30 = 36$

either $\dfrac{12+8}{4} = \dfrac{20}{4} = 5$

or $\dfrac{12+8}{4} = \dfrac{12}{4} + \dfrac{8}{4} = 3 + 2 = 5$

A calculation that involves several multiplications or divisions using the same number and then adding or subtracting the results can be simplified if it is rewritten using brackets with the **common factor** outside. For example:

$$25 + 30 + 35 = 5(5 + 6 + 7)$$

$$3x + 3y + 3z = 3(x + y + z)$$

$$IR_1 + IR_2 + IR_3 = I(R_1 + R_2 + R_3)$$

$$\frac{7}{2} + \frac{3}{2} + \frac{6}{2} = \frac{(7+3+6)}{2}$$

$$\frac{a}{x} + \frac{b}{x} + \frac{c}{x} = \frac{(a+b+c)}{x}$$

3.3 Reciprocals

The value obtained by dividing 1 by a number is called the **reciprocal** of the number (reciprocals can be found using the $1/x$ key of a calculator). Finding the reciprocal of a reciprocal gets you back to the original number. For example:

$$\frac{1}{2} = 0.5 \qquad \frac{1}{0.5} = 2$$

For a wave or oscillation:

$$\text{period } T = \frac{1}{f} \qquad \text{frequency } f = \frac{1}{T}$$

Reciprocals are sometimes written using a negative index:

$$x^{-1} = \frac{1}{x}$$

To find the reciprocal of a fraction, simply turn it the other way up. For example:

$$\frac{1}{\frac{1}{2}} = \frac{2}{1} = 2$$

$$\frac{1}{\frac{2}{3}} = \frac{3}{2} = 1\frac{1}{2}$$

$$\left(\frac{3}{7}\right)^{-1} = \frac{7}{3}$$

This is not just an arbitrary rule. It makes sense if you think in terms of division sums. Consider the second example above. Question: 'How many times does $\frac{2}{3}$ go into 1?' Answer: 'one-and-a-half times.'

Adding and subtracting

One place where you need to add and subtract reciprocals is in calculations of resistors in parallel. To find the net resistance R of several resistors connected in parallel, you must first find the reciprocal of each resistance, then add the reciprocals together (to get $1/R$), then find the reciprocal of $1/R$ to get R.

For example, if $R_1 = 2.0\,\Omega$, $R_2 = 5.0\,\Omega$, $R_3 = 1.0\,\Omega$, then:

$$\frac{1}{R_1} = \frac{1}{2.0}\Omega^{-1} = 0.50\,\Omega^{-1} \text{ (notice the unit of } 1/R\text{)}$$

$$\frac{1}{R_2} = 0.20\,\Omega^{-1}$$

$$\frac{1}{R_3} = 1.00\,\Omega^{-1}$$

(notice that $1/1 = 1$ – the number stays the same but the unit still changes). So:

$$\frac{1}{R} = (0.50 + 0.20 + 1.00)\,\Omega^{-1} = 1.70\,\Omega^{-1}$$

$$R = \frac{1}{1.70}\Omega = 0.59\,\Omega$$

Notice that adding the reciprocals of two numbers is *not* the same as adding the two numbers and then finding the reciprocal of their sum.

Multiplying and dividing

Multiplying by the reciprocal of a number is the same as dividing by that number. For example:

$$7 \times \frac{1}{2} = 7 \div 2 = 3.50$$

Dividing by the reciprocal of a number is the same as multiplying by that number. For example:

$$4 \div \frac{1}{3} = 4 \times 3 = 12$$

$$9 \div \frac{3}{4} = 9 \times \frac{4}{3} = \frac{9 \times 4}{3} = 12$$

For a wave,

$$f = \frac{v}{\lambda}, \qquad \text{time period } T = \frac{1}{f} = \frac{1}{\frac{v}{\lambda}} = \frac{\lambda}{v}$$

We can simplify divisions involving fractions. For example:

$$\frac{3}{4} \div \frac{5}{4} = \frac{3}{4} \times \frac{4}{5} = \frac{3 \times 4}{4 \times 5} = \frac{3}{5} = 0.6$$

3.4 Algebra and elimination

If we have two different relationships that both involve some of the same things, we can combine them to produce a new equation. This allows us to avoid measuring, or calculating, something that is not already known – we can eliminate it (remove it) from the equations. For example, we can take an expression for electrical power

$$P = IV$$

and use the resistance equation

$$V = IR$$

to write IR instead of V:

$$P = I \times IR = I^2 R$$

This enables us to relate P directly to I and R without needing to know or calculate V. Similarly, if we want to eliminate I:

$$P = \frac{V}{R} \times V = \frac{V^2}{R}$$

3.5 Adding and subtracting fractions

You can of course add and subtract fractions on a calculator – you carry out several division sums and add or subtract the results. But for simple fractions it can often be quicker to do the sums 'by hand'.

The trick is to write the fractions so that they have the same **denominator** (the number underneath the fraction). Sometimes it is quite easy to spot how to do this. For example:

$$\frac{3}{4} + \frac{5}{6} = \frac{3 \times 3}{3 \times 4} + \frac{2 \times 5}{2 \times 6}$$

$$= \frac{9}{12} + \frac{10}{12} = \frac{9 + 10}{12} = \frac{19}{12}$$

Otherwise, make a common denominator by multiplying the original denominators together:

$$\frac{1}{17} + \frac{4}{3} = \frac{2 \times 3}{17 \times 3} + \frac{4 \times 17}{3 \times 17}$$

$$= \frac{6}{51} + \frac{68}{51} = \frac{6 + 68}{51} = \frac{74}{51}$$

Another example:

$$\frac{1}{2} + \frac{1}{3} = \frac{3}{6} + \frac{2}{6} = \frac{5}{6}$$

4 Solving equations

It may sound obvious, but the main thing to understand about equations is that the '=' sign means that the two things on either side are *equal* to one another. So whatever you do to one side, you must also do to the other, otherwise they would no longer be equal. (Beware of getting into the bad habit of writing '=' when you really mean 'and so the next step is …'.)

One way to think of an equation is as a 'recipe' for calculating. For example, $F = ma$ tells you how to calculate the net force F if you know the acceleration a that it gives to a mass m. In this example, F is the **subject** of the equation – it is written on its own (usually on the left).

4.1 Rearranging an equation

Quite often, the quantity you want to calculate is wrapped up in the right-hand side of an equation, and you need to make it the subject. When doing this, it helps if you try to understand what you are doing rather than blindly trying to apply a set of rules. It is also wise to write down each step, justifying each one to yourself as you do so. This might sound time-consuming, but it isn't really because it helps you to keep track of what you are doing and, if you do make a slip, it is quite easy to go back and check.

Look at the part of the equation that contains the quantity that you want to know. Think what you need to do to get that quantity on its own, and do the same thing(s) to both sides.

For example, suppose you want to know the acceleration that a force F gives to a mass m:

$$F = ma$$

To get a on its own, you need to divide the right-hand side by m ($ma \div m = a$), so do the same to the left-hand side:

$$\frac{F}{m} = a \qquad \text{or} \qquad a = \frac{F}{m}$$

Another example: suppose you want to calculate internal resistance r from

$$V = \mathscr{E} - Ir$$

It is a good idea first to arrange that the thing you are interested in has a positive sign. You can do this by adding Ir to both sides:

$$V + Ir = \mathscr{E}$$

then to get r on its own you subtract V from both sides:

$$Ir = \mathscr{E} - V$$

and then divide by I

$$r = \frac{\mathscr{E} - V}{I} \qquad \text{or} \qquad r = (\mathscr{E} - V)/I$$

(Notice that you have to divide the *whole* of the right-hand side by I – hence the brackets.)

4.2 Simultaneous equations

Simultaneous equations arise if we have two (or more) different ways of writing a relationship between quantities. If we have two unknown quantities, then they can both be found if we have two simultaneous equations. For three unknown quantities, we'd need three separate equations, and so on.

The trick in solving simultaneous equations is to carry out some algebra and arithmetic to get an expression that involves just *one* of the unknown things, and then use that value to calculate the other one.

MATHS REFERENCE

Brackets and common factors

See Maths note 3.2

For example, the equation $\mathscr{E} = V + Ir$ involves two things that can be measured (V and I). If neither $\mathscr{E}$ nor r is known, then they cannot be found from a single pair of values of V and I. However, if you obtain two *different* pairs of readings (V_1 and I_1, and V_2 and I_2) for the same power supply (using two different external loads), then you can write down two simultaneous equations – two different equations that both describe a relationship between the two unknown things $\mathscr{E}$ and r. These equations let you find both $\mathscr{E}$ and r. So:

$$\mathscr{E} = V_1 + I_1 r$$

$$\mathscr{E} = V_2 + I_2 r$$

Since the right-hand side of each equation is equal to $\mathscr{E}$, then they must also be equal to each other:

$$V_1 + I_1 r = V_2 + I_2 r$$

Subtracting V_1 from each side:

$$I_1 r = V_2 - V_1 + I_2 r$$

Subtracting $I_2 r$ from both sides (and being careful with signs and with the subscripts 1 and 2):

$$I_1 r - I_2 r = V_2 - V_1$$

Now r is a common factor on the left-hand side, so:

$$r(I_1 - I_2) = V_2 - V_1$$

Dividing both sides by $(I_1 - I_2)$ (and using brackets to keep the subtracted things together):

$$r = \frac{(V_2 - V_1)}{(I_2 - I_1)} \qquad \text{or} \qquad r = (V_2 - V_1)/(I_1 - I_2)$$

This value of r can then be used in one of the original equations to find $\mathscr{E}$.

For example: a power supply gives readings of $V_1 = 3\,\text{V}$, $I_1 = 7\,\text{A}$, and $V_2 = 8\,\text{V}$, $I_2 = 2\,\text{A}$. So

$$r = \frac{8\,\text{V} - 3\,\text{V}}{7\,\text{A} - 2\,\text{A}} = \frac{5\,\text{V}}{5\,\text{A}} = 1\,\Omega$$

and

$$\mathscr{E} = V_1 + I_1 r = 3\,\text{V} + 7\,\text{A} \times 1\,\Omega = 3\,\text{V} + 7\,\text{V} = 10\,\text{V}$$

(You would find the same value using V_2 and I_2.)

5 Relationships and graphs

Graphs are extremely useful in physics for giving us a pictorial representation of how one quantity is related to another. Trends in data are not always clear from a table of results, but become immediately evident when viewing a plot of the two quantities involved.

5.1 Graphs and proportionality

Many important relationships in physics involve the idea of direct proportion.

For example, if a conductor obeys Ohm's law, doubling the potential difference produces double the current, tripling the pd triples the current ... and so on. Mathematically, we say that the potential difference is **directly proportional** to the current. In symbols:

$$V \propto I \qquad \text{or} \qquad V = kI$$

The symbol $\propto$ means 'is directly proportional to' and k is called a **constant of proportionality** and has a fixed value for a particular set of values of V and I. (In this example, the constant k is the electrical resistance R.)

If one quantity is directly proportional to another, then a graph of one plotted against the other is a straight line through the origin.

5.2 Linear relationships

The equation

$$V = \mathcal{E} - Ir$$

is an example of a **linear relationship** between two variables, V and I in this case. A graph of V (on the vertical axis, the y-axis) against I (on the horizontal axis, the x-axis) gives a straight line. Linear relationships and graphs are often said to be of the type

$$y = mx + c$$

where y stands for whatever is plotted on the y-axis and x for whatever is plotted on the x-axis, and m and c are constants (they remain fixed when x and y change). This type of graph has two properties that are often useful for doing calculations using experimental results. We can illustrate these with a graph of $y = 2x + 1$, i.e. $m = 2$, $c = 1$ (Figure 2).

On Figure 2, the line cuts the y-axis at $y = 1$ (using the equation, when $x = 0$, $y = c$). The line of such a graph always cuts the y-axis where $y = c$.

If y is directly proportional to x, then the line goes through the origin and $c = 0$.

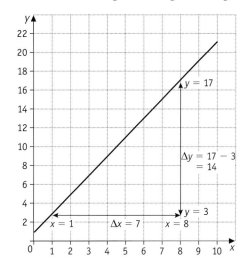

Figure 2 A graph of $y = 2x + 1$

5.3 Gradient of a linear graph

Figure 2 is a graph of the linear relationship $y = 2x + 1$.

The **gradient** (or slope) of the graph is defined as the rise of the graph (the increase in y, Δy) divided by the run (the corresponding increase in x, Δx) found by drawing a right-angled triangle as shown in Figure 2. On Figure 2:

$$\Delta y = 14, \qquad \Delta x = 7,$$

$$\text{gradient} = \frac{\Delta y}{\Delta x} = 2$$

Notice that Δy and Δx are numbers read from the graph scales (*not* lengths measured with a ruler) and that any similar triangle drawn on the graph will give the same value of the gradient.

The gradient of a linear graph of y against x is always equal to the value m in the relationship $y = mx + c$.

The graph in Figure 2 has a positive gradient. If m is negative, then the graph slopes down from left to right.

If two variables measured in an experiment are related by a linear equation, then plotting them on a graph enables you to find the values of the constants relating them. It is helpful if you arrange

MATHS REFERENCE

Error bars and error boxes

See Maths note 7.6

MATHS REFERENCE

Reciprocals

See Maths note 3.3

the relationship so that it looks as much like $y = mx + c$ as possible. For example, by subtracting Ir from both sides you can write $\mathscr{E} = V + Ir$ as

$$V = (-r)I + \mathscr{E}$$

which can be compared directly with

$$y = mx + c$$

If you plot measured values of V on the y-axis against corresponding values of I on the x-axis, the graph will be a straight line that cuts the y-axis at $\mathscr{E}$, and with a gradient $m = -r$.

5.4 Inverse proportionality

If one quantity is **inversely proportional** to another, then as one increases, the other will decrease. For example, the acceleration produced by a given net force is inversely proportional to the mass on which it acts: doubling the mass halves the acceleration, tripling the mass divides the acceleration by three and so on – and vice versa.

Such a relationship is written using reciprocals and the symbol for direct proportion:

$$a \propto \frac{1}{m} \qquad a = \frac{k}{m}$$

or

$$m \propto \frac{1}{a} \qquad m = \frac{k}{a}$$

(In this case, the constant of proportionality is the same as the resultant force F.)

If one quantity is inversely proportional to the other (Table 6), the graph of one plotted against the other is curved as in Figure 3.

But if one quantity is plotted against the *reciprocal* of the other, then the graph is a straight line through the origin, as shown in Figure 4.

m/kg	(1/*m*)/kg^{-1}	*a*/ms^{-2}
1	1.000	5.00
2	0.500	2.50
3	0.333	1.67
4	0.250	1.25
5	0.200	1.00
6	0.167	0.83
7	0.143	0.71
8	0.125	0.63
9	0.111	0.55
10	0.100	0.50

Table 6 Data for Figures 3 and 4

Figure 3 A graph showing how the acceleration a produced by a constant force F (= 5N) depends on mass m (data from Table 6)

Figure 4 The data from Figure 3 plotted as a against $1/m$

5.5 Testing mathematical relationships

Sometimes we are interested in finding a mathematical relationship between two measured quantities. This usually involves some educated guesswork, based on ideas about the underlying physics and/or from looking at the numbers. Plotting graphs provides a way of testing the guesses.

Direct proportion

For example, if both quantities increase together, you might guess that one is directly proportional to the other. Plot a graph of one against the other and see whether you can draw a straight line through all the error boxes.

Examples that give straight-line graphs include:

$s \propto t$ for motion at constant speed

$I \propto V$ for an ohmic conductor.

If the plot does not give a straight line, try something else. For example, motion from rest at constant acceleration is described by the equation

$$s = \frac{1}{2}at^2$$

$$s \propto t^2$$

A graph of distance s against time t is a curve, but a graph of s against t^2 is a straight line with gradient $\frac{a}{2}$ or $\frac{1}{3}a$.

Sometimes you need to use the square root of a quantity to get a straight line. For example, for a simple pendulum a plot of its period T against the square root of its length l gives a straight line:

$$T \propto \sqrt{\ell}$$

Inverse proportion

If one quantity increases as the other decreases, you might guess that you are looking at inverse proportionality, so try plotting a graph using the reciprocal of one quantity.

If this does not give a straight line, try plotting the square, or the square root, of the reciprocal.

For example, suppose you measure the frequency f of the note from a plucked string of mass per unit length μ. Frequency f decreases as you increase μ, but suppose you find that a graph of f against $\frac{1}{\mu}$ is not a straight line.

If a graph of f against $\frac{1}{\mu^2}$ is a straight line, then $f \propto \frac{1}{\mu^2}$.

If you need to plot f against $\frac{1}{\sqrt{\mu}}$ to get a straight line, then $f \propto \frac{1}{\sqrt{\mu}}$.

6 Trigonometry and angular measurements

6.1 Degrees and radians

A **radian**, or **rad** for short, is a unit for measuring angles commonly used in physics instead of degrees. Figure 5 shows how the size of an angle, in radians, is defined.

For a full circle, length of arc = length of circumference = $2\pi r$.

Size of angle = $\frac{2\pi r}{r}$ = 2π radians, i.e. approximately 6.28 rad.

Table 7 lists some useful conversions between radians and degrees.

MATHS REFERENCE

Experimental uncertainty
See Maths note 7.1

Error bars and error boxes
See Maths note 7.6

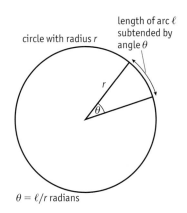

$\theta = \ell/r$ radians

Figure 5 The size of an angle measured in radians

Angle	Size in degrees	Size in radians
Full circle	360°	2π rad = 6.28 rad
Half circle	180°	π rad = 3.14 rad
	114.6°	2.0 rad
Quarter circle	90°	$\pi/2$ rad = 1.57 rad
	60°	$\pi/3$ rad = 1.05 rad
	57.3°	1.0 rad
	45°	$\pi/4$ rad = 0.79 rad
	30°	$\pi/6$ rad = 0.52 rad
	28.6°	0.5 rad

Table 7 Some conversions between radians and degrees

Note that π is a *number* (approximately 3.14) that frequently, but not always, appears in angles measured in radians.

6.2 Sine, cosine and tangent of an angle

Figure 6 shows a right angled triangle. The sides of the triangle are related by Pythagoras's theorem:

$$c^2 = b^2 + a^2$$
$$c = \sqrt{a^2 + b^2}$$

(Careful! You cannot 'cancel' the squares inside the bracket.)

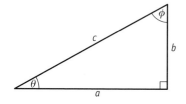

Figure 6 A right-angled triangle

All similar triangles, i.e. those with the same angle θ, will have sides in the same proportion to one another. The ratios of the sides depend only on the angle θ.

The sine, cosine and tangent of the angle θ are known as **trigonometric ratios**.

- Sine of angle θ, $\sin\theta = \dfrac{\text{opposite side}}{\text{hypotenuse}} = \dfrac{b}{c}$

- Cosine of θ, $\cos\theta = \dfrac{\text{adjacent side}}{\text{opposite side}} = \dfrac{a}{c}$

- Tangent of θ, $\tan\theta = \dfrac{\text{opposite side}}{\text{adjacent side}} = \dfrac{b}{a}$

We can combine these to give another useful relationship. Since

$$\frac{b}{a} = \frac{b}{c} \div \frac{a}{c} \quad (c \text{ cancels}),$$

we can write

$$\tan\theta = \frac{\sin\theta}{\cos\theta}$$

Also

$$\sin\phi = \frac{a}{c} = \cos\theta \quad \text{and} \quad \cos\phi = \frac{b}{c} = \sin\theta$$

i.e. if two angles add up to 90°, then the cosine of one is equal to the sine of the other.

Using Pythagoras's theorem leads to another useful result. Dividing $c^2 = a^2 + b^2$ by c^2:

$$1 = \frac{a^2}{c^2} + \frac{b^2}{c^2} = \left(\frac{a}{c}\right)^2 + \left(\frac{b}{c}\right)^2$$
$$1 = (\cos\theta)^2 + (\sin\theta)^2,$$

which is true for any angle and is usually written as:

$$\cos^2\theta + \sin^2\theta = 1$$

6.3 Graphs of trigonometric functions

For angles greater than 90°, Figure 7 shows how sin, cos and tan are defined. For some angles, negative numbers are involved. Figure 8 shows how the sin, cos and tan vary with angle θ. Note that we have labelled the axis in degrees and in radians.

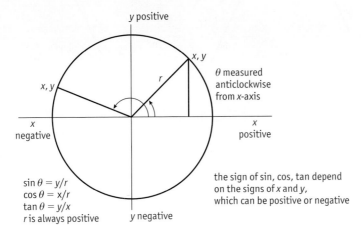

Figure 7 Defining sin, cos and tan for angles greater than 90°

Notice that $\sin \theta$ and $\cos \theta$ are always between +1 and −1, but $\tan \theta$ is infinite for some angles (notice the different scale in Figure 8(c)).

Also notice some useful values, e.g. $\sin 30° = \cos 60° = 0.5$. Look at the values of $\sin \theta$ and $\cos \theta$ when θ is a multiple of 90°.

6.4 Inverse sin, etc.

The angle whose sin is x is written $\sin^{-1} x$. We can write the relationships from Figure 6 as

$$\theta = \sin^{-1}\frac{b}{c} \qquad \theta = \cos^{-1}\frac{a}{c} \qquad \theta = \tan^{-1}\frac{b}{a} \qquad \phi = \sin^{-1}\frac{a}{c}$$

Beware! The index −1 here does *not* indicate a reciprocal:

$$\sin^{-1} x \text{ is } not \text{ the same as } \frac{1}{\sin x}$$

(The reciprocal of $\sin x$ can be written $(\sin x)^{-1}$.)

6.5 Trigonometry on a calculator

You can find the sine, cosine and tangent of an angle on a calculator. For example, to find sin 30°, type 30 and press sin.

Many scientific calculators can be switched between 'degree' and 'radian' modes. The display will indicate which one you are in.

If you switch your calculator to 'radian' mode, you can find sin, etc., of angles in radians without having to convert to degrees. Check that you know how to do this.

With your calculator in radian mode, type π, ÷, 2 (you may need to press = as well) and then press sin or cos. You should get $\sin(\pi/2) = 1$, $\cos(\pi/2) = 0$. If you have your calculator in degree mode by mistake, you will find the sin or cos of 1.57° (3.14° ÷ 2).

Try finding the sin, cos and tan of some angles in degrees and in radians. Check that you get the same values as shown in Figure 8.

If you know the sin, cos or tan of an angle and wish to determine the size of the angle, use the 'inv' key.

For example, to find the angle whose sin is 0.5, type 0.5, press inv and then press sin. You should get 30 if you have your calculator in degree mode. If you do this with your calculator in radian mode, you will get 0.5236 ($\approx \pi/6$).

MATHS REFERENCE

Degrees and radians

See Maths note 6.1

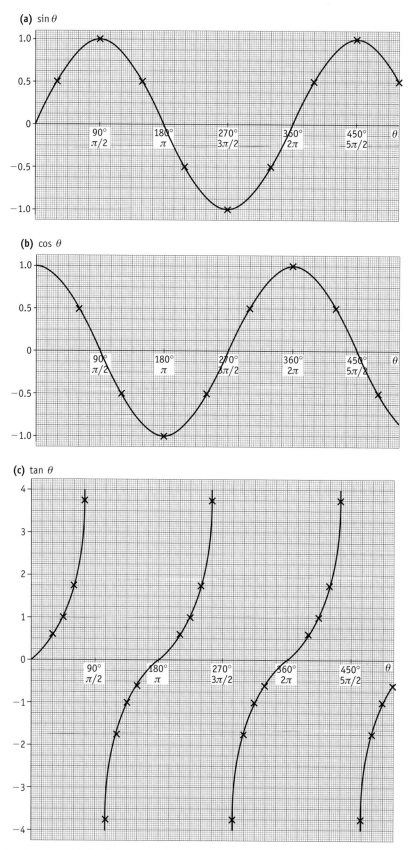

(a) sin θ

(b) cos θ

(c) tan θ

Figure 8 Graphs of trigonometric functions

6.6 The small angle approximations

There are some useful approximations involving the trigonometric ratios of small angles. These become evident when we express the sine and tangent of an angle θ in terms of the right-angled triangles shown in Figure 9.

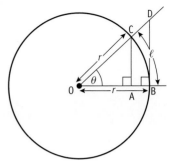

Figure 9 The sine and tangent of an angle

From the triangle OAC:

$$\sin\theta = \frac{AC}{OC} = \frac{AC}{r}$$

and

$$\cos\theta = \frac{OA}{OC} = \frac{OA}{r}$$

From the triangle OBD:

$$\tan\theta = \frac{BD}{OB} = \frac{BD}{r}$$

and

$$\cos\theta = \frac{OB}{OD} = \frac{r}{OB}$$

Figure 9 shows that $\tan\theta$ is always greater than $\sin\theta$ because BD is greater than AC.

As θ is made smaller, the lines AC and BD become closer together and more equal in length, and the lines OA and OD become closer to r, so *for small angles*:

$$\sin\theta \approx \tan\theta$$

and

$$\cos\theta \approx 1$$

With your calculator in degree mode, try finding the sin, cos and tan of the angles listed in Table 7, and some smaller angles. Notice that the approximations get better as the angles get smaller.

Small angles in radians

Comparison with Figure 5 shows that the size of θ *measured in radians* lies between $\sin\theta$ and $\tan\theta$ (the arc length ℓ is longer than AC and shorter than OD):

$$\sin\theta, \ \theta, \tan\theta$$

When θ is small,

$$AC \approx \ell \approx BD$$

and so *for small angles measured in radians* we have some additional approximations:

$$\sin\theta \approx \theta$$

and

$$\tan\theta \approx \theta$$

Switch your calculator into radian mode, and again try finding the sin, cos and tan of various angles. Notice that the approximation gets better at small angles.

7 Measurement and uncertainty

7.1 Experimental measurements

All measurements are subject to systematic and random effects. Such effects are often referred to as 'experimental error', though 'error' here does not mean 'mistake'. However carefully and correctly you make a measurement, there will always be some **random error** (also known as **uncertainty**) and possibly some **systematic error** as well.

A **precise** measurement is one in which the random uncertainty is small. An **accurate** measurement is one that is close to the true value. A precise measurement is not necessarily accurate; there could be some systematic error even if random effects are small.

Systematic errors affect each reading in the same way. For example, a zero error on a micrometer produces an offset so that each measurement is too big or small by the same amount, while a measuring tape that has been stretched leads to readings that are all smaller than the true values by the same proportion. Systematic errors can arise from instrumental effects, or observational methods, or external factors. Sometimes they can be eliminated if their cause is known, for example by checking the zero reading of an instrument.

The uncertainty in a measurement is the range within which you expect the true value to lie. Uncertainty cannot be removed completely but it can sometimes be reduced, e.g. by choice of instrument or method. Do not worry about trying to calculate a precise value for the uncertainty in your experiments – uncertainties are themselves uncertain – but you can usually make reasonable estimates to one significant figure (two at most). There are various ways to estimate the uncertainty in experimental measurements.

MATHS REFERENCE

Significant figures. See Maths note 7.2

Spread in repeated measurements

Suppose you measure a quantity x several times and obtain a series of different values:

$$x_1, x_2, x_3...x_n$$

In your A level physics work, n will usually be a small number, say 3 to 5.

Unless there is reason to suspect that one of the results is seriously out (i.e. it is **anomalous**), the best estimate of the true value of x is the arithmetic mean of the readings:

$$\text{Mean value } \langle x \rangle = \frac{x_1 + x_2 + ...x_n}{n}$$

A reasonable estimate of Δx, the uncertainty in x, is half the range:

$$\Delta x = \frac{(x_{max} - x_{min})}{2}$$

where x_{max} is the maximum and x_{min} the minimum reading of x (ignoring any anomalous readings).

You would write the result of the measurement as:

$$x = \langle x \rangle \pm \Delta x$$

Instrumental uncertainty

Sometimes there may only be a single reading. Sometimes all the readings may be identical. But you cannot therefore assume that there is no uncertainty in the reading(s). The **instrumental uncertainty** is often taken as ± the smallest scale division on the instrument.

When using a digital instrument, the uncertainty cannot be less than half the smallest division of the display. For example, if you are using a digital meter to measure a current and it reads 0.357 A, you can only be sure that the current is closer to 0.357 A than it is to either 0.356 or 0.358 – it could lie anywhere between 0.3565 A and 0.3575 A. So the uncertainty in that reading is 0.0005 A. With analogue instruments, the uncertainty might be smaller if you can interpolate between scale divisions, and you need to consider the particular instrument in order to come up with a realistic estimate.

In many situations, a measurement involves the difference between two readings. For example, the length of an object measured with a ruler is the difference between one mark and another (often zero but not always). There is an uncertainty in each, and both contribute to the uncertainty in the overall measurement. The simplest way to deal with this is to add the two uncertainties and say that the overall uncertainty in the measurement is equal to the smallest scale division. For example, if you were using a ruler graduated in mm, and measured the distance, x, between two markers on a piece of wire to be 864 mm, you would write:

$$x = 864 \, \text{mm} \pm 1 \, \text{mm}$$

Other considerations

Sometimes uncertainty arises because it is difficult to judge exactly what to measure. For example, if you are using a signal generator to produce a sound from a speaker, and adjusting it to give the same pitch as the note from a guitar, it might be hard to judge the 'best' frequency. If you think it is close to 260 Hz, but are unsure by about 10 Hz either way, then (even if the generator had a scale that told you the frequency to the nearest Hz) the uncertainty in your measurement would be $\Delta f \approx 10 \, \text{Hz}$ and you would write:

$$f = 260 \, \text{Hz} \pm 10 \, \text{Hz}$$

7.2 Calculations with uncertainties; significant figures

If you carry out a calculation using a measured value, there will always be an uncertainty in your answer. You can use the uncertainties in the measurements to work out the uncertainty in the calculated value.

For example, suppose you measure a current of $I = 0.24 \, \text{A} \pm 0.01 \, \text{A}$ and a corresponding pd of $V = 0.67 \, \text{V} \pm 0.02 \, \text{V}$.

On a calculator, the resistance found using the 'best' values is:

$$R_{\text{best}} = \frac{V}{I} = \frac{0.67 \, \text{V}}{0.24 \, \text{A}} = 2.791 \, 6667 \, \Omega$$

But, using the largest possible V (0.39 V) and the smallest possible I (0.23 A), the calculated resistance could be as large as:

$$R_{\text{max}} = \frac{0.69 \, \text{V}}{0.23 \, \text{A}} = 3 \, \Omega$$

Or, using the smallest V and the largest I, it could be as small as:

$$R_{\text{min}} = \frac{0.65 \, \text{V}}{0.25 \, \text{A}} = 2.6 \, \Omega$$

There are several things to notice! First, there are quite large differences between the three values. Second, the first value extends to the full length of the calculator display, whereas the others do not.

The large differences show that you *cannot possibly* say that the resistance is precisely 2.791 6667 Ω. This value is close to 2.8 Ω, and the other two differ by 0.2 Ω in either direction, i.e. the uncertainty in R is $\Delta R \approx 0.2 \, \Omega$. The resistance can therefore be written as:

$$R = 2.8 \, \Omega \pm 0.2 \, \Omega$$

The second figure in this answer (the 8 after the decimal point) is uncertain, and so any further figures are meaningless.

Another way of putting this is to say that the answer has (only) two **significant figures** – the one before the decimal point and the first one after it. The rest of the figures in the original 'best' answer are meaningless. They are *not* significant.

7.3 A useful rule of thumb

In a calculation, the answer cannot be known any more precisely than the values used to calculate it. As a useful rule of thumb, the final answer has no more significant figures than the *least* precise value used in the calculation. (The example in Maths note 7.2 illustrates this.)

Suppose you did a calculation to find the frequency f of light whose wavelength is 468 nm (4.68×10^{-7} m). The speed of light is known very precisely: $2.997\,925 \times 10^8$ m s^{-1}.

Using speed ÷ wavelength:

$$f = \frac{2.997\,925 \times 108\,\text{m s}^{-1}}{4.68 \times 10^{-7}\,m}$$
$$= 6.4058 \times 10^{14}\,\text{Hz}$$

However, we only knew the wavelength to three significant figures, so we cannot quote the frequency of this precisely. We must stick to the three significant figures and write:

$$f = 6.41 \times 10^{14}\,\text{Hz}$$

There was no point in using the very precise value for the speed of light. Values listed in data books are often rounded to, say, three significant figures if they are likely to be used only in calculations requiring this precision or less.

7.4 Significant figures and orders of magnitude

The speed of light to seven significant figures is $2.997\,925 \times 10^8$ m s^{-1}; the significant figures are 2997925.

Zeros in front of a number are not significant. The speed of light could be written (rather oddly) as $002.997\,925 \times 10^8$ m s^{-1} or $0.000\,299\,7925 \times 10^{12}$ m s^{-1} without making any difference to its value.

However, zeros at the end of a number are (or at least can be!) significant. If you wrote the speed of light as $299\,792\,500$ m s^{-1}, that would imply that you knew that the last two figures were definitely zeros and not some other numbers. If they are, in fact, not known, it is better to use standard form so that the meaningless zeros can be dropped.

To five significant figures, the speed of light would be 2.9979×10^8 m s^{-1}. To three significant figures, it would be 3.00×10^8 m s^{-1}. Here the zeros *are* significant and should be written down, because 2.997… rounds to 3.00.

To one significant figure the speed of light would be 3×10^8 m s^{-1}.

If a value is rounded to just the nearest power of 10, then we say we are giving just the **order of magnitude**. Two values are said to have the same order of magnitude if one is between 1 and 10 times the other. For example, the wavelengths of red and blue light (about 400 nm and 700 nm) are within the same order of magnitude. But the wavelengths of infrared radiation range from about 10^{-6} m to about 10^{-3} m – they cover three orders of magnitude.

7.5 Combining uncertainties

When calculating a quantity by adding or subtracting measurements, the actual uncertainties are added together. For example, if the starting temperature of a liquid is 18.5 °C ± 0.2 °C, and the final temperature is 30.0 °C ± 0.2 °C, the rise in temperature is 11.5 °C ± 0.4 °C.

When measurements are combined by multiplying or dividing, the *percentage* uncertainty in the result is found by adding the individual percentage uncertainties. For example, suppose a glass stopper has mass m, and volume V (measured by displacement of water) and you want to find the density, ρ, of the glass. This is how you would use the measurements:

$$m = 48.7\,\text{g} \pm 0.5\,\text{g}, \Delta m = \frac{0.5}{487.7} = 0.01 = 1\%$$

$$V = 18\,\text{cm}^3 \pm 1\,\text{cm}^3, \Delta V = \frac{1}{18} = 0.056 = 5.6\%$$

$$\text{density} = \rho = m/V = \frac{48.7\,\text{g}}{18\,\text{cm}^3} = 2.7056\,\text{g cm}^3$$

percentage uncertainty $= 1\% + 5.6\% = 6.6\%$

uncertainty in density, $\Delta\rho = 0.066 \times 2.7056\,\text{g cm}^{-3} = 0.2\,\text{g cm}^{-3}$ (to 1 significant figure)

so density $\rho = 2.7\,\text{g cm}^{-3} \pm 0.2\,\text{g cm}^{-3}$

For formulae involving powers, multiply the percentage uncertainty by the power. This is an extension of the procedure for dealing with multiplication and division. Remember that roots can be expressed as powers, e.g. $\sqrt{x} = x^{\frac{1}{2}}$, so the percentage uncertainty in $\sqrt{x}$ is half the percentage uncertainty in x.

7.6 Error bars and error boxes

When plotting a line graph of experimental data, draw a vertical and a horizontal **error bar** on each point to represent the range of possible values. Use the bars to draw an **error box** around each point. Draw a smooth trend line on your graph which passes through all the boxes, going as close to all their centres as possible. If there is a linear relationship between the variables, you can draw a straight line passing through all the boxes, even if you could not draw one through all the points.

Figure 10 shows graphs of some measurements where the uncertainties were $\Delta y = \pm 0.5$ and $\Delta x = \pm 0.25$. If you only had Figure 10(a) it would be impossible to say whether there is a linear relationship between y and x, but from Figure 10(b) you can say that there is such a relationship within the limits of experimental uncertainty. If x and y were current and pd, Figure 10(b) would indicate that the material obeys Ohm's law.

MATHS REFERENCE

Percentages
See Maths note 3.1

STUDY NOTE

Verify this by finding the maximum and minimum possible densities using values of 49.2 g and 17 cm³, then 48.2 g and 19 cm³

MATHS REFERENCE

Powers that are not whole numbers. See Maths note 1.5.

(a)

(b)

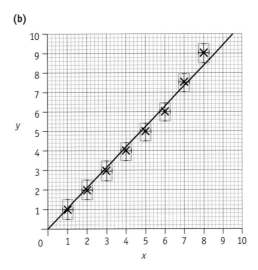

Figure 10 A plot of experimental values (a) without error boxes (b) with error boxes

7.7 Uncertainties and graphs

When you have plotted a graph with error boxes around the points, you can draw the 'best' trend line that passes through all the boxes. The spread of points either side of the line indicates how certain you can be that the 'best' trend line shows the true relationship between the variables – the greater the spread, the greater the uncertainty.

On a linear graph, drawing the steepest and least-steep lines passing through the error boxes allows you to find the uncertainty in the gradient and intercept of the graph, as shown in Figure 11.

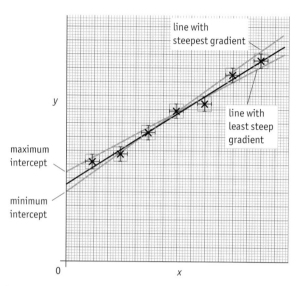

MATHS REFERENCE

Gradient of a linear graph
See Maths note 5.3

Figure 11 Linear graph to show uncertainty

8 Logarithms

8.1 Logs and powers of 10

If a number can be written as *just* a 'power of 10', then the power is the **logarithm** of that number; strictly speaking, it is the **logarithm to base 10**, or **common logarithm**, of the number, but it is often simply called the **log**.

Table 8 lists some examples using whole-number powers.

Number x	$\log_{10}(x)$
$100\,000 = 10^5$	5
$10\,000 = 10^4$	4
$1000 = 10^3$	3
$100 = 10^2$	2
$10 = 10^1$	1
$1 = 10^0$	0
$0.1 = 10^{-1}$	−1
$0.001 = 10^{-2}$	−2

Table 8 Some numbers and their common logarithms

MATHS REFERENCE

Powers that are not whole numbers

See Maths note 1.5

In fact *any positive number* can be expressed as a power of 10, using powers that are not whole numbers. Most whole numbers have logs that are not themselves whole numbers or simple fractions. For example:

$$10^{0.6021} = 4.000$$

so

$$\log_{10}(4.000) = 0.6021$$

All numbers between 1 and 10 have logs that lie between 0 and 1. For example:

$$10^{0.333} = 2.153$$

so

$$\log_{10}(2.513) = 0.333$$

Similarly, all numbers between 10 and 100 have logs that lie between 1 and 2; all numbers between 100 and 1000 have logs between 2 and 3; and so on.

All numbers less than 1 have negative logs. For example:

$$\log_{10}(0.5) = -0.3010$$

$$\log_{10}(0.1) = -1.000$$

8.2 Logs on a calculator

To find the common log of a number using a calculator, type in the number and then press the key marked log or lg.

This process can be reversed to find the **antilog** of a number. Type in the log whose number you want to find, then press the keys marked INV and log (or lg). By doing this, you can show that 4.000 is the antilog of 0.6021, and 2.513 is the antilog of 0.333.

Notice that using the INV and log keys to find the antilog of a number x gives exactly the same result as using the y^x key to find 10^x.

8.3 Logs, multiplication and division

There is a useful relationship between the logs of x and y and the log of their product xy.

Using the definition of a base 10 log:

if $\log_{10}(x) = a$, then $x = 10^a$

and if $\log_{10}(y) = b$, then $y = 10^b$

So we can write:

$$xy = 10^{(a+b)}$$

In other words:

$$\log_{10}(xy) = a + b$$

so

$$\log_{10}(xy) = \log_{10}(x) + \log_{10}(y)$$

You can illustrate this relationship using a calculator to look up the logs of various numbers and see how they relate to one another. For example: $\log_{10}(5) = 0.69897$, $\log_{10}(4) = 0.60206$,

$$\log_{10}(5 \times 4) = \log_{10}(20) = 1.30103$$

$$= 0.69897 + 0.60206$$

$$= \log_{10}(5) + \log_{10}(4)$$

There is a similar relationship for division:

$$\frac{x}{y} = 10^{(a-b)}$$

$$\log_{10}\left(\frac{x}{y}\right) = a - b$$

so

$$\log_{10}\left(\frac{x}{y}\right) = \log_{10}(x) - \log_{10}(y)$$

MATHS REFERENCE

Combining powers
See Maths note 1.3

Logs and powers of 10
See Maths note 8.1

8.4 Logs and powers

We can extend Maths note 8.3 by considering powers. For example:

$$\log_{10}(x^3) = \log_{10}(x) + \log_{10}(x) + \log_{10}(x) = 3\log_{10}(x)$$

A similar line of reasoning says that for any whole number, n,

$$\log_{10}(x^n) = n\log_{10}(x)$$

This rule also works for powers that are not whole numbers, and for negative powers. For *any* value of y we can write:

$$\log_{10}(x^y) = y\log_{10}(x)$$

For example:

$$\log_{10}(9^{1/2}) = \log_{10}(\sqrt{9}) = \log_{10}(3)$$

$$\log_{10}(9) = 0.95424$$

$$\log_{10}(3) = 0.47712 = 0.95424 \div 2$$

and

$$\log_{10}(5^{-1}) = \log_{10}\left(\frac{1}{5}\right) = \log_{10}(0.2)$$

$$\log_{10}(5) = 0.69897$$

$$\log_{10}(0.2) = -0.69897$$

8.5 Logs to other bases; natural logs

We can define logarithms to base 10 using numbers expressed as power of 10. But there is nothing special about the number 10. We can use any number as the base of logarithms.

For example, logarithms to base 2 use the fact that any number can be expressed as a power of 2, as shown in Table 9. Just as with base 10 logs, we are not restricted to whole-number powers.

By trying some examples using numbers from Table 9, you can demonstrate that the relationships set out in Maths notes 8.3 and 8.4 for base 10 logs also work with logs to base 2. (Notice that in *any* base, the log of 1 is zero.)

MATHS REFERENCE

Logs and powers of 10
See Maths note 8.1

Number x	$\log_2(x)$
$16 = 2^4$	4
$8 = 2^3$	3
$4 = 2^2$	2
$2 = 2^1$	1
$1 = 2^0$	0
$0.5 = 2^{-1}$	−1
$0.25 = 2^{-2}$	−2
$1.4142 = 2^{1/2}$	0.5
$5 = 2^{2.32193}$	2.32193
$2.567 = 2^{1.36}$	1.36

Table 9 Some numbers and their logs to base 2

We can go further. The number at the base of a system of logs does not itself have to be a whole number. It can be *any* positive number. For *any* system of logs, it is always true that:

$$\log(xy) = \log(x) + \log(y)$$

$$\log\left(\frac{x}{y}\right) = \log(x) - \log(y)$$

$$\log(x^y) = y\log(x)$$

Natural logs

Apart from common, base 10, logs, the system of logs most widely used in physics is that of **natural logarithms**. This system uses the number e as its base, where e is the **exponential number** ($e \approx 2.718$) which arises in the mathematical description of many naturally occurring changes.

The natural logarithm of a number x is written as $\log_e(x)$ or $\ln(x)$. Most calculators have a natural log key, which is usually labelled $\ln x$.

The inverse of a natural log (its antilog) can be found on a calculator using the INV key and the $\ln x$ key. Finding the natural antilog of a number x is the same as finding e^x.

Table 10 lists some examples of natural logs. By picking suitable numbers from Table 10, you can demonstrate that the relationships for multiplication and division and powers also work for natural logs. Notice that $\ln(e) = 1$, and that $\ln(1) = 0$.

8.6 Using log scales

When we want to plot a graph of numbers that cover a large range of values, we often use a so-called **logarithmic scale** in order to fit the largest numbers on the graph paper while still being able to distinguish between the smallest numbers. Figure 12 shows such a scale. Notice that the powers of 10 are equally spaced – the scale might be better described as a 'powers scale'. The name 'logarithmic' becomes more obvious when we write the (base 10) logs next to the values of x. The logs of x go up in equal steps just like an ordinary graph scale.

Figure 12 A logarithmic scale

When dealing with numbers other than whole-number powers of 10, there are three ways of plotting and reading a log scale – which all amount to the same thing really.

Using logs to plot and read a scale

If you are using ordinary linear graph paper, you need to use logs in order to plot and read a log scale accurately. It can be helpful to write the base 10 logs next to the numbers to be plotted, as was done in Figure 10. Figure 13 shows an expanded view of the part of Figure 12 between $x = 100$ and $x = 1000$, plotted on ordinary graph paper. Table 11 lists some numbers in this range and their base 10 logs.

Figure 13 A log scale covering the range $x = 100$ to $x = 1000$

To plot a number x on the scale shown in Figure 13:

use a calculator to find $\log_{10}(x)$

plot $\log_{10}(x)$ on the side of the scale labelled $\log_{10}(x)$.

To read a number from the scale shown in Figure 13:

Read $\log_{10}(x)$ from the side of the scale labelled $\log_{10}(x)$.

Find its antilog using a calculator.

MATHS REFERENCE

Exponential changes
See Maths note 9.1

Exponential functions
See Maths note 9.2

Number x	$\ln(x)$
0.2	−1.69094
0.5	−069315
1	0.000000
2	0.69315
2.7183	1.0000
3	1.09861
4	1.38629
5	1.60943
9	2.19722
20	2.99573

Table 10 Some numbers and their natural logarithms

x	$\log_{10}(x)$
100	2.000
126	2.100
158	2.200
200	2.300
251	2.400
300	2.477
400	2.602
500	2.698
600	2.778
631	2.800
700	2.845
800	2.903
900	2.954
1000	3.000

Table 11 The numbers plotted on Figure 13 and their logs

Figure 14 A piece of logarithmic graph paper

A short cut

If you do not require values to be plotted or read on a log scale very precisely, there is a useful short cut that avoids having to look up logs and antilogs.

From Figure 13, notice that on the side of the scale labelled *x*:

300 lies roughly halfway between 100 and 1000

200 lies roughly one-third of the way along from 100 to 1000

the distance from 100 to 200 is exactly the same as that from 200 to 400 and from 400 to 800.

This pattern is repeated between *any* two adjacent whole-number powers (e.g. 30 lies mid-way between 10 and 100 on a log scale, 0.3 lies mid-way between 0.1 and 1.0, etc.).

Bearing these points in mind, it is possible to estimate the value of *x* just by looking at where it lies on the scale between two whole-number powers.

Log graph paper

Figure 14 shows a piece of logarithmic graph paper. It covers the range of numbers between two adjacent powers of 10. The grid lines are labelled 1, 2, 3 and so on, corresponding to the positions of 100, 200, 300 on the top scale of Figure 13 – or to 1 ×, 2 ×, 3 × ... any power of 10 that you care to choose.

Plotting and reading values using such a piece of graph paper is just like using ordinary graph paper – the only difference is that the grid spacing varies across the page. There is no need to calculate logs or antilogs, as the paper is already labelled with values of *x*.

8.7 Using log graphs

There are essentially three reasons for plotting graphs using logarithmic scales:

● to represent a large range of values on a compact scale;

● to see whether data are described by an exponential relationship;

● to see whether data are described by a power law.

We will deal with each of the last two in turn.

Exponential relationships

An exponential relationship between two variables *x* and *y* is one that is described as an equation of the form

$$y = Ae^{kx}$$

where *A* and *k* are constants; *k* is sometimes called the **decay constant** (if negative) or the **growth constant**, but often has a particular name according to the situation being described. For example, the attenuation of a signal in a cable or optical fibre is described by the equation

$$I = I_0 e^{-\mu x}$$

where I_0 is the initial intensity of the signal, *I* is its intensity after travelling through a distance *x*, and *μ* the attenuation coefficient.

A plot of *y* against *x* gives a curve, in which *y* changes by equal fractions for equal steps in *x*.

A direct way to determine whether two variables are described by an exponential relationship is to plot a **log–linear graph**. For example, taking logs of the attenuation equation gives:

$$\log(I) = \log(I_0) - \mu x \log(e)$$

This works whichever base of logs we choose. A graph of log (*I*) on the *y*-axis against *x* on the *x*-axis has the form $y = mx + c$; in other words, it is a straight line.

If we choose to take natural logs, the equation becomes simpler:

$$\ln (I) = \ln (I_0) - \mu x \ln (e)$$

i.e.

$$\ln (I) = \ln (I_0) - \mu x$$

A graph of $\ln (I)$ against x has a gradient equal to $-\mu$.

This is an example of a general rule: a log-linear graph of an exponential relationship is a straight line whose gradient is equal to the growth or decay constant (see Figure 15).

MATHS REFERENCE

Gradient of a linear graph
See Maths note 5.3

Logs, multiplication and division
See Maths note 8.3

Logs and powers
See Maths note 8.4

Logs to other bases; natural logs
See Maths note 8.5

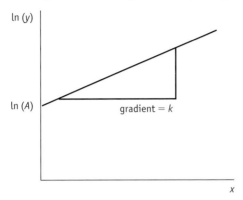

Figure 15 A log-linear graph of the relationship $y = Ae^{kx}$

Power-law relationships

Two variables x and y are related by a **power law** if they obey an equation of the form

$$y = Ax^p$$

where A and p and constants; p is called the **exponent** (*not* to be confused with the exponential number e).

An example of such relationship is Coulomb's law

$$F = \frac{kqQ}{r^2}$$

which describes how the magnitude of the electrostatic force, F, between two charges, q and Q, depends on their separation r^p, where $p = -2$.

A graph of x against y gives a curve whose shape depends on the exponent p.

A power-law curve can superficially look like an exponential growth or decay curve, but it does *not* follow the 'equal fractions in equal steps' pattern that characterises exponential curves.

One direct way to determine whether variables obey a power law is to plot a **log-log graph**. For example, taking logs (to *any* base) of the Coulomb's law equation gives:

$$\log (F) = \log (kqQ) - 2 \log (r)$$

A graph of $\log (F)$ against $\log (r)$ has the form $y = mx + c$. It is a straight line with gradient -2.

This is an example of a general rule. A log-log graph of a power-law relationship is a straight line whose gradient is equal to the exponent (see Figure 16).

MATHS REFERENCE

Gradient of a linear graph
See Maths note 5.3

Logs, multiplication and division
See Maths note 8.3

Logs and powers
See Maths note 8.4

Logs to other bases; natural logs
See Maths note 8.5

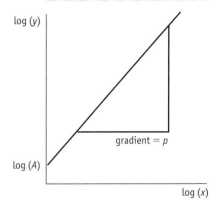

Figure 16 A log-log plot of the relationship $y = Ax^p$

9 Exponentials

9.1 Exponential changes

Many naturally occurring changes follow a pattern in which the rate of change is proportional to the amount, or number, of something present. Such a change is called an **exponential change**. Such changes are represented by equations of the form

$$\frac{dy}{dt} = kt \quad \text{or} \quad \frac{dy}{dx} = kx$$

depending on whether the change takes place over time, t, or over distance, x.

One example of exponential change is radioactive decay, where the number of radioactive disintegrations per second is proportional to the number of unstable nuclei present in a sample. Mathematically, this decay can be represented by an equation

$$\frac{dN}{dt} = -\lambda N$$

where N is the number of unstable nuclei and λ the **decay constant** for the particular isotope involved.

Another example is population growth

$$\frac{dN}{dt} = kN$$

where now N is the number of organisms in a population and k the **growth constant**.

Figure 17 shows graphs of exponential decay and growth. These graphs have certain characteristics that are unique to exponential changes (and which are related to one another):

- the gradient at any point is proportional to the value of y at that point (this is described by the equations stated overleaf)

- equal intervals of time or distance result in equal fractional changes in y

- the time or distance taken for y to halve, or double, is independent of the initial value of y; when dealing with decay over time, this time is called the **half-life** $(t_{1/2})$ of the decay.

9.2 Exponential functions

Exponential changes can always be described by an equation of the form

$$y = y_0 a^{bt} \quad \text{or} \quad y = y_0 a^{bx}$$

where a and b are constants and y_0 is the value of y when $x = 0$ or $t = 0$. For growth over time or distance b is positive, for decay b is negative.

For example:

- if $N = 2^t$, then when t increases by 1, N doubles.

- if $y = 3 \times 10^{-x}$, then when x increases by 1, y is divided by 10.

- radioactive decay can be described by the equation

$$N = N_0 \times 2^{-(t/t_{1/2})}$$

where $t_{1/2}$ is the half-life of the decay.

All mathematical functions of the type $y = y_0 a^{bt}$ are called **exponential functions**.

One very commonly used exponential function uses the **exponential number** e ($e \approx 2.718$). This number arises from the mathematical description of situations where the constant k in the growth or decay equation is numerically equal to 1.

(a)

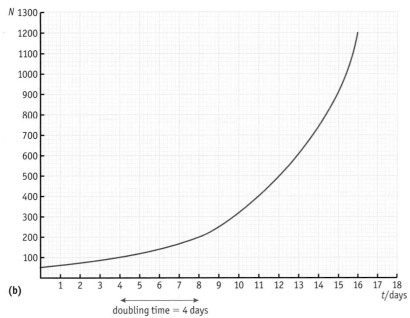

(b)

Figure 17 Graphs of exponential change; (a) decay, (b) growth

If

$$\frac{dy}{dx} = y$$

then the change is described by

$$y = y_0 e^x$$

Similarly, if

$$\frac{dy}{dt} = y$$

then

$$y = y_0 e^t$$

The number e can be used to describe *all* exponential changes – and this is the most common way that we write them. If

$$\frac{dy}{dx} = ky$$

then

$$y = y_0 e^{kx}$$

and, similarly, if

$$\frac{dy}{dt} = ky$$

then

$$y = y_0 e^{kt}$$

For example, if a radioactive decay is described by the equation

$$\frac{dN}{dt} = -\lambda N$$

then it is also described by the equation

$$N = N_0 e^{-\lambda t}$$

Some texts write e^x as exp (x), and so on, so our last example would be written

$$N = N_0 \exp(-\lambda t)$$

9.3 Exponentials and logs

MATHS REFERENCE

Logs to other bases; natural logs
See Maths note 8.5

The exponential number e ($e \approx 2.718$) is used as the base of so-called natural logarithms. By taking natural logs of exponential growth and decay equations, we arrive at some useful relationships. If

$$y = y_0 e^{kt}$$

then

$$\ln(y) = \ln(y_0) + kt$$

This form of the equation is useful as it shows how we can use a log-linear graph to determine whether a change is exponential and, if it is, to determine the value of k.

MATHS REFERENCE

Using log graphs
See Maths note 8.7

It also enables us to find the value of k if we know y_0 and y at a given time, t:

$$kt = \ln(y) - \ln(y_0) = \ln\left(\frac{y}{y_0}\right)$$

Similarly, in radioactive decay

$$N = N_0 \exp(-\lambda t)$$

so

$$\ln(N) = \ln(N_0) - \lambda t$$

$$\lambda t = \ln\left(\frac{N_0}{N}\right)$$

This expression leads to a useful relationship between λ and the half-life, $t_{1/2}$, of the decay. If $t = t_{1/2}$, then by definition $N = N_0/2$,

MATHS REFERENCE

Logs, multiplication and division
See Maths note 8.3

so

$$\lambda t_{1/2} = \ln(2)$$

$$t_{1/2} = \frac{\ln(2)}{\lambda}$$